Symbiosis in Cell Evolution

Microbial Communities in the
Archean and Proterozoic Eons

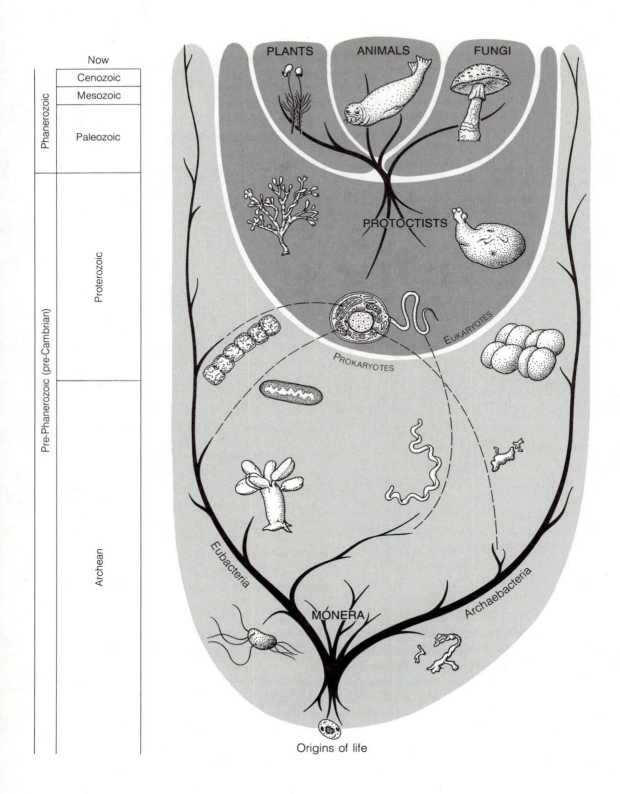

Origins of life

SYMBIOSIS IN CELL EVOLUTION

Microbial Communities in the Archean and Proterozoic Eons

SECOND EDITION

LYNN MARGULIS
UNIVERSITY OF MASSACHUSETTS

W. H. Freeman and Company

NEW YORK

This book is dedicated to the memory of my mother,
Leone Wise Alexander
1914–1977

QH581.2
.M372
1993

Frontispiece based on a drawing by Christie Lyons.

Library of Congress Cataloging-in-Publication Data

Margulis, Lynn, 1938–
   Symbiosis in cell evolution : microbial communities in the Archean
and Proterozoic eons / Lynn Margulis. -- 2nd ed.
      p.   cm.
   Includes bibliographical references (p.   ) and index.
   ISBN 0-7167-7028-8 (hard). -- ISBN 0-7167-7029-6 (soft)
   1. Eukaryotic cells--Evolution.   2. Life--Origin.
3. Symbiogenesis.   4. Paleontology--Precambrian.   I. Title.
QH581.2.M372   1993   92-14565
577--dc20   CIP

Printed in the United States of America

1 2 3 4 5 6 7 8 9 0   VB   9 9 8 7 6 5 4 3

CONTENTS

LIST OF
ILLUSTRATIONS

LIST OF TABLES

FOREWORD

By her own account, Lynn Margulis has chosen audacity over prudence. She would prefer risking her professional reputation to being ignored. Following this paradigm throughout her scientific career, she has already done great service to biological science by bringing "symbiogenesis" to critical attention, and she has been vindicated by the consensual acceptance of the symbiotic origin of mitochondria and chloroplasts. She continues to press us to take a similar view of the organelles of eukaryote mitosis and motility, that these are derived from symbiotic spirochetes.

In this work, Dr. Margulis has amassed an incredible range of information, belying her despair that no important text could any longer come from a single pen. It is hard to find many counterexamples, but I did think of Lubert Stryer's *Biochemistry* as a rule-exemplifying exception. She also tells a story, with no punches barred, of macroevolutionary events that will transfix attention in the very process of critical evaluation.

My own style might be inclined to be more agnostic about evolutionary explanation, particularly about the temporal sequences. How could we ever be sure that undulipodia are derived from spirochetes, rather than vice versa by gene transfer from a eu- to a prokaryote? Does the contemporary living world really retain well-ordered historical relics, from which we might conclude that motility preceded mitosis, or are they booby traps set by Nature to confuse us? Such questions might be the path of scientific caution, but — I admit — at the price of a nihilism that paralyzes the motive to develop evolutionary ideas to their logical culmination and eventual experimental test.

Nevertheless, I do have to record a few points where a still broader range of hypotheses might well be given attention. Above all, I would urge against leaping directly from symbiotic associations among discrete cells to the engulfment of intact genomes. There appear to be innumerable examples of DNA segments, fragments of genomes, being exchanged among disparate organisms — outstandingly by viral transduction, or by uptake of raw DNA. Whatever the primordial genotype of the uhr-mitochondrion, much of the genetic information for the mitochondrial structure now appears in the nucleus, some of it surely having been transposed there, other parts likely newly evolved. If many kinetosomes now lack DNA, it is hard to aver whether this, or the controverted alternatives, are the more primitive condition. Well, there I go again: of course it's "hard to be sure," but let us be grateful that someone will set up clear targets for us to aim at.

As to the ultimate chemical origin of life, Dr. Margulis, in accord with the common wisdom, has emphasized terrestrial history. But I am glad to note that she is also willing to go even further out into deep space. I can then share with her some contemporary revival of the idea that cometary fragment infall might account for a substantial part of terrestrial organic matter. This would leave a wider range of possibilities about the primordial chemistry than if we are confined to the early Earth's atmosphere. Again, I am troubled about whether the germinal matter of contemporary life has not totally obscured the primordial: witness the current fad about "the RNA world."

In conclusion, while I have reserved judgment about the state of proof of many of her detailed conclusions, I have no doubt that we would be much poorer if she had repressed them, or if others were to ignore them. About the central theme, that evolution is a reticulum more than an ever-diverging arborization, she has made a compelling case that is wonderfully narrated and pictured in this book.

May 1992

Joshua Lederberg
Professor
Rockefeller University

FOREWORD TO THE
FIRST EDITION

Fifty years ago, the basic ideas explored in this book appeared "too fantastic for present mention in polite biological society," as E. B. Wilson wrote, or so out of harmony with the established point of view that they "could not be discussed at respectable scientific meetings," as more recently A. Takhtajan put it. Though some may still reject these ideas, they have now obviously become polite and respectable subjects of scientific discourse. This quiet revolution in biological thought is primarily due to the insight and enthusiasm of Lynn Margulis, the author of this book. Though her earlier books and papers are now deservedly well known, an enormous number of relevant new facts has been discovered in recent years. In the present volume she evaluates these findings and concludes that most fit neatly into her theoretical scheme.

The range covered is enormous: pre-Phanerozoic geology, molecular genetics, and the taxonomy of all the major groups of lower organisms. In discussing what she calls the mix-match principle, she points out that, as in cookery one can take the primary ingredients for granted, so too in assembling a eukaryote the component simple organisms are integrated, in successful cases, to produce new and unexpected results. The same principle is clearly applied in the writing of the book, which turns out to be a very rich dish indeed. It is of immense interest to anyone concerned, as most biologists are, with its main topic. It is a truly important contribution to the reintegration of our terribly fragmented science into a coherent discipline. Beyond all this, it provides a glimpse of the working of one of the most constructively speculative minds, immensely learned, highly imaginative, and occasionally a little naughty, that is now engaged in the study of the great problems posed by living organisms.

July 1980
G. Evelyn Hutchinson (1903–1991)
Sterling Professor Emeritus of Zoology
Yale University

PREFACE

Food, enemy, poison, fiber, or source of scent? All peoples and all languages have ways of systematizing knowledge about the millions of living organisms with which we people share the Earth. Yet biology, the science of life, which organizes consciously the knowledge of life, is a social activity only one century old. A goal of biology is depiction of the genetic relations between extant organisms through diagrams of their evolution from common ancestors. The science of genetics, which so illumines evolutionary thought, began as the analysis of the transmission of traits from parents to offspring. It absolutely required study of reproductive behavior of live beings. Genetics has been transformed in the second half of this century by a peculiar integration of observations of living organisms, their descendants, and the materials (nucleic acids and proteins) which travel between generations. Genetics has recombined arcane studies of macromolecular chemistry with the controlled growth, in the laboratory, of a select group of rapidly growing organisms: the colon bacterium *Escherichia coli*; the green alga *Chlamydomonas*; the yeast of beer and bread, *Saccharomyces*; the fruit fly *Drosophila*; and the roundworm *Caenorhabditis*.

The ways in which scientists strive to reconstruct the evolutionary history of life have transformed. Large teams of investigators, mainly in Europe, North America, and Japan, work in laboratories that resemble more emergency rooms of hospitals than the groves of academe; hundreds of co-workers paid from large government grants have replaced the contemplative scholar at the edge of his botany pond surrounded by grasshoppers. No major textbook is authored by a single scientist anymore, nor generally does any investigator who participates in molecular biology and genetics enter nature's fields. Almost no serious student of molecular biology or genetics, at any stage of her professional life, is introduced formally to any aspect of atmospheric or geological science. Paleontologists concerned with the bodily remains of ancient organisms employ a language with hardly a word in common with that of the biochemist. In this worsening climate of mutually suspicious specialists and rampant incompletions, the renovation of my story of the role of symbiogenesis, the ultimate intimacy, in the evolution of life is at once an act of civil disobedience and unremitting arrogance. Furthermore, it is an undertaking even more audacious and risky than before.

In the decade since the first edition of this book appeared, biology has undergone staggering changes. Perhaps most marked is the immense accumulation of macromolecular sequence data for both proteins and nucleic acids from diverse sources. Crucial to the enterprise are techniques for "cloning" (replicating) the molecule of interest, whether it is the gene for interferon — to produce a protein for therapeutic use — or the gene for the 16S RNA of an obscure amoebomastigote — to place it on the next great phylogeny, the definitive grand scheme relating all life on Earth. Cognizant of the rate of influx of new molecular data, I find the risk is less the loss of my scientific reputation and more of not being read at all. Audacity, another name for this narrative, is the requisite antidote.

There may be more than 30 million species of organisms on Earth, each averaging over one billion bases of DNA in each of its cells. To understand, from sequence data, all life on Earth, we would need to know at least 3×10^{16} details. If we add to this the probabilities that 99.9 percent of all organisms once alive are now extinct and that perhaps 99.9 percent of all microorganisms are uncatalogued in the literature of science, we collect six more orders of magnitude of ignorance. We calculate that, in principle, so many zillions of details underlie the past history of life that its reconstruction is probably intrinsically unknowable. But the exigencies of curiosity are unsatisfied with long, ordered lists of adenine and cytosine nucleotides and such calculations. Evolution is the history of life, the machinations of the ancestors, and the antecedents of the inhabitants of today's living world. What alternative paths present themselves? Hefting any massive textbook of biochemistry, molecular genetics, or cell biology that purports to present modern truths, I slyly deride both authors and readers for self-deception: it is apparent from the team effort alone that no one human can master the miscellany of bald protein and nucleic acid sequences, names of organisms, and other unrelated "facts." Either I despair or decide — as I have — that a radically different approach is needed. I deliberately replace a "pass-the-test" mentality, the kind that has ritualized memorization since the earliest priesthoods indoctrinated novitiates, with the literary device known well to mythologists: I tell a story. Consistent with the findings of modern science, I attempt to develop overtly the principles of biological knowledge that are deeply embedded in the rush of information that daily floods the professional literature.

Accordingly, this book cannot claim to "cover the field" of cell evolution with equanimity, if it were even possible. For reasons of background, nationality, date, and style, I suspect that to announce such an effort would be self-deluding, indeed as impossible as determining macromolecular sequences from entirely silicified fossil wood. But I attempt to provide a narrative, a possible retelling of 3500 million years of evolution from the perspective of the microcosm, the subvisible world of life too small to be perceived directly by humans.

In his remarkable little book, Mereschkovsky (1920) wrote: "I call this process symbiogenesis, which means: the origin of organisms through combination and unification of two or more beings, entering into symbiosis." By symbiosis, Mereschkovsky meant simply the living together of different kinds of organisms. The importance of the Mereschkovskian concept of symbiogenesis as a mechanism of evolution has fared well during these turbulent years of the molecular biological revolution. It is precisely the plethora of sequence data generated by the new science of molecular biology itself that substantiates the role of symbiosis in the origins of classes of cell organelles and hence symbiogenesis in the polyphyletic origins of eukaryotes.

The new data have led to two major revisions in this second edition of the book: the first is the idea that microtubule systems preceded symbiotic acquisition of mitochondria in the origin of the earliest eukaryotes. The second, the spectacular discovery of centriole-kinetosome DNA in *Chlamydomonas* (Hall, Ramanis, and Luck, 1989), constrains enormously the possible narratives of the origins of these motility organelles. The earlier appearance of nuclei and centriole-based, intracellular motility systems is deduced from the distribution of cilia, division centers, and other microtubule-based organelles in many obscure protoctists. The documentation of genera, families, and even orders of nucleated, undulipodiated anaerobic protoctists in anoxic zones in which no members have mitochondria but many contain other sorts of cytoplasmic symbionts suggests that animal, plant, and fungal mitochondria are a special case of a general phenomenon of bacteria-to-organelle transitions in early eukaryotes. This reinterpretation of data led to the reordering of Chapters 8, 9, and 10 in this edition. The discovery by David Luck's group at Rockefeller University of some six megabases of DNA in the centriole-kinetosome of the green alga *Chlamydomonas*, although challenged by at least two teams of investigators, has greatly modified the discussion of centriole-kinetosome origin in Chapter 8. Besides these major changes, I have made every attempt to incorporate the most recent scientific literature into the text. Of course, this means facing the lonely futility of the magnitude of the task and suffering accompanying waves of depression. The source of the compensating joy is the intrinsic fascination of the story itself and the challenge of sensible, direct communication with you, the reader.

May 1992   *Lynn Margulis*

ACKNOWLEDGMENTS

I am most grateful to my students, both informal and formal (for example, those who have taken Environmental Evolution, Symbiosis, Evolution, Protists, and a miscellany of graduate seminars at Boston University or the University of Massachusetts, Amherst), for their embarrassing questions. More than anything else, such free inquiry leads to the posing of important scientific problems. W. Ambrosino, D. Ashendorf, S. Awramik, S. Banerjee, D. Bermudes, S. Campbell, D. G. Chase, R. Chesselet, P. E. Cloud, Jr., P. Collins, J. O. Corliss, R. Dickerson, W. F. Doolittle, Betsey Dyer, M. Enzien, S. Fracek, S. Francis, I. D. Gharogozlou, S. Giovannoni, J. Giusto, S. Golubic, E. Gong, A. Grosovsky, R. Guerrero, H. O. Halvorson, E. Hoffman, A. Hollande, J. Howard, G. E. Hutchinson, J. K. Kelleher, G. Kline, A. Knoll, A. Lazcano-Araujo, C. Limoges, A. López-Cortés, J. E. Lovelock, FRS, D. Lovley, H. A. Lowenstam, T. N. Margulis, M. McElroy, H. McKhann, E. G. Merinfeld, L. Meszoly, C. Monty, K. E. Nealson, K. H. Nealson, L. Olendzenski, Wm. Ormerod, J. Oró, T. C. Owen, C. Ponnamperuma, M. Rambler, H. Ris, M. San Francisco, J. Wm. Schopf, K. V. Schwartz, G. Small, D. C. Smith, Wm. Solomon, M.-O. Soyer, J. F. Stolz, P. Strother, T. Swain, I. Taylor, Wm. Taylor, G. Thorington, L. P. To Isaacs, N. Todd, R. K. Trench, J. C. G. Walker, R. H. Whittaker, R. S. Young, and M. Zavitkovsky all aided. Without the generous lessons in every aspect of this work given me by Elso S. Barghoorn for more than twenty years, this book would never have been written.

The research was supported, directly or indirectly, by the NASA Life Science Office (National Aeronautics and Space Administration Grant NASA-NGT-004-025), the Boston University Graduate School, the University of Massachusetts Botany Department and College of Natural Sciences and Mathematics (Dean Frederick Byron), and the Division of Geology and Planetary Science of the California Institute of Technology. Some research on spirochetes and microtubules was supported, if reluctantly, by the National Science Foundation. My debt to the Sherman Fairchild Distinguished Fellowship program of the California Institute of Technology (and thus to Barclay Kamb and Lea Sterrett) is too great for words. In the last decade the generosity of the Richard Lounsberry Foundation (New York) permitted the work to continue. The peace and quiet provided by the Guggenheim Foundation and T. N. Margulis's fierce defense of my privilege to work was a prerequisite to development of the first edition. I thank J. R. Williams, Wm. Bennett, Gary Carlson, Kirk Jensen, I. Krohn, Arthur Bartlett, and especially G. Hamilton and J. Tannenbaum for editorial work.

Dorion Sagan, Jeremy Sagan, Zachary Margulis, Jennifer Margulis (my children), and Morris Alexander (my father) all helped with the first edition. My debt to Stephanie Hiebert is overwhelming. My chairman, James Walker, and others at the Botany Department of the University of Massachusetts — J. B. Ashen, R. Charron, G. Hinkle, R. Fester, J. Klenz, S. Klinginer, T. Lang, D. Munson, L. Nault, L. Olendzenski, J. A. Stricker, D. Reppard, D. Sagan, M. Solé, L. Stone, and O. West — fostered the new research included here.

Residency in the calm beauty and luxurious accommodations of the Rockefeller Foundation's Bellagio Study Center in the foothills of the Alps during the summer of 1991 was essential to the preparation of this second edition. My gratitude is to Antonella Ancafora, Gianna Cellini, Susan Garfield, Frank and Jackie Sutton, and the others of the dedicated staff and remarkable visitors at Lake Como's shores. As a visiting professor at the autonomous University of Barcelona (Bellaterra) on my "año sabatico en España" I was aided by scientist-helpers and scientists R. Amils, C. Chica, I. Esteve, N. Gaju, both Jordis Mas, C. Pedros-Alío, M. Piqueras, and especially Ricardo Guerrero. The index was prepared in collaboration with S. Hiebert, D. Sagan, and L. Stone, to whom I continue to be grateful. In Bellagio, Barcelona, and elsewhere, Professor Ricardo Guerrero generously donated both general knowledge and specific information to this project. Without Bellagio and Stephanie Hiebert's devotion to the Bellagio manuscript, the book would never have been completed.

L. M.

PREFACE TO THE
FIRST EDITION

Various "symbiosis theories" have appeared from time to time, and it is appropriate to recall them here, not for their own sake, since all are certainly defunct, but because they preface other matters. Those theories by Portier (1918), Wallin (1927), and Schanderl (1948) are the main ones. Common to all three theories was the claim that bacteria occur among the cytoplasmic inclusions of cells in general, and that intracellular symbiosis with bacteria is a universal characteristic of plants or animals! Momentous significance was variously attached to this supposed symbiosis. . . . It is hardly surprising that each of these theories in turn was ignored or summarily rejected by most authorities.

R. LANGE, 1966

This book presents an old idea in a modern and precise focus: that nucleated or eukaryotic cells — those cells comprising the bodies of animals, fungi, and plants — evolved from bacterial ancestors by a series of symbioses. Thus, the origin of eukaryotic cells is perceived as a special case of a general phenomenon, the evolution of microbial associations. The emergence of partner species and their coevolution began at least 3500 million years ago and has persisted until the present. Furthermore, microbial interactions not only played a major role in the evolution of cells but they have had profound effects both on the surface sediments of the Earth and on its atmosphere.

The attitude of the scientific community toward the importance of symbiosis and cooperation in effecting evolutionary change has altered in the decade since my book on the origin of eukaryotic cells was published (Margulis, 1970). Most biologists now agree that symbiosis has led to innovation, although there still is no consensus on details. Furthermore, virtually all biologists agree that differences between the cells of eukaryotes (protoctists, fungi, animals, and plants) and the non-nucleated cells of prokaryotes (bacteria) are far more profound than differences between animal and plant cells. Unlike prokaryotes, all eukaryotic cells are polygenomic; they contain several kinds of organelles that harbor distinct genetic systems. However, the origin of the polygenomic cells of eukaryotes is still warmly debated. This book presents one view — the "extreme" version of the serial endosymbiosis theory of the origin of organelles.

The sophistication with which the interaction of the nucleocytoplasm and organellar genetic systems can be studied has led to inquiry into the biogenesis of cell organelles. Recent discoveries about organellar development can be interpreted in an evolutionary context. Many phenomena can be understood as the consequence of growing intimacies between once free-living microbial partners. The biochemical cytology of the 1950s has become the discipline called cell biology. Adolescence has given way to fat maturity; thousands of scientists, the vast majority of whom are not concerned with evolution, practice this new profession. Unwittingly, they have provided much data for the evolutionist. The past decade has also seen the rebirth of the nineteenth-century field of microbial ecology and the conception of another discipline, microbial evolution.

We have been to the Moon and have seen our planet from space; the extent to which the Earth is now and has been an abode for life, yet modified by life, has awakened consciousness and stirred consciences. A consequence of these activities is that the study of microbiology and the evolution of cells is no longer an obscure, exclusively biological undertaking. Microbiological data once thought to be the concern of only the health or soil scientists is becoming integrated into the conceptual framework of the rest of evolution; in the next decade it will be integrated into an even newer science, comparative planetology.

Cellular chemistry can no longer be viewed as merely a "data base" for research in human diseases, food preservation, sewage treatment, or crop production. The history of the cell is inextricably tied to the history of the Earth's surface in ways that we are only now beginning to comprehend. The ecosystems that compose the biosphere change with time; evolution is not only the change in gene frequencies in populations of organisms but the change in the Earth's environments as well. This book attempts to weave those often isolated threads of science into the whole cloth of our most ancient history. It traces the history of cells, the units of life, from their obscure beginning through their present diversity and ubiquity to their influence on the planet itself.

Concepts of cell evolution that can be examined experimentally and can be expected to generate new lines of research are discussed here. The twentieth-century myth of man's place in nature, drawing from all branches of science and having the virtue of testability, may be better able to fill the cracks of man's curiosity than all the traditional myths before it. Based on the detailed research of many dedicated people, this scientific origin myth has the advantage of generating testable predictions and revealing the outlines of the unknown in a realistic way. Not only is the story fascinating, but, unlike preceding folk tales, if in error it can be modified, rearranged, updated, and amended by many members of the tribe.

This book was written for two audiences, science students having some experience with biology, geology, and chemistry, and professionals engaged in untangling subplots of the epic drama of the early evolution of life on Earth. The influx of new data will lead to continuous renovation of the script. I hope that this version of the narrative will be questioned and discussed in seminars and classrooms, that it will be challenged and modified as new information requires. For, unlike many scholarly works, this book is not so much a collection of truths as a presentation of a point of view concerning the likely course that evolution took and the role symbiosis played in the emergence of eukaryotic cells.

September 1980   *Lynn Margulis*

CHAPTER 1

SERIAL ENDOSYMBIOSIS THEORY

OVERVIEW

Just as reproduction insures the *perpetuation of existing species*, the author believes that Symbionticism insures the *origin of new species*.

I. E. WALLIN, 1927

The symbiotic theory of the origin and evolution of cells rests on two concepts of biology. The first is that the most fundamental division in the organization of the living world is between the prokaryotic and the eukaryotic organisms, between the bacteria and those organisms composed of cells with nuclei—the protoctists, animals, fungi, and plants. The second concept is that some parts of eukaryotic cells resulted directly from the formation of permanent associations between organisms of different species, the evolution of symbioses. Three classes of organelles—cilia, mitochondria, and photosynthetic plastids—were once free-living bacteria that were acquired symbiotically in a certain sequence by other, different bacteria. (In contrast, the traditional view, direct filiation, holds that cell organelles—centrioles, mitochondria, and plastids—evolved by compartmentalization inside cells.) The symbiotic theory depends heavily on evolutionary thought as developed by geneticists, ecologists, and cytologists, including scientists who welded Mendelian genetics to Darwin's ideas of natural selection. It also relies on new or revitalized scientific fields: molecular biology, especially the study of nucleic acid and protein sequence and structure; micropaleontology, which studies the earliest evidence of life; and even atmospheric physics and chemistry, insofar as they are concerned with biologically generated gases. The theory is summarized in the later sections of this chapter and explained more fully in the following chapters.

Chapter 2 reviews the diversity of life from the point of view of the symbiotic theory. All organisms composed of cells are grouped into five kingdoms: the prokaryotic kingdom (Prokaryotae or Monera, comprising all archaebacteria and eubacteria) and the eukaryotic kingdoms (Protoctista, Animalia, Fungi, and Plantae). Protoctists are eukaryotic organisms that are not animals, fungi, or plants. Algae,* protozoa,† slime molds, slime nets, and other obscure eukaryotic organisms are members of the kingdom Protoctista. Protists, the smaller eukaryotes, are defined more restrictively as few-celled or single-celled nucleated organisms. Thus, the kingdom Protoctista includes more than just protists; it also includes their immediate multicellular descendants, such as red and brown seaweeds, and many fungus-like microbes, such as slime molds and oomycetes. (Descriptions of criteria for classification in the five kingdoms, including examples of the best-known genera, are in the Appendix.)

The remarkable history, primarily European (including Russian), of the concepts of symbiogenesis is included in Chapter 3, which compares the symbiotic and traditional theories of cell evolution and traces their fluctuating fortunes.

Chapter 4 attempts to define life by looking at its minimal manifestations. Presenting the geological context in which cells must have originated, it also contains a status report on the unsolved and tricky question of the origin of life itself. The next two chapters (5 and 6) present cell evolution from 3500 million years ago onward, more or less chronologically as evinced from the fossil record.

Chapter 5 traces the modern microbiology and evolution of major anaerobic metabolic pathways in bacteria (see Brock and Madigan, 1991, for a comprehensive discussion). Chapter 6 explores the evolution of aerobic metabolism and its relationship to the accumulation of atmospheric oxygen. Chapter 7 presents criteria for distinguishing organelles that originated as symbionts from those formed by intracellular differentiation; extant symbioses illustrate the power of symbiosis as a mechanism of evolutionary innovation.

---

*In this book, the terms **algae** and **fungi** apply exclusively to organisms having eukaryotic organization.

†"Protozoa" is an obsolete phylum in the two-kingdom system that classifies bacteria and fungi as plants and a miscellany of motile heterotrophs as animals (see Appendix for current five-kingdom classification). All traditional "protozoa" are accommodated in one of the fifty or so phyla of the kingdom Protoctista (Margulis, Corliss, Melkonian, and Chapman, 1990).

The next four chapters consider the history of the major classes of eukaryotic organelles. Nuclei and centrioles, undulipodia and their underlying kinetosomes, microtubules, and the sexual systems based on mitosis and meiosis are considered in Chapter 8. The most controversial—and perhaps the most interesting—concept in the book, the origin of undulipodia (centrioles, kinetosomes, cilia and other microtubular organelles) from motility symbioses between archaebacteria (such as *Thermoplasma*) and free-living spirochetes (such as *Spirosymplokos*), is described in Chapter 9. Such motility symbioses preceded the origins of other organelles (mitochondria, plastids). The integration of disparate symbionts is identical to the evolution of the mitotic-meiotic-Mendelian genetic systems of the pore-studded nuclei of eukaryotes. Reasons for these assertions, coupled with new information, are detailed in Chapter 9. Origins of mitochondria in protoctists are taken up in Chapter 10. The symbiotic acquisition of cyanobacteria that converts heterotrophic protists to algae as they acquire photosynthetic organelles—the plastids—is the subject of Chapter 11. Finally, Chapter 12 discusses evolution during the Phanerozoic Eon—the most recent 600 million years. It develops a perspective on dramatic planet-wide alterations of the Earth's surface by the biota. What has been thought of as the passive environment of life, the Earth's surface and lower atmosphere, can no longer be considered an inert physicochemical system; the misnamed "environment" has been part of life's sphere of influence, and aspects of it have been regulated by life for at least two billion years.

AUTOPOIESIS AND MANY GENOMES

The eukaryotic "cell" is a multiple of the prokaryotic "cell."

F. J. R. TAYLOR, 1974

Cells, minimal autopoietic systems, are membrane-bounded entities, the smallest capable of self-reproduction (Fleischaker, 1990). Even the tiniest cells today contain genes in the form of DNA molecules, single or in several copies, and protein-synthesizing machinery composed of several types of RNA and many proteins. All cells contain ribosomes, bodies about 0.02 μm in diameter composed of at least three kinds of RNA and about fifty different proteins.

Prokaryotic (non-nucleated) organisms are the smallest units that fit the definition of a cell. On the other hand, single eukaryotic (nucleated) cells are, in a way, not units at all. They are collections of different protein-synthesizing units—for example, nucleocytoplasm, mitochondria, and plastids—the kind and number of which vary with species. Thus, prokaryotic cells are single protein-synthesizing systems, whereas eukaryotic cells are multiple ones. Eukaryotic cells have nuclei containing most of the DNA, packaged as chromosomes. Generally they also contain many, even hundreds, of mitochondria. Plant and algal cells contain not only the same organelles that fungal and animal cells contain, but also one, several, or even thousands of plastids. Mitochondria and plastids contain enzymes for special functions: respiration in mitochondria and photosynthesis in plastids. Plastids also contain pigments, colored compounds that work with complex arrays of enzymes to capture solar energy and convert it into chemical energy. Both classes of organelles can be considered deficient cells: although they contain genes and protein-synthesizing machinery and are bounded by membranes, they lack the complete set of components needed for their own duplication.

The theory set forth in this book states that mitochondria developed efficient oxygen-respiring capabilities when they were still free-living bacteria and that plastids derived from independent photosynthetic bacteria. Hence, the functions now performed by cell organelles are thought to have evolved long before the eukaryotic cell itself existed. Photosynthesis evolved in anaerobic bacteria in the absence of molecular oxygen very early in the history of life. The type of photosynthesis that releases gaseous oxygen appeared later. Oxygen-respiring organisms evolved only after photosynthetically produced oxygen was available to them. Many different metabolic mechanisms evolved to cope with and eventually utilize oxygen, a highly reactive and potentially poisonous gas. The nucleocytoplasmic part of eukaryotic cells, the "host," evolved independently of the organelles. It was neither photosynthetic nor adept at utilizing oxygen. However, ancestors of the nucleocytoplasm could withstand high temperatures and acidic conditions. Furthermore, the whiplike cilia that are absent in prokaryotes and nearly omnipresent in eukaryotes are thought also to have derived from still another group of free-living bacteria. Eukaryotic cells are thus considered to have originated as communities of interacting entities that joined together in a definite order; with time, the members of the consortium, already specialized for motility, oxygen respiration, and photosynthesis—in that order—became organelles of a fermenting archaebacterial host. This particular story of the separate origins of parts of eukaryotic cells and their merging to form the whole has been dubbed by Taylor (1974) the **serial endosymbiosis theory** (SET). It is diagrammed in Figure 1-1.

The nucleocytoplasm became the largest and least specialized entity; however, hot acidic conditions had caused it to evolve a special class of

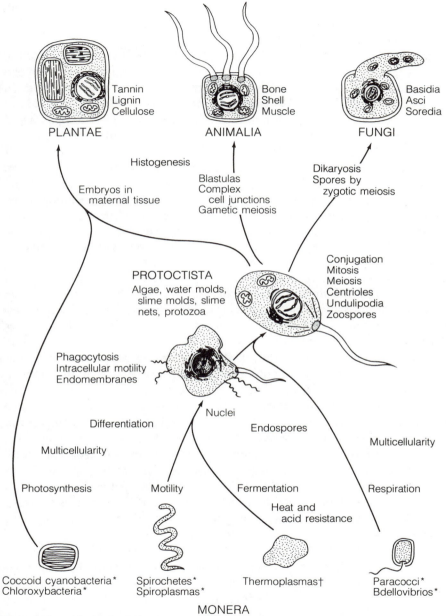

PLANTAE — Tannin / Lignin / Cellulose

ANIMALIA — Bone / Shell / Muscle

FUNGI — Basidia / Asci / Soredia

Histogenesis

Embryos in maternal tissue

Blastulas / Complex / cell junctions / Gametic meiosis

Dikaryosis / Spores by zygotic meiosis

PROTOCTISTA
Algae, water molds, slime molds, slime nets, protozoa

Conjugation / Mitosis / Meiosis / Centrioles / Undulipodia / Zoospores

Phagocytosis / Intracellular motility / Endomembranes

Nuclei

Differentiation

Endospores

Multicellularity

Multicellularity

Photosynthesis

Motility

Fermentation

Respiration

Heat and acid resistance

Coccoid cyanobacteria* / Chloroxybacteria*

Spirochetes* / Spiroplasmas*

Thermoplasmas†

Paracocci* / Bdellovibrios*

MONERA

*Eubacteria   †Archaebacteria

FIGURE 1-1

Model for the origin of eukaryotic cells by symbiosis.

proteins, namely histones, to protect its DNA (Searcy and Stein, 1980; Searcy and Delange, 1980). The mitochondria were virtuosos at breaking down three-carbon compounds to carbon dioxide and water in the tricarboxylic acid cycle. These three-carbon compounds, the waste products of fermentative metabolism of sugars by the nucleocytoplasm, became food for the mitochondria; thus, more energy was wrested from the original sugars by the consortium. First the symbiotic association of nucleocytoplasm and undulipodia, and then that of nucleocytoplasm-undulipodia and mitochondria, persisted and changed, becoming mutually dependent.

The near-identity of eukaryotic cilia and flagella and their unrelatedness to bacterial flagella has resulted in terminological confusion. The term **undulipodium** is therefore used for the eukaryotic flagellum and cilium, each of which develops invariably from a structure called the **kinetosome**; the term **flagella** is reserved for bacterial structures (Kuznicki, Jahn, and Fonseca, 1970; Corliss, 1979; Margulis, 1980). Hence, since only bacteria can be flagellated, the traditional term **mastigote** refers to the eukaryotic, undulipodiated microbes (formerly, **flagellated protozoa**).

The need for appropriate terminology is imperative. Confusion reigns even among professional biologists because the term **flagella** refers to two vastly different structures (i.e., bacterial rotary flagella and eukaryotic undulipodia), while cilia and eukaryotic "flagella" are identical structures composed of microtubules in a nine-doublet $[9(2) + 2]$ array. (The $[9(2) + 2]$ refers to the pattern of 24-nm-diameter tubulin-rich microtubules as seen in electron micrographs in transverse section of undulipodia. A $[9(3) + 0]$ pattern is typical of kinetosome-centrioles.) As recognized by Harold (1986) in his superb introduction to biological energetics and urged by my colleagues and me (Hinkle, 1991; Margulis, Corliss, Melkonian, and Chapman, 1990), the appropriate solution is to adopt unambiguous (yet traditional) terms: **kinetosome-undulipodium** for the eukaryotic motility organelle, and the traditional term **flagellum** for the rotating structure composed of flagellin proteins in bacteria. **Mastigote**, a term that has always referred to eukaryotes, retains its meaning, whereas the confusing word **flagellate** (with its sadomasochistic overtones) should be retired (along with **Protozoa**, an obsolete phylum in the animal kingdom).

The least accepted, most original, and most questionable aspect of the SET is the hypothesis that undulipodia originated from spirochete bacteria. Spirochetes originally attached to archaebacterial hosts (protoeukaryotes) and fed on small organic compounds that leaked across the membranes of their hosts. Natural selection transformed spirochete-host symbioses into highly motile complexes—ancestors of today's anaerobic (or at least mitochondria-less) mastigotes, for example, *Trichomonas* or *Giardia*. This transformation was followed by morphogenesis, the evocation of form, in the development of impressive cellular asymmetries. The consequences were profound; eventually, the remnants of spirochetes became internal cell structures. Today there is only the barest evidence that undulipodia are remnants of motile bacterial symbionts.

Individual bionts interact on behavioral, metabolic, gene product (RNA, protein), and genic levels in the formation of symbionts. They develop common new structures, their membranes fuse, and genes transfer among them as individual bionts become symbionts. The integrated symbionts (holobionts) become new organisms with a greater level of complexity. With the integration of mutually dependent nucleocytoplasm, undulipodia, and mitochondria, the polygenomic structure of the eukaryotic cell was complete. The first eukaryotic cell, with its multiple ancestry, was anaerobic and heterotrophic, feeding on preformed organic compounds. Mitosis, and subsequently meiosis, evolved in such early eukaryotes. I suspect that, more than once during the evolution of meiosis, aerobic bacteria were acquired symbiotically in protist lineages and integrated to become mitochondria.

The most recent event in the saga, the acquisition of photosynthesis by eukaryotes, occurred during and after the establishment of the fundamental eukaryotic organization. Photosynthetic eukaryotes are most easily described by the equation **heterotrophic eukaryote + photosynthetic prokaryote = alga** or **plant**. This acquisition of plastids also occurred at least several times and involved distinct phototropic partners.

Although the nucleus, as a membrane-bounded sphere universal in eukaryotes, is the defining characteristic of the nucleated cell, the origin of this body and its relation to symbiosis is unclear. My current thought favors the following scenario, for reasons mentioned here and discussed in Chapter 8: The nucleus probably evolved more than once (polyphyletically), both a product of direct filiation and symbiotic integration. Nuclear equivalents, as membrane-bounded spheres, evolved in prokaryotes—some, like *Gemmata*, remained prokaryotic, whereas others, archaebacteria like *Thermoplasma*, acquired other symbiotic genomes and became eukaryotes. Extant descendants of the earliest polygenomic microbes are today, of course, protoctists. The range of variation in nuclear structure and genomic organization even in extant protoctists is astoundingly large. *Pelomyxa*, a huge, multinucleate protist, harbors at least three types of endosymbiotic bacteria, including two different kinds of methanogens; the numerous small nuclei of *Pelomyxa* bear no necessary genealogical connection to the diploid micronucleus of ciliates or the endosomal nucleus of euglenoids.

The impetus to form a membrane-bounded structure that segregates DNA is apparent in bacteria; it was stimulated, probably more than once, by the acquisition of motile symbionts in the origin of the karyomastigont system of zoomastigina even prior to the symbiotic acquisition of mitochrondria in at least some of their members (such as *Trichomonas* or *Giardia*). Insinuating former parasites, each with its own original genome, led to membrane hypertrophy as a natural defense.

In today's world a milliliter of wood-eating-cockroach hindgut fluid can contain up to 30 sorts of protoctists (zoomastigina and chytrids) with even more different sorts of motile and stationary epi- and endosymbiotic bacteria. Complex microbial associations are rampant even in free-living organisms of the sapropel. I find it increasingly likely that many more different consortia toward eukaryosis existed in the past, and that of these, a few enjoyed spectacular success in their peculiar descendants. Not only are protoctist nuclei polyphyletic, by this reckoning, but so are the microtubule systems that form the mitotic spindle apparatus.

Among protoctists the cell structures are so varied, the diversity of bacterial symbionts so impressive, that a compelling case can be made for the origin of the nuclear and endomembrane systems in several lineages of beleaguered archaebacteria whose metabolic products leaked and whose phosphate-rich genomes were under siege by predaceous spirochetes and other hungry attackers. Such polygenomic, anaerobic mastigotes are still with us as various groups of zoomastigina (for example, pyrsonymphids and diplomonads). Some were threatened further by daptobacters, bdellovibrios, or other invasive respiring eubacteria that became mitochondria, whereas others, after developing intracellular motility systems as a result of uneasy truces with their former spirochete invaders, probably engulfed oxygen-respiring bacteria. Thus, to me, polyphyly of protoctist mitochondria also seems far more plausible than a single, paradigmatic event of eukaryosis. More than one origin of microtubule systems from attached spirochetes was detailed by Szathmary (1987), who emphasized the implications of his scheme for the polyphyly of mitochondria as well.

The symbiotic complexes that became plant and animal cells have never stopped evolving. Metabolic and developmental innovations created new levels of organization impossible to the partners individually. The symbioses became more integrated as the partners became more dependent on each other. The dependence between each organelle and the metabolic products of others is now so complete that the metabolic pathways of the original partners can be traced only with difficulty.

The evolution of biochemical pathways largely preceded the evolution of animals and plants. The eukaryotes show remarkable variation in structure, but they retain their metabolic uniformity; most, if not all, metabolic talents of eukaryotes were perfected in the different bacteria before they came together to become nucleocytoplasm and organelles. Prokaryotes, although structurally simpler than eukaryotes, are now and always have been far more diverse metabolically.

Prokaryotic cells were the products of that irritatingly mysterious event, the origin of life. They diversified in their ways of obtaining energy and in biosynthetic and degradative virtuosity. They entered extreme environments of all kinds and formed many associations. Bacterial communities altered both lithosphere and atmosphere long before eukaryotic cells formed larged organisms such as animals with skeletons.

ORIGINS

It must be admitted that no one yet knows how life began.

S. L. MILLER and L. E. ORGEL, 1974

The common ancestors of life today arose more than three billion years ago from interacting organic compounds formed in comets, in meteorites, and on the Earth's surface. The appeal of this concept is based in part on its utility. Panspermia, the idea that Earth was inoculated with life from other bodies in the universe, is far more difficult, perhaps impossible, to test. By contrast, the formation of organic compounds from inorganic precursors on the young Earth may be simulated in the laboratory.

"Chemical evolution" was suggested by the Russian biochemist Oparin in the 1920s but not tested in the laboratory until 1953, when Miller and Urey showed that amino acids and other organic molecules are formed under atmospheric conditions thought representative of those on the early Earth (Lazcano, 1992). This line of inquiry founded **probiotic chemistry**, the experimental study of the origin of life. Since then, many organic compounds have been synthesized from inorganics like water, methane, ammonia, and hydrogen cyanide. The amino acids found in proteins, and even nucleotides, i.e., hydrolytic products of nucleic acids, can, in the absence of enzymes and cells, be produced in the laboratory from inorganic precursors. Different organic molecules, including many found in all of today's living systems, probably formed spontaneously during the Earth's first billion years. Organic compounds have also been found in interstellar space and in meteorites; inevitably, organic compounds form by the interaction of lightweight elements with appropriate energy sources. Likely sources of energy for the prebiotic production of organic compounds on Earth were ultraviolet and visible light from the sun.

Organic compounds are not life. But many different interacting organic compounds were produced and concentrated in a way that led to minimal self-replicating entities. Prebiotic organic compounds became organized into macromolecular systems with properties recognizable as "living." How this occurred is not known and may never be known. The universality of the biochemistry of reproduction implies that all life on Earth derives from common ancestors: all life today has descended from cells that contained information-transmitting systems based on DNA replication and messenger RNA–directed protein synthesis. We have only the barest glimpses of the steps leading to the first cells; however, the path of evolution from the earliest bacteria to the polygenomic cells of protoctists, animals, fungi, and plants is becoming more apparent.

Toward phototrophy

La vie sans air.

L. PASTEUR, 1866

The best studied of the oldest sedimentary rocks on Earth that have not suffered metamorphosis come from the Swaziland System near Barberton, South Africa. These rocks are cherts—siliceous rocks, some of which preserve biological structure. They contain a large quantity of organic carbon, which was probably produced by microbial photosynthesis (Reimer, Barghoorn, and Margulis, 1979). In cherts from the Swaziland System, objects interpreted as fossils of microorganisms have been described, some even undergoing cell division (Knoll and Barghoorn, 1977). The direct evidence for life in the Archean Eon has been reviewed extensively by Schopf (1983, 1992). Stromatolites, sedimentary structures recognized as the remains of lithified microbial communities, have been observed in some of the most ancient sediments. They are found on all major continents and are a major part of the early fossil record. Stromatolites about 3000 million years old have been found in the Pongola Group of rocks in Africa; the Warrawoona Group in the Pilbara region of Western Australia contains even older stromatolites (3500 million years old; Walter, 1976) and microfossils (Awramik, Schopf, and Walter, 1983). These bona fide stromatolites and fossil microorganisms tell us that bacteria must have originated well before 3000 million years ago.

From the beginning, nucleic acids may have been isolated from their environment by lipid–protein membranes. At some point in the transition to life — the stage known as the **RNA world** — RNA replication and information systems must have prevailed. Later, in cells, genetic information became coded in the nucleotide sequences of DNA molecules that determined the amino acid sequences of polymerases. These proteins can catalyze the polymerization of nucleotides to make more DNA. Even in the earliest ancestral cells, information transfer must have been based on a form of the triplet code because the order of the three-nucleotide units in DNA determines the order of amino acids in the proteins of all modern organisms. DNA-directed protein synthesis is so conservative that it is assumed to have been present in the earliest ancestors of life today.

What was the energy source for the earliest cells? Organic compounds formed in laboratory simulations are rich in energy. Furthermore, these simulations have produced, under many conditions, the compounds needed for cell reproduction, such as adenosine triphosphate (ATP) and the protein amino acids. The earliest cell, a network of biochemical processes bounded by a membrane of its own making, probably used such compounds both as direct sources of energy and as structural components. Thus, early life was most like heterotrophic bacteria in that it obtained food and energy from abiotically produced organic compounds.

Three phenomena are necessary for evolution: reproduction, heritable variation (mutation), and environmental selective pressures. Reproduction is the sine qua non of biological evolution; if the fidelity of reproduction is assured, mutation and natural selection are unavoidable. To explain the origin of self-maintaining metabolic systems capable of reproduction is to explain the origin of life itself. That life originated on Earth is assumed.

As soon as cell reproduction evolved, evolution was under way. The ordinary hazards of the environment — variations in temperature, in quantity and quality of sunlight, and in concentration of salts in water — must have been among the many selective pressures leading, in different environments, to distinct populations of organisms varying from a common ancestor. The environment became depleted of abiotically produced organic compounds. It is logical to suspect that, early in cell evolution, the fermentation of small molecules, universal metabolites such as acetate and lactate, provided energy and carbon for the synthesis of nucleic acids. Any organism that could convert some available compound to a compound required for cell reproduction could then survive in the absence of the formerly necessary compound. Organisms that developed pathways for the synthesis of required cell components reproduced. Many different biosynthetic pathways evolved, including the synthesis of amino acids and nucleic acid derivatives. From the present distribution of metabolic pathways in prokaryotes, the selective advantages of their end products, and the fact of biochemical conservatism, a plausible sequence of the evolution of metabolism in bacteria can be inferred. As the supply of abiotic organics became limiting, cells evolved other ways of obtaining carbon and energy.

Anaerobic methanogenic microbes were among the earliest autotrophs, organisms capable of generating their organic requirements and energy from inorganic sources. There are three reasons for believing this. First, methanogens grow by using atmospheric carbon dioxide and hydrogen. Both gases were probably abundant 3000 million years ago, owing to volcanic emissions (Walker, 1980). Second, methanogens differ markedly from most other bacteria in their ribosomal RNA, suggesting that they evolved independently, perhaps even before ribosomal function was optimized and stabilized (Woese and Fox, 1977). Third, because methanogens have no defenses against atmospheric oxygen, they probably have avoided it ever since it appeared in abundance. Methanogenesis did not feed the biosphere indefinitely—a more steady, abundant, and direct source of energy was put to use to produce food in the most consequential metabolic innovation in the history of life on our planet.

Photosynthesis evolved originally in oxygen-intolerant bacteria. The first photosynthetic bacteria used hydrogen, hydrogen sulfide, or small organic compounds as reducing agents; using sunlight absorbed by bacterial chlorophyll, they converted atmospheric carbon dioxide into organic compounds. Descendants of these bacteria still thrive in anaerobic muds, swamps, marshes, tidal flats, and other environments in which light is available and oxygen is depleted. The evolution of anaerobic photosynthesis altered the Earth's surface and atmosphere permanently. Carbon dioxide was removed from the atmosphere and converted to organic matter according to the general equation for photosynthesis:

$$2H_2X + CO_2 \xrightarrow[\text{(Light)}]{hv} CH_2O + 2X + H_2O$$

Hydrogen donor   Cell material   Waste   Water

The new sources of carbon and energy led eventually to spectacular increases in organismic diversity and in the number of idiosyncratic evolutionary strategies.

Photosynthetic bacteria that used hydrogen sulfide to reduce carbon dioxide deposited elemental sulfur. They gave rise to immensely important descendants: bacteria capable of oxygen-releasing (oxygenic) photosynthesis. Photosynthesis that releases oxygen has traditionally been called "green plant" or "algal" photosynthesis. No prokaryotic "algae" exist; thus, in recognition of their bacterial nature, cyanophytes, cyanophyceans, cyano-phyta, or blue-green bacteria are called here **cyanobacteria**, rather than **blue-green algae**, an obsolete term. However, there is no doubt that oxy-gen-releasing photosynthesis evolved neither in plants nor in algae. Proba-bly it originated by accumulations of mutations in photosynthetic sulfur bacteria; these mutations culminated in the evolution of the cyanobacteria (Chapter 6).

A group of oxygen-releasing photosynthetic prokaryotes, discovered by Lewin of the Scripps Institution of Oceanography, are grass-green in color. Their pigment, like that of green algae and land plants, consists of chloro-phylls *a* and *b* only; they lack the phycobiliproteins that give the cyanobac-teria their bluish color. In ultrastructure these chloroxybacteria are nearly indistinguishable from coccoid cyanobacteria. The genera include the coc-coid *Prochloron*, filamentous *Prochlorothrix*, and extremely abundant, un-named, tiny coccoids in the ocean (Chisholm et al., 1988). They evolved polyphyletically from cyanobacterial ancestors (Urbach, Robertson, and Chisholm, 1992). Such green and blue-green oxygen-releasing photosyn-thetic microbes must have been abundant prior to the origin of eukaryotes.

Cyanobacteria, including their grass-green relatives like *Prochloron*, are hypothesized to be ancestors of the plastids of algal and plant cells. Water is the hydrogen donor in carbon dioxide fixation by cyanobacteria. Oxygen is released as a waste product. For every four hydrogen atoms used to reduce carbon dioxide into organic molecules, a molecule of oxygen is formed. It may have taken less than thousands or more than millions of years after the evolution of oxygenic photosynthesis for surface rocks and atmospheric volcanic gases to be oxidized. Eventually — probably more than 2000 mil-lion years ago — net quantities of oxygen began to escape from coastal waters, lakes, and soil into the atmosphere. As oxygen accumulated from microbial photosynthesis, the oxygen of the Earth's atmosphere ap-proached its present concentration of 20 percent.

The increase in oxygen provoked a crisis — prior to this time the world had been populated by anaerobic bacteria. Now survival required that microbes either avoid oxygen, by crawling into the mud or finding special anaerobic environments, or develop a metabolism able to cope with the increasing amount of oxygen. The responses of members of the kingdom Monera to gaseous oxygen are remarkably diverse, especially compared with the nearly uniform aerobiosis of eukaryotes. Groups of microbes that had diverged from each other much earlier, such as actinobacteria, bacilli, spirochetes, and cyanobacteria, developed different but analogous ways of coping with the planet-wide oxygen increase. Oxygen tolerance probably evolved first in photosynthesizers producing the gas. Oxygen-detoxifying mechanisms were followed by respiratory metabolism and other ways of using oxygen directly. Some members of nearly every major group of anaerobic prokaryotes became oxygen-tolerant and even oxygen-utilizing. At least one of them was ancestral to the bacteria that became mitochondria.

The carbon compounds produced by photosynthetic microbes must have supported complex communities of microorganisms. An abundance of stromatolites, especially in sedimentary sequences between 2400 million and 500 million years old, is direct evidence of such past microcosms. By 2000 million years ago, metabolic diversity was probably widespread and genetic exchange rampant (Sonea, 1991). Nearly any organic compound produced by prokaryotes can be metabolized by other prokaryotes, either as individuals or as groups of interacting species. The DNA-based ability to produce specific enzymes can be transferred between prokaryotes by viruses and plasmids. These genetic exchanges permit the rapid spread of metabolic innovations through bacterial populations even in the absence of mutations. Even aromatic or fibrous organic materials, such as toluene, cellulose, chitin, or industrial polymers, made later by plants and animals, can be degraded by bacteria, usually in communities.

The biosynthetic ability of prokaryotes is equally astounding: from carbon dioxide, a few salts, and reduced inorganic compounds that serve as energy sources, microbes called **chemoautotrophs** can synthesize in unlit oxygenated water all of the complex macromolecular requirements for their growth and reproduction. Methane or ammonia can be oxidized as a sole source of energy by some and are produced as metabolic end products by others. Methane is oxidized by some bacteria even in the absence of gaseous oxygen: sulfate is the oxidizing agent (Panganiban et al., 1979). Hydrogen, nitrogen, oxygen, and carbon dioxide can all be fixed — converted into organic compounds. Hydrogen sulfide can be oxidized to sulfur; sulfur deposited by anaerobic photosynthesizing sulfur bacteria can be further oxidized aerobically to sulfate, which in turn can be reduced back to hydrogen sulfide and returned to the muds by sulfate reducers. These biochemical processes, first in the absence and then in the presence of oxygen, led to element-cycling systems in the environment like those of today, long before animals, fungi, plants, or even protoctists evolved.

ASSOCIATIONS AND EUKARYOTES

The case for the symbiotic origin [of chloroplasts] appears overwhelming.

P. H. RAVEN, 1970

Ris (1962) has demonstrated the presence of DNA in the chloroplasts of several algae. He concluded that this DNA represents "the genetic system of the plastid" and suggested that the plastid of *Chlamydomonas* . . . may be an endosymbiont of an original blue-green alga. This bad penny has been circulating for a long time. . . . Clearly there is no chemical, structural, or phylogenetic basis for this belief.

R. KLEIN and A. CRONQUIST, 1967

By the time oxygen had accumulated in the atmosphere, many types of bacteria were inhabiting the seas and lakes, covering the soil, and releasing spores into the air. Stromatolites and the microfossils in cherts, abundant and widely separated in time and space, are incontrovertible testimony to the golden age of prokaryotes long before the appearance of animals. Like organisms in every period, prokaryotes of different kinds interacted with each other. The entire biosphere depended on proteins and carbohydrates produced by purple, green, and blue-green photosynthetic and other autotrophic bacteria. The accumulation of atmospheric oxygen put an end to the abiotic production of organic compounds that are quickly destroyed by free oxygen. Furthermore, ozone now absorbed much of the ultraviolet radiation that had been instrumental earlier in the production of amino acids, nucleic acid derivatives, and other organic compounds.

The ions of many elements (zinc, copper, and arsenic, for example) are toxic and tend to accumulate in cells because the ions chemically resemble those used metabolically. Many microbes, perhaps all, expel such poisons as their gaseous methylated derivatives, such as trimethyl arsene. Bacteria were directly and intimately dependent on others for gas supply and ventilation or gaseous removal of waste products. Interdependence for nutrition and protection also existed. The biosphere became its patchy, recognizable self. Many different kinds of relations arose, including symbiosis, parasitism, and predation. One of these, a certain series of symbioses, led to the formation of new kinds of cells (Chapter 7).

According to what Taylor (1979) has called the "extreme version" of SET, all eukaryotes were derived from symbioses between extremely different kinds of prokaryotes: nucleocytoplasm evolved from fermenting thermoplasmas, undulipodia from spirochetes, mitochondria from aerobic respiring eubacteria, and plastids from cyanobacteria. Mitotic cell division evolved in the early protoctists only after internalization of the spirochetes and redeployment of their parts (Chapter 9).

Thus, the history of eukaryotes began with wall-less, pleiomorphic microbes that could ferment glucose to three-carbon end products by the

Embden-Meyerhof pathway. These microbes, which were to become the nucleocytoplasm, acquired spirochete endosymbionts. The hosts were presumably microaerophilic, that is, adapted to less than the present 20-percent concentration of oxygen in the air. The most similar living microbes are microaerophilic, wall-less archaebacteria such as *Thermoplasma*, considered good candidates for hosts because they have arginine- and lysine-rich, histone-like proteins that protect their DNA from attack by hot acid. The microbes that became the nucleocytoplasm may have lacked the circular DNA genomic organization common to most other bacteria.

The first step toward full eukaryotic status was the symbiotic acquisition of motility. Since many groups of known protists have typical [9(2) + 2] undulipodia but lack mitochondria, the acquisition of motile symbionts and their integration probably preceded the acquisition of mitochondria. Full integration of motile bacteria preadapted ancestral eukaryotes for phagocytosis and other intracellular motility systems. Some of the thermoplasma-like microbes that had formed associations with attaching, still-motile spirochetes greatly increased their swimming rates. Luckily for evolutionists, these particular attached spirochetes contained at least single or double microtubules in their protoplasmic cylinders. The microtubules, composed of neat arrays of protein, may even already have formed a [9(2) + 2] pattern. Alternatively, some such ninefold symmetry in transverse section evolved after the integration of the symbionts. Whenever these patterns arose, their universal presence in undulipodia argues strongly that these organelles are all evolutionary homologs, whether they are the tails of human sperm or moss sperm, the tracheal cilia of vertebrates or the body cilia of ciliates. Although motility by means of enigmatic gliding movement and the rotary motors of bacterial flagella had evolved far earlier, undulipodia are products of symbiosis. They evolved as intracellular organelles only after spirochetes (associated with the archaebacteria that became the nucleocytoplasm) underwent permanent modification.

The symbiotic spirochetes, or **protoundulipodia**, became more and more intimately associated with their hosts. As in all stable symbiotic associations, their rate of reproduction adjusted itself closely to that of the hosts; attachment sites became kinetosomes. Thus, new complexes appeared, containing two heterologous genomes; host and undulipodium. These digenomic complexes speciated into many different protoctist types. The original spirochete genome differentiated eventually to form many sorts of structures, including, for example, the locomotory and feeding cortical appendages of ciliates. Protoundulipodia dedifferentiated to become fibrogranular complexes (Dirksen, 1991); in many cases, they were drawn inside their hosts, so that their property of motility was lost. Yet they retained their basic replicative ability; nucleic acids and their products, formerly those of spirochetes, differentiated by mutation and gave rise with time to the replicating centrioles, the attachments to chromosomal kineto-chores, and the mitotic spindle. In short, symbiotic bacteria that became undulipodia (Chapter 9) were preadaptations for mitosis (Chapter 8).

Mitosis itself evolved by a long, indirect series of mutations in protoctists—it may have taken the better part of a billion years. The evolution of mitosis, certainly complete before 570 million years ago, led to an adaptive radiation that formed myriad species of eukaryotic microorganisms, the Protoctista and their descendants. Eventually, heterotrophic eukaryotic microorganisms spawned animals and fungi. Soft-bodied protoctists and some animals were widespread and diverse before the appearance of forms with skeletons 500 million years ago (Glaessner, 1968). The earliest metazoans were soft-bodied; hard parts developed apparently more or less concurrently in many remotely related animals and protoctists. Did worldwide late Proterozoic environmental change lead to more active precipitation of carbonate and silicate (McMenamin and McMenamin, 1990)? Or was predation, which required protection from predators, an evolutionary motive force (Hutchinson, 1959; Stanley, 1976a, b)? Regardless of the reasons for extensive deposition of skeletal materials in the metazoan fossil record, mitotic and eventually meiotic sexual systems were prerequisites (Chapter 12).

In the next step in eukaryotic cell evolution, undulipodiated protoctists acquired mitochondria in the form of respiring bacteria. From where did the respiring endosymbionts come? These **protomitochondria** were oxygen-respiring, rod-shaped organisms related to certain living bacteria, purple bacteria such as *Paracoccus denitrificans*. They oxidized fermentation products completely to carbon dioxide and water. The associations may have begun as bacterial predation on undulipodiated hosts. No known prokaryote engulfs other living cells (that is, they all lack phagocytosis). Perhaps some mastigotes, with developed undulipodia, were capable of phagocytosis. But little is known about intracellular motility in obligate anaerobic mitochondria-less mastigotes; they may be incapable of phagotrophic nutrition. Thus it is possible that promitochondria invaded their hosts just as the modern predatory *Bdellovibrio* bacteria invade their larger prey bacteria (Chapter 10).

Reproduction of symbionts inside the osmotically controlled nucleocytoplasm led to selection against their cell walls and against biosynthetic pathways that duplicated those in the host. The endoplasmic reticulum and its derivatives, such as the Golgi apparatus and the nuclear membrane, differentiated from internal cytoplasmic membranes to accommodate metabolic capabilities coded for by host and symbiont genomes. DNA segregation in the host had been either on the plasma membrane or on internal membranes, as apparently it still is in prokaryotes. Thus, the nuclear membrane may have evolved first to distribute newly synthesized DNA molecules deep inside the cell before the evolution of chromosomes and mitosis. Protomitochondria increased the surface area of their membranes for oxidation-reduction reactions. Natural selection led to the differentiation of cristae — the folded membranes of modern mitochondria. These accommodations to the symbiosis culminated in trigenomic cells: aerobic, mitochondria-containing, nucleated amoebomastigotes and other amoeba-like protists. Here, for simplicity, they will be called **amoeboids**.

Another series of symbiotic acquisitions resulted in photosynthetic eukaryotic organisms. As natural selection optimized the processes of mitosis and meiosis, various kinds of photosynthetic prokaryotes (**protoplastids**) became associated with more than one kind of heterotrophic protist. The microbial photosynthesizers were ingested but not digested. Green photosynthetic prokaryotes (chloroxybacteria, or prochlorons) were probably independently ancestral to the plastids of euglenoids, chlorophytes, and others. Coccoid cyanobacteria like the genera *Gloeocapsa* and *Synechocystis* of today were ancestral to the red plastids of algae of the phylum Rhodophyta and to a few obscure protists, such as cryptomonads, *Glaucocystis*, and *Cyanophora*. Close associations between photosynthetic endosymbionts and their several hosts became progressively more obligate, leading to the evolution of different algal phyla. Although several protist groups, such as the peranemids, evolved secondarily by loss of plastids from photosynthetic forms like *Euglena*, only these can be considered "apochlorotic" algae. According to the SET as presented here, the vast majority of eukaryotes evolved directly from heterotrophic microbes that never acquired photosynthesis (Chapter 11).

All of the great innovations in the evolution of cells occurred before any animals, plants, or fungi appeared. The major biochemical pathways had been established, and patterns of mitosis, meiosis, and fertilization had developed in some, but not all, protists. Some heterotrophic eukaryotes had ingested photosynthetic prokaryotes and become algae. That even this final event occurred more than 750 million years ago is deduced from the existence of *Bangia*-like red algal fossils from the Proterozoic of Arctic Canada (Butterfield, Knoll, and Swett, 1990) and Greenland (Enzien, 1990). Fossil fungi in plant root tissue are well preserved in the Rhynic chert of Scotland. By about 400 million years ago, then, the major animal, plant, and fungi phyla had been established (Chapter 12 and Appendix).

CHAPTER 2

DIVERSITY

CLASSIFICATION AND EVOLUTION

1st chapter, which telleth of the four footed animals;
2nd chapter which telleth of all the different kinds of
birds, of whatever sort; 3rd chapter which telleth of all
the animals which dwell in the water; 4th chapter
which telleth of still other animals which live in the
water, which are inedible; 5th chapter which telleth of
the various serpents, and of still other creatures which
live on the ground; 6th chapter which telleth of the
various trees, and of the various properties which
correspond to them, such as their strength; 7th
chapter, which telleth of all the different herbs . . .

    FRAY BERNARDINO DE SAHAGÚN, *Florentine Codex*,
General History of the Things of New Spain, Sixteenth Century

The diversity of the living world is bewildering; it is estimated that over 30 million species are alive today. Organization of the knowledge of living things—some classification scheme—is essential. Early schemes for classifying larger organisms tended to be arbitrary and immediately practical. Detailed properties of more than a thousand plants, animals, and minerals were known to the Nahuatl-speaking and other civilized North American peoples before the arrival of the first Europeans, and many thousands of kinds of organisms have been known to societies of the Far East and Eurasia. Aristotle recognized about 540 animals, which he classified into groups such as red-blooded, bloodless, viviparous, hair-coated, and plant-like. He also recognized an inanimate world of plants with "more vitality" (Coonen, 1977). Unfortunately, his botanical works are lost, and to today's reader his classification system may seem childishly inadequate.

In European civilization before the eighteenth century, biological knowledge generally was organized much like the Aztec classification quoted above from the Florentine Codex—living things were grouped according to commonly perceived properties. Especially important were characteristics relevant to human needs: the provision of food, fiber, and poison. In nearly all accounts, small, stationary organisms were herbs; motile beings, regardless of size, were animals. Even in the twentieth century, when Darwinian evolutionary concepts led to the realization that organisms are related by ancestry, and after the development of microscopy, a single dichotomy, animal/plant, was considered logical and sufficient.

Modern biological classification began with Carl von Linné (1707–1778), known more commonly as Linnaeus. In his great work, *Systema Naturae*, first published in 1735, this Swedish naturalist grouped all organisms known to him—some 36,000—into distinct categories called species, and similar species into categories called genera. He gave each species a generic and a specific name, for example *Homo sapiens* (human) and *Pseudoplatanus acerifolium* (sycamore maple). On the basis of observable characteristics, he grouped genera into families, families into orders, and orders into larger, even more inclusive groups called classes.

In 1789, in *Genera Plantarum*, Antoine Laurent de Jussieu (1748–1836) of Paris extended Linnaeus's classification to more plants and stressed the significance of the internal organization of organisms. Georges Cuvier (1769–1832), the founder of comparative anatomy, applied the new Jussieu approach to animals, some known only from the fossil record. He introduced a more inclusive grouping based on the overall functional organization of animals, the **embranchement**, into which he placed related classes. Cuvier recognized four embranchements of the animal kingdom: vertebrata, mollusca, articulata (joint-footed forms, including arthropods and segmented worms), and zoophyta (radially symmetrical forms, echinoderms, and everything excluded from the other groups) (Coleman, 1971). To Cuvier, although nearly every fossil animal differed from its modern counterparts, the correspondence between living and extinct was usually so obvious that he could assign fossils unambiguously to one of his four embranchements. He played a pivotal role in relating biological to geological knowledge, thus helping to establish the field of paleontology.

The Swiss-French botanist Augustin-Pyramus de Candolle (1778–1841) invented the word **taxonomy**—the science of identifying, naming, and classifying—and expanded Cuvier's analysis by creating larger inclusive groupings of the plant kingdom, divisions. These botanical divisions, some of which are still in use (at least informally), correspond to the phyla of zoologists. Candolle distinguished thallophytes, cryptogamic plants (nowadays called algae), phanerogamic plants, nonflowering vascular plants, and flowering vascular plants.

These early workers believed in the fixity of species created by God and in the division of the living world into two great *regna* (reigns): the plant and animal kingdoms. In the second half of the nineteenth century, several workers proposed third kingdoms. One of the first was J. Hogg (1861), who introduced *Regnum Primogeneum*, Kingdom Protoctista, for organisms clearly neither animal nor plant (he included the sponges).

Darwin's powerful generalization, that all organisms are related by ancestry, greatly impacted subsequent classification schemes. The great German philosopher-biologist Ernst Haeckel (1834–1919) was the first major evolutionist to attempt a classification based on evolutionary history. He, too, attempted to replace the plant/animal two-kingdom system with a three-kingdom one (Haeckel, 1878). His third kingdom was primarily for unicellular organisms (protists), in which he sometimes included fungi and sponges (Haeckel, 1866, 1878). Haeckel recognized among the protists a major group lacking nuclei: the monera. In this group he included bacteria, blue-green algae, and some amoebae (which were later found to be nucleated).

Gregor Mendel (1822–1884) discovered rules by which certain discernible traits were transmitted from parents to offspring. Underlying these Mendelian rules for the reassortment of heritable traits was the behavior of chromosomes. In the 1920s, the "chromosomal theory of heredity" became accepted widely as fact. Soon after, in the 1930s and 1940s, the relationship between genetics, evolution, and taxonomy was formalized with the founding of the science of systematics. This was a "new synthesis" of the relationships among organisms, their morphology, physiology, genetic systems, fossil histories, and ecological interactions. Most biologists today are neo-Darwinists. They acknowledge Darwin's basic insight into the mechanism of evolution, and they have the advantage of knowing how genetic changes—point mutation, duplication, and chromosomal recombination —generate heritable variations in populations of organisms. Developing a mathematical language to describe the behavior of genes in populations, they founded the science of population genetics.

Systematists estimate now that over 30 million species of living organisms can be grouped into some one hundred phyla. The number of individuals per species varies from a few hundred, for rare large vertebrates, to enormous numbers—for example, 10^{10} per milliliter—for small microbes in nutrient-rich waters. An additional 130,000 fossil species have been identified on the basis of distinguishable morphology; most can be assigned to modern phyla. These fossilized organisms are thought to represent only a minute sample of those that have existed.

How are species defined? Although the members of a species may be different, each has traits in common with others in the group. Eukaryotes that look alike, especially if they mate and form viable offspring, are considered members of the same species. Species are then grouped into more inclusive taxa, which are grouped into larger taxa in a hierarchical classification system of units that increases in inclusiveness and decreases in number from each level to the next. From more to less inclusive, the taxa generally recognized by botanists and zoologists are **kingdom, phylum** (still called **division** by botanists), **class, order, family, genus** (plural, **genera**), and **species**. Species comprise populations of organisms that are diverging from each other; they may be divided into interfertile groups called subspecies, varieties, or races. The concepts of the cell-symbiosis theory have little effect on taxonomic level below that of class or order. However, the higher taxa, kingdoms and phyla, are profoundly affected.

Over 20,000 types of bacteria have been recognized, as distinguished by their nutrition, metabolic products, spore formation, habitat, morphology, and other traits (Balows, Trüper, Dworkin, et al., 1992). All are capable of reproduction in the absence of a sexual partner. Many types, perhaps even all, are capable of shedding genes into the environment and receiving them from outside sources in the form of viruses, plasmids, conjugation factors and other DNAs. Bacterial genetic exchange has led Sonea (1991) to question the validity of the species concept in bacteria, while detailed studies of bacterial chemistry have led Woese and his colleagues (1990) to recognize two great groups of prokaryotes on the basis of ribosomal RNA sequences (16–18S) and lipid and cell wall composition. These two groups, called **kingdoms** or **domains** by the authors, are Archaebacteria or Archaea (including methanogenic, halophilic and thermoacidophilic bacteria) and Eubacteria (all other bacteria). Primarily on the basis of nucleocytoplasmic sequence data, Woese and colleagues place all nucleated organisms together in the Eukarya domain. Because all Eukarya have more than a single type of 16S ribosomal RNA as a consequence of their polygenomic ancestry, the three-domain system proposed by these molecular biologists privileges one (the nucleocytoplasmic) over the others (e.g., plastid or mitochondrial 16S ribosomal RNA). Thus, the scheme which has been so useful for systematizing the bacteria—leading to the recognition of two subkingdoms (Archaebacteria and Eubacteria)—is invalid as a taxonomic tool to organize eukaryotes, all of which have more than a single genetic system because of their derivation from symbionts.

Since the founding of systematics as a branch of evolutionary science, many have continued the attempt to classify the diversity of life. Even though each biologist is familiar with only certain groups, many strive to devise systematic classifications that reflect evolutionary history. Classifications not explicitly based on evolutionary relationships are called "artificial"—for example, organisms are often grouped according to usefulness (economic plants), size (trees, shrubs), color (flowers), and so forth. Many find such nonsystematic classifications useful.

Some classification systems published since 1956 and currently in use are shown in Table 2-1; Table 2-2 compares the classification of five organisms in the traditional two-kingdom system, the newer five-kingdom one, and the molecular biologists' new three "domains."

Evolutionists agree that there is a direct relationship between time and taxa. As Simpson and Beck (1965) have written,

> A systematic unit of organisms in nature is a population or a group of related populations. . . . Its anatomical and physiological characteristics are simply the total of those characteristics in the individuals making up the population. The pattern of characteristics is neither a real individual nor an idealized abstraction of the character of an individual. A "systematic unit" is a frequency distribution of the different variants of each character actually present at any given time. Species are populations of individuals of common descent living together in similar environments in a particular region, with similar ecological relationships and tending to have a unified and continuing evolutionary role distinct from that of other species. . . . Systematic units more inclusive than a species are groups of one or more species of common descent.

The phrase "populations of individuals of common descent" refers to a time relationship. The lowest taxon into which two organisms both belong should somehow reflect how long it has been since the two populations of organisms diverged from their most recent common ancestor. Usually, two organisms that belong to the same genus (for example, the beans *Phaseolus vulgaris* and *Phaseolus lunatus*, or the frogs *Rana pipiens* and *Rana catesbiana*) diverged more recently from their common ancestor than did two organisms that belong only to the same family (for example, *Phaseolus vulgaris* and *Vicia americana*, purple vetch, or *Rana pipiens* and the toad *Bufo terrestris*). This is not an absolute rule, however, because rates of evolution and criteria for classification differ in different groups.

A remarkable discovery of molecular biology in the last three decades is the commonality of the fundamental chemical processes underlying the most essential biological phenomenon: the ability to maintain and reproduce a membrane-bounded system and, thus, to evolve. The genetic code, the rules relating the sequence of nucleotides in RNA to the sequence of amino acids in protein, is essentially the same in all organisms studied — from monera to monkey. That messenger RNA is complementary to DNA and binds to the ribosomes in protein synthesis is universal. Members of all kingdoms use adenosine triphosphate (ATP) in energy-yielding reactions. The details of molecular biology provide direct evidence that all living organisms are ultimately related.

TABLE 2 - 1

Some recent classification systems.

2 KINGDOMS	3 KINGDOMS			4 KINGDOMS
TRADITIONAL (ALTMAN AND DITTMER, 1972)	CURTIS, 1968	STANIER ET AL., 1970	WOESE ET AL., 1990	COPELAND, 1956
Plantae	**Protista**	**Protista**	**Bacteria**[a]	**Monera**
Bacteria	Bacteria	Bacteria	Thermotogales	Bacteria
Blue-green algae	Blue-green algae	Blue-green algae	Flavobacteria	Blue-green algae
Green algae	Protozoa	Protozoa	Cyanobacteria	**Protoctista**
Chrysophytes	Slime molds	Green algae	Purple bacteria	Protozoa
Brown algae	**Plantae**	Chrysophytes	Gram-positive	Green algae
Red algae	Green algae	Brown algae	bacteria	Chrysophytes
Slime molds	Chrysophytes	Red algae	Green nonsulfur	Brown algae
True fungi	Brown algae	Slime molds	bacteria	Red algae
Bryophytes	Red algae	True fungi	**Archaea**[a]	Slime molds
Tracheophytes	True fungi	**Plantae**	Crenarchaeota	True fungi
Animalia	Bryophytes	Bryophytes	(thermoacidophils,	**Plantae**
Protozoa	Tracheophytes	Tracheophytes	some	Bryophytes
Multicellular	**Animalia**	**Animalia**	methanogens)	Tracheophytes
animals	Multicellular	Multicellular	Euryarchaeota	**Animalia**
	animals	animals	(some	Multicellular
			methanogens,	animals
			extreme halophils)	
			Eukarya	
			Animals	
			Ciliates	
			Green plants	
			Fungi	
			Flagellates	
			Microsporidia	

One goal of evolutionary studies is the construction of phylogenies that represent history accurately. Often diagrammed as family trees, with twigs and branches converging into a single trunk, phylogenies are summaries of evolutionary information. They relate taxa to each other graphically, in time or according to degree of similarity in some trait. An organism placed at a fork in the evolutionary line is the most recent common ancestor of those on the branches growing from the fork. Phylogenies may be eclectic, using information from many sources. For example, the number of mutations that have occurred since two organisms diverged from a common ancestor may be estimated from the number and kind of differences between the amino acid sequences of proteins that the organisms have in common, such as cytochrome c or hemoglobin. Nucleotide sequences of ribosomal RNAs (Figure 2-1) from the same two organisms can be analyzed in the same way. Computer programs can now compute composite phylo-

5 KINGDOMS		8 KINGDOMS	13 KINGDOMS
WHITTAKER, 1969	WHITTAKER AND MARGULIS, 1978; THIS BOOK[b]	EDWARDS 1976	LEEDALE, 1974
Monera	**Monera**	**Cyanochlorobionta**	**Monera**
Bacteria	Bacteria	Blue-green algae	Bacteria
Blue-green algae	Cyanobacteria	**Erythrobionta**	Blue-green algae
and gliding	Actinobacteria	**Chlorobionta**	**Red algae**
bacteria	**Protoctista**[c]	Green algae	**Plantae**
Protista	Dinomastigotes	Bryophytes	Green algae
Protozoa	Ciliates	Tracheophytes	Stoneworts
Chrysophytes	Amoebomastigotes	Euglenids	Bryophytes
Euglenids	Slime molds	**Myxobionta**	Tracheophytes
Hyphochytrids	Slime nets	Slime molds	**Euglenoids**
Plasmodiophorans	Chytrids	**Fungi 1**	**Myxomycetes**
Plantae	**Fungi**	Nonflagellated fungi	**Fungi**
Green algae	Zygomycetes	Chytrids	True fungi
Brown algae	Ascomycetes	**Fungi 2**	(nonundulipodiated)
Red algae	Basidiomycetes	Slime nets	**Heterokonts**
Bryophytes	Lichens	Hyphochytrids	Oomycetes
Tracheophytes	**Plantae**	Oomycetes	Brown algae
Fungi	Bryophytes	**Chromobionta**	Diatoms
Slime molds	Tracheophytes	Brown algae	Yellow-green algae
Oomycetes	**Animalia**	Chrysophytes	**Eustigmatophyta**
Chytrids	Poriferans	**Animalia**	**Haptophyta**
True fungi	Metazoans	Metazoans	**Cryptomonads**
Animalia			**Dinoflagellates**
Multicellular			**Mesozoans**
animals			**Animalia**

[a]Both groups are prokaryotes. Domains are equivalent to kingdoms in this system.

[b]The Whittaker and Margulis classification (1978) has two alternatives—a Kingdom Protista or a Kingdom Protoctista. In the first alternative, undulipodiated "fungi" would be included in Kingdom Fungi and the seaweeds in Kingdom Plantae; in the second, used in this book, these organisms are included in Kingdom Protoctista.

[c]See *Handbook of Protoctista* (Margulis, Corliss, Melkonian, and Chapman, 1990) for the many other protoctist phyla and Margulis and Schwartz (1988) for other moneran, plant, and animal phyla.

genies (see Figure 3-9) consistent with all the protein and RNA sequences known for the organisms in question (Woese and Fox, 1977). Certain judgments and assumptions must be made in designing phylogenies. Partial phylogenies derived from macromolecular data should be compared with those based primarily on skeletal structures, external morphology of living and fossil organisms, or chromosomal cytology.

Classifications based on phylogenies integrate immense quantities of information, although they do require revision as new evidence appears.

TABLE 2-2

Five kingdoms and two kingdoms: The classification of five organisms.

TAXON	ORGANISM				
	OSCILLATORIA	STENTOR	FIELD MUSHROOM	HUMAN	MAIZE
FIVE KINGDOMS					
Kingdom	Monera	Protoctista	Fungi	Animalia	Plantae
Grade	—	Mitotica	—	Coelomata	Tracheophyta
Phylum	Cyanobacteria	Ciliophora	Basidiomycota	Chordata	Angiospermophyta
Class	Hormogoneae	Polyhymenophora	Homobasidiomycetae	Mammalia	Monocotyledonae
Order	Nostocales	Heterotrichales	Agaricales	Primates	Graminales
Family	Nostocaceae	Stentoridae	Agaricaceae	Hominidae	Gramineae
Genus	*Oscillatoria*	*Stentor*	*Agaricus*	*Homo*	*Zea*
Species	*marinus*	*coeruleus*	*campestris*	*sapiens*	*mays*
Common designation	Blue-green bacterium (cyanobacterium)	Ciliate	Edible mushroom	Person	Corn
TWO KINGDOMS (traditional)[a]					
Kingdom	Plantae	Animalia	Plantae	Animalia	Plantae
Phylum	Schizophyta	Protozoa	Eumycetes	Chordata	Spermatophyta
THREE KINGDOMS (Woese et al., 1990)[a]					
Domain[b]	Bacteria	Eukarya	Eukarya	Eukarya	Eukarya
Phylum	Cyanobacteria	Ciliates	Fungi	Animals	Plants

[a]Lower taxa are the same as in the five-kingdom system.
[b]None of the five organisms are Archaea. Domain is equivalent to kingdom.

Acceptable phylogenies have been drawn up by both botanists and zoologists. Paleontology, comparative anatomy, physiology, and biochemistry have converged to a consensus on the broad outlines of the last 600 million years of the evolutionary history of animals and plants. However, the systematics of the so-called "lower" organisms (bacteria, protoctista and fungi) has been in turmoil. The confusion stems in part from the professional organization of microbiologists and botanists: their traditions poorly reflect the relationships between the organisms they study (Table 2-3). Furthermore, identifying and classifying practices have differed greatly in botanical, zoological, and bacteriological literature (Table 2-4).

However, tradition cannot be blamed for everything. Only since the early 1960s has it been known that the fossil history of microbial life extends back for at least two billion years. Many microbes have no obvious sexual stages, so the definition of species depends solely on the frequency distribution of morphological and physiological traits. Many, indeed most, microbes have not been grown and studied in pure culture; hence, crucial information is often lacking. The attempt to apply criteria of morphology

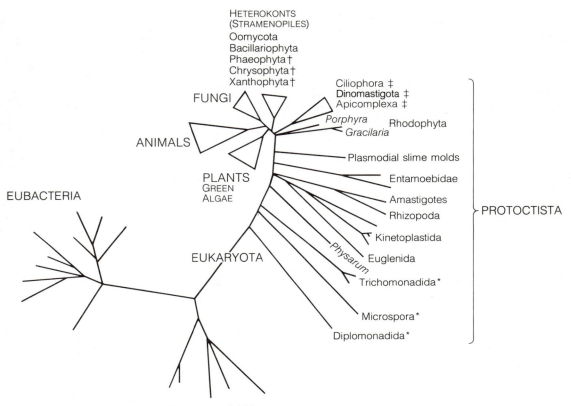

HETEROKONTS
(STRAMENOPILES)
Oomycota
Bacillariophyta
Phaeophyta†
Chrysophyta†
Xanthophyta†

Ciliophora ‡
Dinomastigota ‡
Apicomplexa ‡

FUNGI

Porphyra   Rhodophyta
Gracilaria

ANIMALS

Plasmodial slime molds

PLANTS
GREEN
ALGAE

Entamoebidae

EUBACTERIA

Amastigotes

Rhizopoda

Kinetoplastida

EUKARYOTA

Physarum   Euglenida

Trichomonadida*

PROTOCTISTA

Microspora*

Diplomonadida*

ARCHAEBACTERIA

†Chromophyta   ‡ Alveolates

FIGURE 2-1

A three-domain/five-kingdom unrooted tree inferred from 16S-like ribosomal RNA sequence similarities by using distance matrix methods. The depth of branching for the eukaryotes, indicated by relative line-segment lengths, is even greater than that observed in the members of the eubacteria or archaebacteria. The earliest diverging eukaryotic lineages are protists that lack mitochondria (asterisks). The five major eukaryotic assemblages (thick branches) separated nearly simultaneously and relatively late in the evolutionary history of eukaryotic cells. [Courtesy of G. Hinkle and M. Sogin; based on an original drawing by Kathryn Delisle.]

and nutritional mode, a tactic successful in the systematization of animals and plants, has not been particularly useful for devising microbial phylogenies. With the exception of dinomastigotes, radiolarians, foraminiferans, diatoms, cyanobacteria, and a few other groups of shelled or sediment-trapping microbes, well-preserved microfossils have been so rare that the con-

TABLE 2-3
Organisms and traditional fields of study.

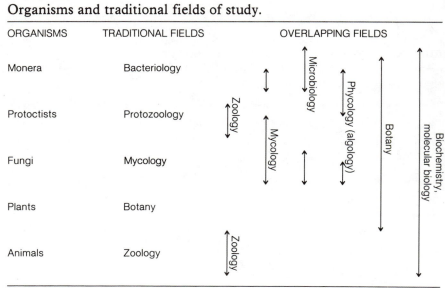

TABLE 2 - 4
Contradictory classification of a photosynthetic protist.

TAXONOMICAL LEVEL	BY BOTANISTS		BY ZOOLOGISTS	
	TAXON	CRITERION	TAXON	CRITERION
Kingdom	Plantae	Photosynthesis	Animalia	Motility
Phylum	Chrysophyta	Golden yellow pigmentation	Protozoa	Single-celled animal
Class	Chrysophyceae	Golden yellow pigmentation	Phytomastigophorea	Bears plastids
Order	Chrysamoebida	Amoeboflagellate motile stages[a]	Chrysomonadida	Golden-yellow plastids
Genus and species	*Ochromonas danica*	Morphology	*Ochromonas danica*	Morphology

[a]Undulipodated.

struction of phylogenies and the taxonomy and systematics of most bacterial and protoctist groups have been avoided or based primarily on comparisons among living forms. In summary, insights into the metabolism, ultrastructural morphology, life cycles, nutrition, genetics, and fossil record of the microbial world have been gained only recently. For this reason, phylogenies for the bacteria (Woese and Fox, 1977) and protoctists (Sogin, 1991) have not been constructed until recently, and the listings in the Appendix are provisional. In spite of these difficulties, one agreement has certainly been reached: the prokaryote/eukaryote difference in cell organization is far more profound than the traditional animal/plant distinction.

PROKARYOTES AND EUKARYOTES

The numerous and fundamental differences between eukaryotic and prokaryotic organisms . . . have been fully recognized only in the past few years. In fact, this basic divergence in cellular structure which separates the bacteria and the blue green algae from all other cellular organisms, probably represents the greatest single evolutionary discontinuity to be found in the present-day living world.

R. Y. STANIER, M. DOUDOROFF, and E. A. ADELBERG, 1970

Prokaryotic organisms include bacteria—those traditionally regarded as bacteria, the blue-green algae (cyanobacteria), and the grass-green *Prochloron* organisms (chloroxybacteria)—and some multicellular organisms, such as actinobacteria (actinomycetes) and gliding myxobacteria. All are microbes. The name prokaryote comes from the Greek words pro and karyon, meaning "before" and "seed" or "nucleus"; the word eukaryote from the Greek eu, "good" or "true." Most prokaryotic cells are smaller than eukaryotic cells (Figures 2-2 and 2-3) (Clements and Bullivant, 1991). The prokaryotic gene-bearing structure, often misnamed "bacterial chromosome," should be called the **genophore** (Ris, 1975). It is made up of a circular thread of DNA, which is not contained in a membrane-bounded nucleus; under the electron microscope, the genophore (or linkage group) appears as a relatively transparent region called the **nucleoid**. In a eukaryotic cell, the gene-bearing structures are chromosomes contained in a membrane-bounded nucleus. In exceptionally fine translucent preparations, live chromosomes can be seen with a light microscope; more often,

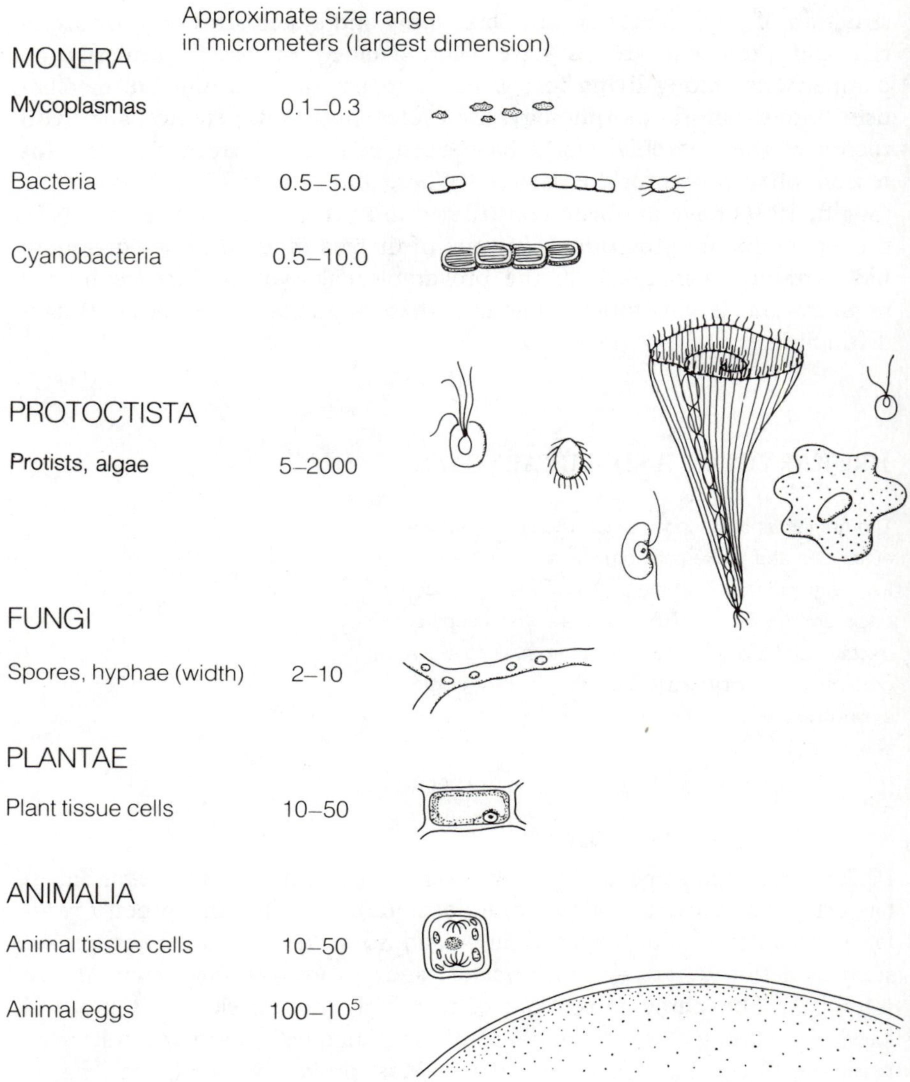

MONERA

Approximate size range
in micrometers (largest dimension)

Mycoplasmas   0.1–0.3

Bacteria   0.5–5.0

Cyanobacteria   0.5–10.0

PROTOCTISTA

Protists, algae   5–2000

FUNGI

Spores, hyphae (width)   2–10

PLANTAE

Plant tissue cells   10–50

ANIMALIA

Animal tissue cells 10–50

Animal eggs $100–10^5$

FIGURE 2-2
Comparison of some cell sizes.

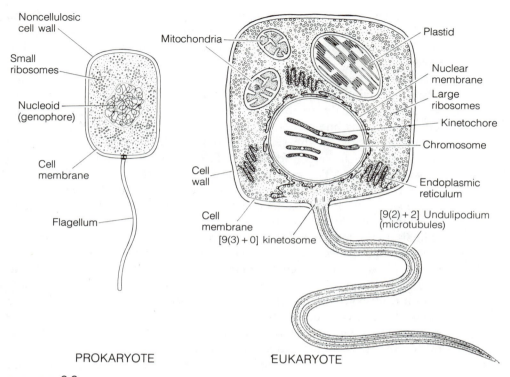

Noncellulosic cell wall

Small ribosomes

Nucleoid (genophore)

Cell membrane

Flagellum

Mitochondria

Plastid

Nuclear membrane

Large ribosomes

Kinetochore

Chromosome

Cell wall

Cell membrane

[9(3) + 0] kinetosome

Endoplasmic reticulum

[9(2) + 2] Undulipodium (microtubules)

PROKARYOTE

EUKARYOTE

FIGURE 2-3

Comparison of prokaryotic and eukaryotic cell structure. Composite diagrams based on observations at the light- and electron-microscopic levels.

they are studied in fixed and stained cells. (Unlike prokaryote genophores, chromosomes stain red with the Feulgen reagent.) Chromosomes are made of DNA complexed with some five histones; these are arginine- and lysine-rich proteins that, in most eukaryotes, form a large fraction—more than half—of the mass of chromosomes. They give chromosomes some of their characteristic properties, such as elasticity, tight packing, and stainability. However, they are not responsible for the ability of chromosomes to move, which depends on the mitotic spindle or on related microtubule systems. In most prokaryotes, the genophore is not complexed with histone or other proteins.

All familiar organisms—seaweeds, ciliates, molds, mushrooms, animals, and plants—are composed of eukaryotic cells. Except for some protoctists, the cells of these organisms divide by mitosis, the indirect cell division in which chromosomes "split" longitudinally and move centrifugally in two groups to opposite cell poles. The word **mitosis** in this book pertains to chromosomes and the mitotic apparatus. It does not include the precise direct distribution of genes of the genophore of bacteria. Prokaryotic cells may divide equally (by fission) or unequally (by budding), but never by mitosis.

Sex is defined as any process that leads to the formation of an individual offspring having more than a single parent—usually two. Prokaryotes reproduce asexually. In some, no sexual processes at all are known: offspring have only one parent. In prokaryotes capable of sex, sexual systems are unidirectional in the sense that donor cells ("males") transfer their genes to recipient cells ("females"). The donor cell retains a copy of the transferred genes. The number of genes transferred varies in the mating process. The genes take the form of a long DNA molecule, and generally only a small fraction of the total genome (but sometimes almost all of it) is transferred. In bacterial mating, cytoplasm does not fuse as it does in all animals, in the hyphae of fungi, and in many plants and protoctists; the new prokaryotic organism, called the **recombinant,** consists of the recipient cell itself with some of its genes replaced by those of the donor (Jacob and Wollman, 1961). Thus, in prokaryotes, there is almost never equal contribution from both parents. In the sexually formed eukaryote, on the other hand, the contributions of the parents are equal, or nearly so: the new eukaryotic individual (the zygote) generally takes half the genes and some of the nucleoplasm and cytoplasm from each parent.

Chromosomes are made of DNA, histones, and other proteins, but isolated preparations often also contain a large percentage of RNA from elsewhere in the nucleus. This RNA, probably both messenger and ribosomal, tends to adhere to isolated chromosomes. The eukaryotic nucleus also contains nucleoli, composed of several sizes of RNA and many proteins; nucleoli are made up of precursors of the cytoplasmic ribosomes. Other organelles unique to eukaryotic cells include mitochondria, plastids, centrioles, and kinetosomes and their undulipodia. These organelles lie outside the nuclear membrane.

Eukaryotic organelles of motility, which are all about 0.25 μm in diameter, have traditionally been called **flagella** if they are from 10 to about 15 μm long and if there are only a few per cell, and have been called **cilia** if they are shorter and more numerous. The electron microscope has shown remarkable structural similarity among all cilia and eukaryotic flagella; in cross section, they show the same [9(2) + 2] array of proteinaceous microtubules, each about 0.024 μm (24 nm) in diameter. These organelles are far more complex than and entirely different in structure and in protein composition from bacterial flagella (Figure 2-4). Their names should reflect new knowledge; thus, in this book, an old term, **undulipodia,** replaces cilia, flagella, and related organelles of eukaryotes (for example, shafts of

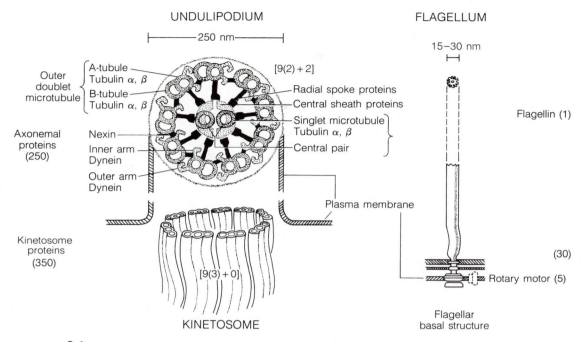

UNDULIPODIUM

├─── 250 nm ───┤

Outer doublet microtubule
{ A-tubule Tubulin α, β
B-tubule Tubulin α, β

[9(2) + 2]

Radial spoke proteins
Central sheath proteins
Singlet microtubule Tubulin α, β
Central pair

Axonemal proteins (250)
Nexin
Inner arm Dynein
Outer arm Dynein

Plasma membrane

Kinetosome proteins (350)

[9(3) + 0]

KINETOSOME

FLAGELLUM

15–30 nm
├─┤

Flagellin (1)

(30)
Rotary motor (5)

Flagellar basal structure

F I G U R E 2-4

Comparison of undulipodium underlain by kinetosome and flagellum moved by rotary motor. The approximate number of different proteins comprising each structure is indicated in parentheses. [Based on an original drawing by Kathryn Delisle.]

sperm tails, the units of cirri, and other [9(2) + 2] structures and their derivatives that develop from kinetosomes—which themselves show a [9(3) + 0] structure in cross section). The term **flagella** is reserved for the solid, thin, bacterial flagella and homologous structures, such as spirochete axial fibrils. Usually flagella are too small to be seen with the ordinary light microscope. The earliest use of "undulipodium" that I can find is Doflein (1916); see also Shmagina (1948). Both use the term, without definition, to refer to "*schwingende*" organelles, "waving feet,"—in retrospect, cilia and eukaryotic "flagella" (Kuznicki, Jahn, and Fonseca, 1970; Corliss, 1979).

Table 2-5 lists major differences between prokaryotic and eukaryotic organisms, whereas Table 2-6 summarizes the surprising chemical abilities of prokaryotes, members of at least two huge groups of distantly related organisms (Archaebacteria and Eubacteria).

TABLE 2-5
Differences between prokaryotes and eukaryotes.

PROPERTY	PROKARYOTES	EUKARYOTES
Cell size	Most are small cells (1–10 μm); some are larger than 100 μm.[a]	Most are large cells (10–100 μm); some are larger than 1 mm.
General characteristics	All are microbes. Unicellular or colonial. The most complex morphologically are filamentous or mycelial with fruiting bodies. With the exception of *Gemmata obscuriglobus*,[b] the nucleoid is not membrane-bounded.	Some are microbes; most are large organisms. Unicellular, colonial, mycelial, multicellular. The most complex morphologically are the vertebrates and the flowering plants. All possess a membrane-bounded nucleus.
Cell division	No mitosis; direct, mainly by binary fission or budding. Genophore contains DNA but no protein; it is not responsive to Feulgen staining. No centrioles, mitotic spindle, or microtubules.	Various forms of mitosis. Generally many chromosomes containing DNA, RNA, and proteins, bright red Feulgen-staining. Centrioles present in many; also, mitotic spindle or some arrangement of microtubules.
Sexual systems	Where present, unidirectional transfer of genetic material from donor to recipient; not required for reproduction.	Present in animals, plants, and fungi. Variable in protoctists. Male and female participate equally in fertilization.
Development	Both cyclical (e.g., spores) or terminal (e.g., heterocyst) development; no alternation of haploid and diploid generations; no extensive tissue differentiation; individual and colonial forms. Lack complex cell junctions. Multicellularity.[c]	Haploid forms produced by meiosis, diploid from zygotes; multicellular organisms show extensive tissue differentiation. Plasmodesmata, desmosomes, septate or gap junctions, or other complex structures between cells. Metamorphosis common.
Oxygen toleration	Strictly or facultatively anaerobic, microaerophilic, aerotolerant, aerobic.	Aerobic. (Exceptions, except in some protist groups, are clearly secondary modifications.)
Metabolism	Various metabolic patterns; no packaged membrane-bounded enzyme sacs for oxidation of organic molecules (lack mitochondria, plastids, and other heritable organelles).	Same oxidation pattern within all kingdoms: membrane-bounded sacs (mitochondria) containing enzymes for oxidation of 3-carbon organic acids.
Photosynthesis	If photosynthetic, enzymes for photosynthesis bound to cell membranes (chromatophores) but not packaged separately. Anaerobic and aerobic photosynthesis; sulfur, sulfate, or oxygen elimination; hydrogen donor may be H_2, H_2O, H_2S, or $(H_2CO)_n$. Lipids: vaccenic and oleic acids, hopanes; steroids very rare. Aminoglycoside antibiotics formed.	If photosynthetic, enzymes for photosynthesis packaged in membrane-bounded plastids. Mostly oxygen-releasing photosynthesis, in which the hydrogen donor is always H_2O. Lipids: linoleic and linolenic acids; steroids common (ergosterol, cycloartenol, cholesterol). Alkaloids, flavonoids, acetogenins, and other secondary metabolites common, especially in plants.
Means of motility	Some have simple bacterial flagella composed of flagellin; some glide. Intracellular motility rare or absent; no phagocytosis, pinocytosis, or cyclosis.	Except fungi, most have undulipodia: [9(2) + 2] "flagella" or cilia; 9 + 0 or 6 + 0 structures represent developmental modifications of the [9(2) + 2] plan. Pseudopods containing actin protein are common. Intracellular motility (pinocytosis, phagocytosis, cyclosis) by means of motility proteins: actin, myosin, tubulin.
Cell wall	Diaminopimelic acid, muramic acid glycopeptides; glycoproteins rare or absent; ascorbic acid not required.	Chitin or cellulose; glycoproteins with hydroxylated amino acids common; ascorbic acid required.
Propagules	Spores (endo- and exo-) resistant to desiccation; heat-resistant endospores contain calcium dipicolinate; actinospores.	Cysts, seeds, etc.; vary with phylum; no calcium dipicolinate; sporopollenin in spores; lack endospores. Less resistant to heat and desiccation than bacteria.

[a]Clements and Bullivant, 1991.
[b]Fuerst and Webb, 1991.
[c]Dworkin, 1992. See text.

TABLE 2-6
Bacterial virtuosities.

METABOLISM			
ENERGY	SOURCE OF ELECTRONS (OR HYDROGEN DONORS)	SOURCE OF CARBON	ORGANISMS AND THEIR HYDROGEN OR ELECTRON DONORS[a]
Photo- (light)	Litho- (inorganic compounds and C_1)	Auto- (CO_2)	Chlorobiaceae, H_2S, S (e) Chromatiaceae, H_2S, S (e) Rhodospirillaceae, H_2 (e) Cyanobacteria, H_2O (e) (including chloroxybacteria, H_2O) (e)
		Hetero- $(CH_2O)_n$	None
	Organo- (organic compounds)	Auto-	None
		Hetero-	Chromatiaceae, org. cpd. (e) Chloroflexaceae, org. cpd. (e) Halobacteriaceae (a) Heliobacteriaceae, org. cpd. (e) \ *Rhodomicrobium*, C_2, C_3 (e)
Chemo- (chemical compounds)	Litho-	Auto-	Methanogens, H_2 (a) Hydrogen oxidizers, H_2 (e) Methylotrophs, CH_4, CHOH, etc. (e) Ammonia, nitrite oxidizers, NH_3, NO_2 (e)
		Hetero-	Sulfur bacteria, S Manganese oxidizers, Mn^{2+} Iron bacteria, Fe^{2+} Sulfide oxidizers (e.g., *Beggiatoa*) Sulfate reducers (e.g., *Desulfovibrio*)
	Organo-	Auto-	Clostridia, etc., grown on CO_2 as sole source of carbon (H_2, $-CH_2$) (e)
		Hetero-	Many (including NO_3^{2-}, SO_4^{2-}, O_2, and $PO_4^{2-(b)}$ as terminal electron acceptors) (e) Thermoacidophils (a)

MORPHOLOGY		GENETICS
Spores Cysts Myxocyst sporophores Flagella Sheaths	Heterocystous filaments Matting and branching trichomes Parenchymatous organization	Small and large replicon maintenance Conjugation Transduction Transformation Transfection

[a](a) = Archaebacteria; (e) = Eubacteria; org. cpd. = organic compounds, e.g., acetate, propionate, pyruvate.
[b]Detection of phosphine (Dévai et al., 1988).

FIVE KINGDOMS

There are those who consider questions in science which
have no unequivocal experimentally determined answer
scarcely worth discussing. Such feeling, along with
conservatism, may have been responsible for the long
and almost unchallenged dominance of the system of two
kingdoms—plants and animals. . . . The unchallenged
position of these kingdoms has ended, however;
alternative systems are being widely considered.

R. H. WHITTAKER, 1969

Classification schemes attempt to reflect evolutionary history, but didactic
and ecological factors must be considered as well. Evolutionary considerations require the grouping of organisms with their most recent common
ancestors into the least inclusive taxa. If strict monophyly were adhered to,
the members of each taxon, including the multiple genomic system which
comprises it, would have to be more related by common ancestry to other
members of the taxon than to any organism outside the taxon. The number
of groups, especially in higher taxa, would be overwhelming, and taxon
definition would be so technical as to be useless to all except devoted
specialists. Definable, manageable, teachable groups are needed. Such
didactic considerations force the abandonment of the principle of strict
monophyly.

With respect to ecological considerations, when Whittaker first suggested five kingdoms (1959), he emphasized that, in the interacting webs of
nature, large organisms are producers, ingesters, or decomposers. These
styles of nutrition correspond approximately to the life styles of the three
kingdoms of macroscopic organisms. That is to say, plants are producers,
animals are ingesters, and fungi are decomposers. All producers are either
photo- or chemoautotrophs: they derive food from inorganic sources either
by using energy from light or by direct oxidation of reduced inorganic
compounds. However, the latter, chemoautotrophy, is used only by some
prokaryotes. Thus, eukaryotic autotrophy is always photoautotrophy. Ingesters and decomposers are both heterotrophs, requiring preformed organic compounds as food, but in different ways. Ingesters require organic
compounds in the form of other organisms: if they eat photoautotrophs,
they are herbivores; if they eat organisms that eat photoautotrophs, they are
predators. All ingesters take food into their bodies, where it is digested.
Decomposers are heterotrophs that do not ingest; they secrete digestive
enzymes into the environment to break down their food externally, and
thus they are osmotrophs that absorb the resulting small molecules from
solution.

To complement Whittaker's characterization by nutritional mode, the three kingdoms can be classified unambiguously by development: in all animals, the zygote, formed by fertilization of the female by the male gamete, develops into a ball of cells called a blastula; plants develop from multicellular diploid embryos that are nursed by sterile photosynthetic tissue; fungi, although capable of producing sexual structures, lack undulipodia at all stages of their development — they also produce hyphae from haploid spores.

With these three kingdoms thus neatly defined in this modified Whittaker scheme, only the nonanimal, nonplant, and nonfungal organisms remain to be distinguished. These fall into two groups: all prokaryotic organisms and the remaining eukaryotes which, lacking embryogenesis, fail to fit the definition of animal, plant, or fungus. Kingdom Monera, which is classified in this book into sixteen, admittedly tentative, phyla, comprises all prokaryotic organisms. The eukaryotic microorganisms and their direct descendants make up Kingdom Protoctista. Protoctists are nutritionally diverse: there are osmotrophs that absorb, producers that photosynthesize, and phagotrophs that ingest. Prokaryotes exhibit all possible nutritional patterns except the ingestive; on the scale of a single cell, ingestion depends on phagocytosis, which prokaryotes lack. The distinguishing features of the five kingdoms are summarized in Table 2-7, as well as the approximate time of their appearance in the fossil record. Expansion of the five-kingdom system to the level of class is shown in the Appendix and more extensively in Margulis and Schwartz, 1988.

Viruses are omitted from the kingdom classification because they are not autopoietic systems. The minimum number of enzymes, nucleic acids, and other macromolecular constituents necessary for maintenance and reproduction (i.e., autopoiesis) of a theoretical "minimal cell" is estimated to be about 50 (Morowitz and Wallace, 1973), corresponding to about 50 genes. Viruses contain less genetic information than this; they all require host cells. Probably they originated many times from different hosts after cell replication evolved. They are also probably more closely related to their hosts than to each other (Joklik, 1974), although some of the largest and most complex viruses, such as the vaccinias, may have descended directly from bacteria by continued reduction of function. Table 2-8 contains a classification of viruses.

A major innovation of Whittaker's five-kingdom system is the recognition of the great differences between plants and fungi. Although traditionally classified as plants, fungi have little ecologically or physiologically in common with plants besides a sedentary habit. Green plants are photoautotrophic primary producers, whereas fungi are absorptive decomposers. However, Whittaker — like Copeland (1956), Dodson (1971), and Leedale (1974), among others — wrestled with the problem of "blurred boundaries"

TABLE 2-7

Five-kingdom summary.

KINGDOM	EXAMPLES	GENETIC ORGANIZATION	EARLIEST APPEARANCE (EON) AND APPROXIMATE TIME OF DIVERSIFICATION[a] (MILLIONS OF YEARS)
SUPERKINGDOM PROKARYOTA			
Monera	Bacteria: cyanobacteria, mycelial bacteria, gliding bacteria	Prokaryotic genophore; merozygotes only; sex unidirectional	Archean Eon (3500)
SUPERKINGDOM EUKARYOTA			
Protoctista	Eukaryotic microorganisms: undulipodiated "fungi," slime molds, slime nets	Eukaryotic chromosomes, some lacking histones; mitotic cell divisions absent, idiosyncratic, or standard; various ploidy levels; various meiotic sexual systems	Proterozoic or Phanerozoic Eon (1200–700)
Animalia	Metazoa: anisogametous diploids developing from blastula embryos	Diploid; meiosis precedes gametogenesis	Late Proterozoic or Phanerozoic Eon (750)
Plantae	Metaphyta: anisogametous spore-formers developing from maternally-retained embryos; multicellular reproductive organs	Alternation between haploid and diploid phases	Phanerozoic Eon rhyniophytes found in Downtonian rocks in Wales, Czechoslovakia, and New York State (450)
Fungi	Amastigomycota: conjugation fungi, sac fungi (molds), club fungi (mushrooms), yeasts	Haploid and dikaryotic; zygote formation followed by meiosis; haploid spore formation	Phanerozoic Eon *Palaeomyces asteroxyli* found in Rhynie Chert (450)

[a]The geochronology used in this book is outlined in Table 4-2.

between plants and fungi and the protoctists from which they presumably evolved. Multicellularity and nutritional style are inadequate to distinguish plants, fungi, and protoctists. Permanently nonphotosynthetic plants are well known; multicellular organization has evolved in all five kingdoms, including the bacteria. Many, if not most, bacteria in nature are multicellular (Balows, Trüper, Dworkin, et al., 1991). Some, like *Gomphosphaera*, never display any single-cell stage. In fact, no classification scheme regards

CHARACTERISTICS	SIGNIFICANT SELECTIVE FACTORS	REFERENCES
Ultraviolet photoprotection; photosynthesis; motility; geochemical cycling of elements	Amount of ultraviolet and visible solar radiation; increasing concentration of atmospheric oxygen; depletion of nutrients	Schopf, 1983, 1992 Knoll, 1990
Mendelian genetic systems; mitosis and meiosis; obligate recombination each generation; intracellular motility (phagocytosis and pinocytosis)	Depletion of nutrients; feeding strategies	Vidal and Knoll, 1983; Knoll, 1990
Tissue; desmosomes, gap junctions, or other specialized intercellular connections for heterotrophic specializations; ingestive nutrition	Transition from aquatic to terrestrial and aerial environments; feeding strategies	McMenamin and McMenamin, 1990
Tissue development for autotrophic specializations; photosynthetic nutrition	Transition from aquatic to terrestrial environments; resistance against predation	Banks, 1972a, b; Richards in Schopf, 1992
Complex mycelial development; absorptive nutrition	Transition from aquatic to terrestrial environments; nutrient sources; nature of hosts	Pirozynski and Malloch, 1975; Tiffney and Barghoorn, 1974

all protists as single-celled (consider the volvocalean algae or colonial ciliates). In recognition of this difficulty, Whittaker, Schwartz, and I developed a systematic classification straightforward enough to be generally useful (Whittaker and Margulis, 1978; Margulis and Schwartz, 1988). The three kingdoms that contain large organisms are fungi, plants and animals. The others, Monera and Protoctista, are the eukaryotic and prokaryotic microbes and their descendants, respectively.

TABLE 2-8
Viruses.[a]

TYPE	STRUCTURE	EXAMPLES
Ribovira (RNA viruses)	Ribohelica (nucleocapsid with helical symmetry)	Tobacco mosaic virus, myxovirus (influenza A), pea mosaic virus, ribgrass virus
	Ribocubica (nucleocapsid with cubical symmetry)	Napoviridae (turnip yellow mosaic virus, poliovirus, Coxsackie virus 1, echovirus, RNA phage), Reoviridae (reovirus), Arboviridae (arbovirus)
Deoxyvira (DNA viruses)	Deoxyhelica (nucleocapsid with helical symmetry)	Smallpox, fowlpox, rabbit myxoma, viruses, cyanophages (phage ϕX174)
	Deoxycubica (nucleocapsid with cubical symmetry)	Shope papillomavirus, polyomavirus, herpes simplex viruses, adenovirus
	Deoxybinala (with head and tail)	Bacteriophages (T, λ, 434, ϕ80, and P22)

[a]After Whitehouse (1973).

The fungal kingdom includes haploid or dikaryotic mycelial organisms reproducing by spores and excludes all of the water molds and other organisms that have undulipodia at any time during their life cycles—organisms often grouped with fungi. Kingdom Fungi includes here only molds and mushrooms (for example, zygomycota, ascomycota, and basidiomycota). The plant kingdom includes only embryophytes—plants that display alternation of generations: they develop from diploid embryos and from haploid spores—they primarily live on land. By exclusion, Kingdom Protoctista then becomes what some have called a ragbag: that remaining miscellaneous assemblage of phyla in which profound evolutionary experimentation occurred on the themes of multiple genomes, varied nutritional strategies, mitosis, and meiosis (Margulis, 1988; Margulis, Corliss, Melkonian, and Chapman, 1990). Thus defined, the kingdom includes brown and red algae (all the seaweeds), slime and water molds, and many other descendants of eukaryotic microbes. Most of these groups have multicellular descendants. It is because the term **protist** implies a single cell that the term **protoctist** is used, recognizing Hogg's (1861) precedent, as Copeland (1956) suggested.

In the classification of extant life presented in the Appendix, there are fewer than a hundred fundamentally different body or metabolic plans represented by living phyla. Paleontologists estimate that perhaps another three or four phyla are extinct (for example, archeocyathids and chitinozoa). It is remarkable that, although no species that existed 500 million years ago is still alive today, by about that time nearly all of the phyla still with us had already evolved. Why? What is the basis for this lack of stability of species coupled with extreme conservation of phyla? Whatever the reason, it is profitable to assume that the rapid turnover of lower taxa and the conservation of higher is true also in the microbial world. The complexes of traits that define phyla are conserved, forming the basis for the construction of phylogenies and for the reconstruction of the early history of life.

CHAPTER 3

CELL EVOLUTION IN PERSPECTIVE

DIRECT FILIATION

The classic view logically assumed that all primary
eukaryotic organelles were produced by the
differentiation of the protoeukaryote's own cell
substance, i.e., autogenously.

F. J. R. TAYLOR, 1976

The origin and evolution of nucleated cells has been assumed to have occurred in a unique ancestral population by the accumulation of various kinds of mutations acted upon by natural selection. Proponents of this idea think that point mutations, deletions, duplications, and other types of mutations were responsible for the differentiation of eukaryotic cells from prokaryotic cells. Defenders of **direct filiation,** or the nonsymbiotic view of cell origins, include Allsopp (1969), Raff and Mahler (1972), Uzzell and Spolsky (1974), Cavalier-Smith (1975), Reijnders (1975), and Taylor (1976).

Direct filiation as an explanation of the origin of eukaryotes from prokaryotes is inadequate, in my opinion. There is an immense hiatus of form both in the fossil record and among living organisms. The discontinuity between the nonmitotic cyanobacteria and the fully mitotic red algae, for example, is difficult to understand by direct filiation, but does follow from the symbiotic theory. The direct filiation and symbiotic theories are contrasted in Figure 3-1. In a critical discussion of the symbiotic theory, Taylor (1974) rejected symbiosis as the origin of undulipodial motility and

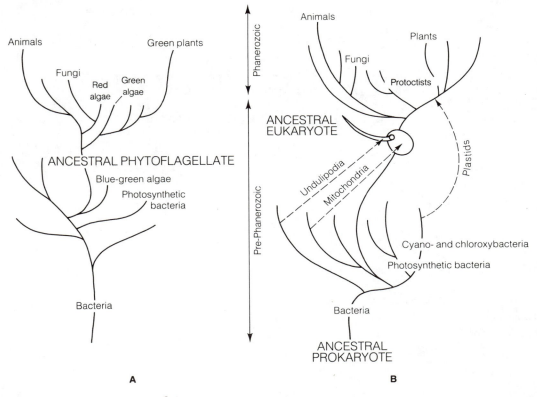

FIGURE 3-1

Comparison of direct-filiation (**A**) and symbiotic (**B**) phylogenies.

dubbed it the **SET (Serial Endosymbiosis Theory)**, a name we continue to use.

The theory of direct filiation and the SET both assume that all organisms on Earth derived from bacterial ancestors with common biochemistry and that plants and metazoans arose from some—still unknown—protoctist. However, the two theories differ entirely in their concepts of the process by which prokaryotes gave rise to eukaryotic cells. Among themselves, the direct-filiation theorists differ on details and in the degree to which they have developed their schemes (Bessey, 1950; Dougherty and Allen, 1960; Klein and Cronquist, 1967; Raff and Mahler, 1972; Cavalier-Smith, 1975; Reijnders, 1975; Taylor, 1976). However, they share the concepts that photosynthetic bacteria gave rise to algae and, eventually, to plants, and that some algae lost their plastids and evolved into ancestors of fungi and animals. They also agree that plastids and other eukaryotic organelles, including the nucleus, evolved by differentiation within the cells themselves — that organelles evolved by compartmentalization.

The "lower" plants, since the time of Jussieu, have been called **thallophytes** (algae, including blue-greens, and fungi), and the "higher" plants **bryophytes** (mosses, liverworts) and **tracheophytes** (ferns, cone- and flower-bearing plants). According to direct-filiation concepts, the common ancestors of photosynthetic bacteria and blue-green algae (cyanobacteria) gave rise directly to all the other algae. Whatever the details of any particular direct-filiation idea, a direct evolutionary connection between photosynthetic prokaryotes and photosynthetic eukaryotes is required conceptually. The groups of organisms whose phylogenies are affected most by the direct-filiation concept have fallen into the province of botany — bacteria, algae, fungi, and plants. The few in the zoological domain — photosynthetic protists and their nonphotosynthetic relatives — are but a tiny fraction of the constituency of scrutiny of zoologists, who, for this reason, have not concerned themselves much with the earlier stages of evolution. On the other hand, various ancestors of algae have been suggested by outstanding botanists (Fritsch, 1935; Bessey, 1950; Dougherty and Allen, 1958). Many different, partial phylogenies have been called the **ancestral phytomonads** or the **uralgae.** Some authors claim that *Cyanidium* or *Cyanophora,* eukaryotes with pigment systems like cyanobacteria, make good candidates for the uralgae (Klein and Cronquist, 1967; Seckbach, Fredrick, and Garbary, 1983). However, *Cyanophora* probably originated as a symbiosis between mastigotes and cyanobacteria (Trench et al., 1978), whereas *Cyanidium* may be a red algae highly specialized and adapted to hot, acid waters (Ragan and Chapman, 1978). In fact, the search for the common ancestor between blue-greens and eukaryotes led to the unequivocal classification of blue-greens as bacteria.

The idea of a common ancestor linking the cyanobacteria with mitotic algae led to the search for the origin of most features of eukaryotes — for example, plastids, mitochondria, and the mitotic spindle — in plants. Early cytologists soon found that cells of many algae divide by standard mitosis, but they continued to seek more primitive cell-division systems among the microscopic forms of life. Because they thought that single-celled algae gave rise to plants and, by the loss of photosynthesis, to fungi and animals, they expected to find in algae evidence of trends from "primitive" cell division to the complex mitotic and meiotic divisions of vascular plants. The microscopists of the early twentieth century studied in this context the nuclear "division figures" of a wide assortment of algae, protozoans, and fungi. However, their observations did not fall into an easily interpretable pattern (Wilson, 1959).

The "thallophytes" did not show a trend from the nonmitotic division of cyanobacteria to the fully developed standard mitoses of green algae and

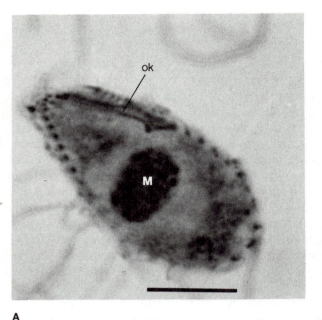

A

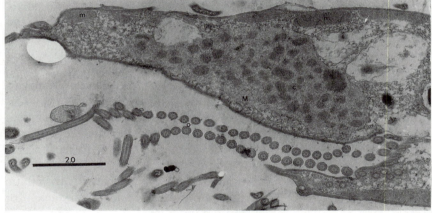

B

FIGURE 3-2

Idiosyncratic nuclei of protists.
(**A**) The ciliate *Pseudocohnilembus salinus*. Macronucleus (M), oral kinetids (ok), and body kinetids are visible. Light micrograph of Protargol-stained preparation, × 2500; bar = 10 μm. [Courtesy of L. Olendzenski.] (**B**) Anterior end of *Pseudocohnilembus salinus*. M = macronucleus; m = mitochondria; c = condensed chromatin; o = oral cilia. Transmission electron micrograph, × 20,000, reproduced at 62%; bar = 2 μm. [Courtesy of L. Olendzenski.] (**C**) *Pelomyxa palustris*. The nuclei of this giant multinucleate amoeba divide directly; no chromosomes, kinetochores, centrioles, asters, or mitotic spindle have ever been seen. This organism, the only species in Phylum Karyoblastea (see Appendix), is a most distinctive eukaryote, since it lacks mitochondria but contains methanogenic and nonmethanogenic bacteria. N = nucleus; B = endosymbiotic bacteria; PN = perinuclear bacteria. Transmission electron micrograph, × 4900; bar = 1 μm. [Courtesy of E. W. Daniels.]

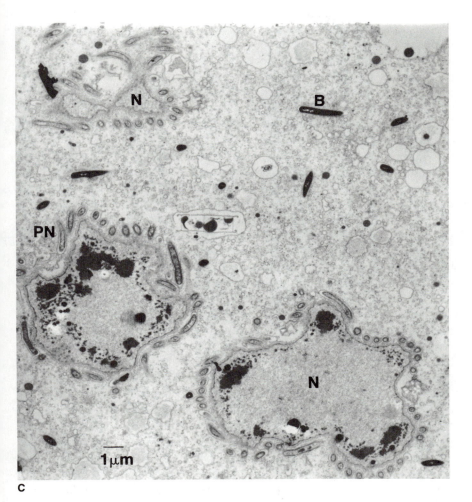

c

plants. Investigations led rather to internally consistent but baffling observations. Mitosis was fundamentally dualistic: the mitotic apparatus (that is,
the non–Feulgen-staining "achromatic" apparatus—spindle, centrioles,
and so on) moved independently of the chromosomes. Furthermore, many
algae and protozoans had neither mitosis nor a genetic system like that of
animals and plants. For example, the ciliates, the phylum to which *Paramecium, Stentor, Euplotes*, and *Tetrahymena* belong, have an idiosyncratic
genetic system involving two types of nuclei, often many of each per cell:
macronuclei and smaller micronuclei (Figure 3-2; Beale, 1954; Sonneborn,
1959; Grell, 1967). Micronuclei are dispensable; yet they store copies of the
genes, and in the sexual cycles, they undergo meiosis. Macronuclei are the

sites of transcription of RNA and thus are indispensable to the physiology of the cell.

Even more unexpected nuclear behavior was found in other groups — for example, in dinomastigotes, eukaryotic planktonic marine organisms with both photosynthetic, algalike, plastid-containing members and protozoanlike, plastid-lacking members (Dodge, 1966; Soyer, 1972; Ris and Kubai, 1974; Soyer and Haapala, 1974; Taylor, 1978). The chromosomes of dinomastigotes do not coil but remain dark-staining (condensed) throughout the cell cycle. In general, electron-microscopic, genetic, and molecular-biological studies from the early 1960s to the present have confirmed that unconventional mitotic and sexual processes occur in protoctists, but no trend toward standard plant and animal mitosis can be related to the "plant-like" nature of the cell, i.e., to the presence of red, yellow, or green plastids. No data supported a scheme of development of eukaryotic features, especially mitosis, in cyanobacteria as they evolved to become algae. In fact, unusual nuclear organization and genetic systems were observed primarily in nonphotosynthetic protists.

One problem in relating cyanobacteria to algae was the lack of evidence for the origin of undulipodia: no intermediate stages were found between the cyanobacteria, which always lack undulipodia, and the many algae that bear them. Another problem was the absence of meiotic sexuality in cyanobacteria: no intermediates were discovered between cyanobacteria and the primarily sexual green and brown algae. After 1963, the use of glutaraldehyde fixatives in electron-microscopic studies became common, and unexpected relationships were established between undulipodia and mitosis (Figure 3-3). Many algae were found to have undulipodia at some stage of their life cycle. In these algae, in ciliates, and even in anaerobic mastigotes, animals, and plants, the motile structure was the same: all were composed of microtubules 0.024 μm (24 nm) in diameter arranged in the [9(2) + 2] pattern in cross section or related in some obvious way to the [9(2) + 2] theme. Furthermore, the mitotic spindle, even in fungi, which lack undulipodia, was discovered to be composed of protein microtubules of the same diameter as those in undulipodia. Electron microscopy and protein biochemistry revealed striking homologies between the undulipodia and microtubules of all eukaryotes.

Although fungi and flowering plants lack undulipodia, their cells undergo mitotic and meiotic processes that require spindles composed of microtubules. Botanists have assumed that fungal and plant undulipodia were lost in the evolution of these organisms, during their transition from aquatic to terrestrial environments. Whether the lack of undulipodia is secondary in the red algae — the rhodophytes, which are nearly all marine organisms — has been debated, with no convincing evidence. It is generally

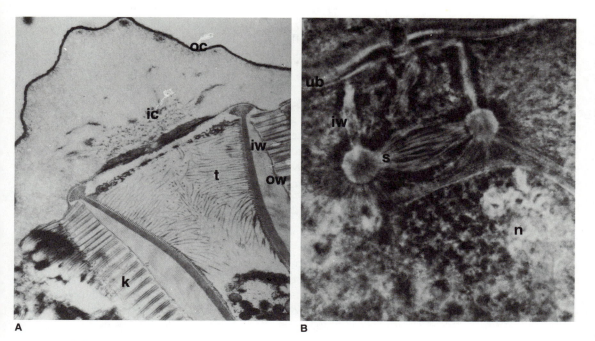

FIGURE 3-3

The relationship of kinetosomes to the centrioles and mitotic spindle of hypermastigotes.

(**A**) *Pseudotrichonympha* in early division showing how the entire anterior (rostral) end of the cell is involved in formation of the spindle. *oc* = outer cortex; *ic* = inner cortex; *iw* = inner wall, which Cleveland (1963) interpreted to be a "long centriole." The walls function as mitotic poles as spindle microtubules (*t*) assemble between. The rows of kinetosomes (*k*) are aligned along the outer wall (*ow*). Electron micrograph, × 7300. [From Grimstone and Gibbons (1966).] (**B**) A slightly later stage in division of a related species, *Barbulonympha*, showing why Cleveland called the inner wall a "long centriole." The proximal sides of the walls may be thought of as microtubule-organizing centers (MCs). *ub* = undulipodial bands; *iw* = inner wall; *s* = mitotic spindle; *n* = nucleus. Light microscope, × 450. [Photograph from film by L. R. Cleveland, *Sexuality and Gametogenesis in Barbulonympha*. See Grimstone and Gibbons (1966).]

agreed that microtubule-containing organisms derive from undulipodiated protoctists, but from which, no one knows. Yet no algae, or any other organisms, contain structures interpretable as stages in the evolutionary development of undulipodia. What selective pressures led to these microtubule-based structures? Motility is assumed to be the factor selected in the origin of cilia and other undulipodia. Whenever it appears in evolution, motility is associated with heterotrophic nutrition—the pursuit of food. Cheetahs, paramecia, and flagellated bacteria are motile by means of analogous organs—legs and feet, undulipodia, flagella—because they seek out their food. Trees, mosses, and algae, on the other hand, are autotrophs dependent on water, minerals, sunlight, and atmospheric carbon dioxide. Photoautotrophs tend to settle permanently in one place; except for the gliding cyanobacteria, few motile organisms, if any, are obligate photoautotrophs. How, then, can one hypothesize the appearance of undulipodia in cyanobacteria? If cyanobacteria evolved into ancestors of the undulipodiated algae, why do they all lack mitosis and other eukaryotic traits?

The complexity and antiquity of photosynthesis lends support to the idea of an early evolution of this nutritive process which sustains the biosphere, as does the fact that many organisms (for example, bacteria, euglenoids, and parasitic plants) lose photosynthesis by mutation both in the field and in the laboratory. Because photosynthesis is an anaerobic process upon which all organisms ultimately depend, it must have developed very early in the history of life on the planet, before mitosis. Photosynthesis involves many pigments, lipids, and enzymes arrayed precisely to form highly structured membranes. In fact, the details of the photosynthetic patterns of algae and plants are remarkably similar to those of cyanobacteria. All release gaseous oxygen from water, which is the source of hydrogen atoms for the reduction of carbon dioxide to organic cell matter. All use ribulose biphosphocarboxylase in the same steps to reduce carbon dioxide to organic compounds (Bassham, 1980). This precise photosynthetic pattern with all its complexities did not evolve independently in separate lines of organisms. Therefore, the photosynthetic apparatus of cyanobacteria must be homologous with that of algae and plants. But cell division in cyanobacteria is always of the bacterial type. (Claims in the 1950s that cyanobacteria performed mitotic cell division have been withdrawn.) Nuclear membranes, chromosomes, mitochondria, undulipodia, and spindle microtubules have never been found in any of the thousands of widely distributed, highly diversified species of cyanobacteria. Their structure, chemical composition, genetic organization, and lack of mitosis show unquestionably that blue-green algae are bacteria (Figures 3-4 to 3-7).

An impasse was reached. The more that morphological and physiological trends from asexual cyanobacteria to nucleated photosynthetic organisms, especially the sexual red and green algae, were sought, the more firmly the discontinuity between the two groups was established. The differences in cell division between cyanobacteria and all photosynthetic eukaryotes were particularly striking (Figures 3-7 and 3-8; Table 3-1). At the same time, the evolutionary literature on classification and possible relationships among "lower" plants was so fraught with confusion and contradiction that it was ignored by most biologists. Classification of algae

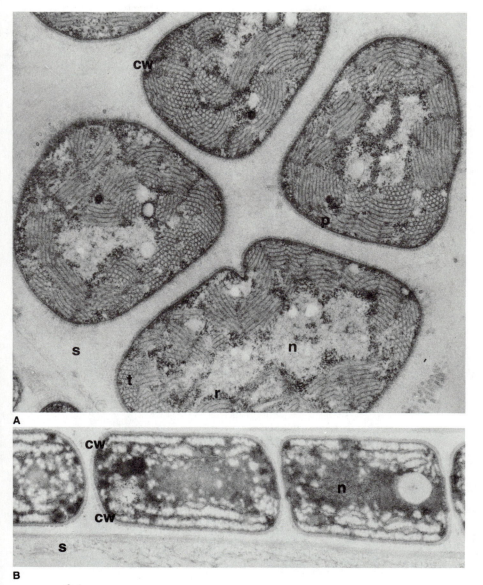

FIGURE 3-4

Prokaryotic cell structure.

(**A**) *Thiocapsa pfennigi*, a photosynthetic purple sulfur bacterium from microbial mats at Laguna Figueroa, Baja California Norte. *cw* = cell wall; *p* = plasma membrane; *s* = sheath; *t* = thylakoids; *n* = nucleoid; *r* = ribosomes. Transmission electron micrograph, × 27,500. [Courtesy of D. Chase] (**B**) Green filamentous prokaryote from Laguna Figueroa, Baja California Norte. Transmission electron micrograph, × 23,000. [Courtesy of D. Chase]

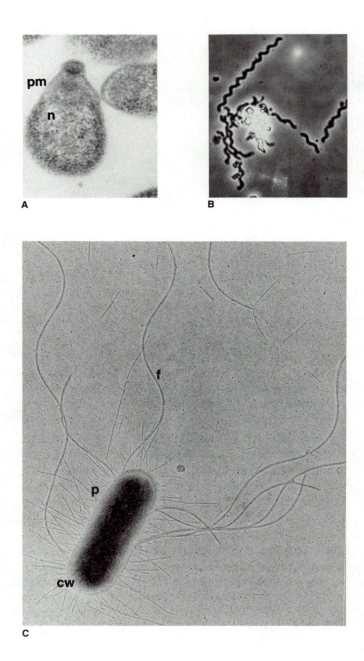

at the level of higher taxa (class or order) traditionally has been based on the distinctly colored pigments of photosynthetic plastids. Depending on authority, the ancestors of plants are to be sought among green algae, red algae, golden-yellow algae, or hypothesized uralgal organisms yet to be discovered. Yet several naturalists studying the live organisms have been struck by the fact that the position of the undulipodium in algae or their gametes is often a reliable taxonomic criterion: it is highly correlated with other aspects of biology (Copeland, 1956; Lund, 1964). Why? In Chapter 8, we shall see that the nonphotosynthetic portion of algal cells, and especially undulipodial arrangement (the kinetid) because it correlates with mitotic pattern and sexuality, provides far better clues to the relationships among algae than does the color of their plastids.

For half a century, it seemed reasonable to believe that cyanobacteria evolved into nucleated algae, an unstated assumption that provided both the impetus for research and a framework for the evaluation of results. According to the concept, both undulipodia and mitosis must have evolved in algae — somewhere between cyanobacteria and green algae. Hence, variations of mitosis in heterotrophs, for example in parasitic protists, must be viewed as irrelevant because they are "off the main line" of evolution from cyanobacteria to plants. However, if cyanobacteria are ancestral to algae, in what population of organisms did undulipodia and the related mitotic spindle arise? One would not expect this structure for the distribution of chromosomes to be lost after it had evolved. If cyanobacteria are not ancestral to nucleated algae, why is photosynthesis in both so similar? In what organisms and in what environments did selective pressures operate for the origin of both photoautotrophic nutrition and undulipodial motility? When?

The claim that links between cyanobacteria and eukaryotes have all died out, leaving neither fossil record nor living relics (Klein and Cron-

FIGURE 3-5

Prokaryotic cell structure.
(A) *Mycoplasma gallisepticum*, a heterotrophic wall-less bacterium. pm = plasma membrane; n = nucleoid. Transmission electron micrograph, $\times$ 75,000. [Courtesy of J. Maniloff.] (B) *Saprospira albida*, a flexibacterium whose motility and form resemble those of the cyanobacterium *Spirulina*. Phase-contrast light micrograph of live cells, $\times$ 600. [Courtesy of R. A. Lewin.] (C) *Escherichia coli*, the colon bacterium, the best known living organism. f = peritrichous flagella; cw = cell wall; p = shorter thinner pili, which may function in attachment. Special pili are used in the sexual process, by which compatible bacteria recognize each other and the donor (male) transfers genes to the recipient (female). Negative-stained electron micrograph, $\times$ 11,150. [Courtesy of D. Chase.]

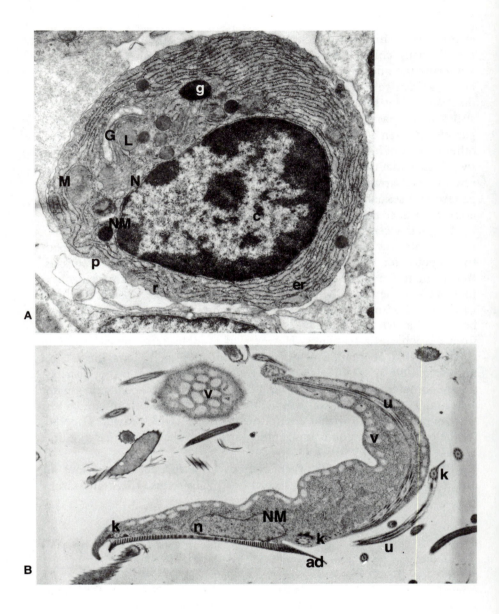

quist, 1967; Dougherty and Allen, 1960) has become less satisfactory with the availability of new information about living microbes and the fossil record. Major intermediates between photosynthetic prokaryotes and eukaryotes did not die out without a trace; the thesis of this book is that they never existed. Biological systems are remarkably conservative, and evidence for intermediate forms between these two great cell types is plentiful; it needs only to be organized around new paradigms; for example, the SET. According to the SET, cyanobacteria are indeed ancestral, but only to plastids — not to the rest of the eukaryotic cell.

Yet, the SET is consistent with direct filiation regarding the origin of the nucleus. As Taylor (1978) pointed out, the SET and direct filiation are not mutually exclusive. Conceivable modifications of the extreme version of the SET include, for instance, symbiotic origin of plastids but not of mitochondria, or symbiotic origin of plastids and mitochondria but not undulipodia. Even the origin of the nucleus by symbiosis has been suggested by Hartman (1975) and by Pickett-Heaps (1974). Still other permutations of the concept have been evaluated by Taylor (1974, 1978). These theories for the origin of eukaryotes lead to many deductions that are subject to experimental analysis. The mutually exclusive views are testable, and to some extent they have been tested. One example will suffice here: if photosynthetic prokaryotes are ancestral to the plastids of photosynthetic eukaryotes but not to their nucleocytoplasm, then amino acid sequences of specific proteins from living prokaryotes ought to show more homology with eukaryotic plastid proteins than the plastid proteins do with other proteins in the same cell. The opposite prediction, namely a closer relationship of the plastids to their own cells than to free-living prokaryotes, is deduced from the direct-filiation idea. Data bearing on this issue point to plastid origin by symbiosis (see Figure 3-9 **C**). Likewise, by homology

FIGURE 3-6

Eukaryotic cell structure.
(**A**) Animal: a plasma cell from the intestinal lining (lamina propria of the intestinal mucosa) of a rat. N = nucleus; NM = nuclear membrane; c = chromatin; p = plasma membrane; r = ribosomes; er = endoplasmic reticulum; M = mitochondrion; G = Golgi body; L = lysosomes; g = granule. (**B**) Protoctist: *Giardia muris* (Phylum Zoomastigina, Class Diplomonadida), taken from the small intestine of a rat. These eukaryotes attach to the intestines of their mammalian hosts, including people, by means of their adhesive disks (*ad*). They have two nuclei (one shown, *n*), eight undulipodia (*u*) underlain by kinetosomes (*k*), vesicles (*v*) of unknown function (the one seen above the *Giardia* shown is from another *Giardia*), and no mitochondria. NM = nuclear membrane. Transmission electron micrograph, × 8750. [Courtesy of D. Chase.]

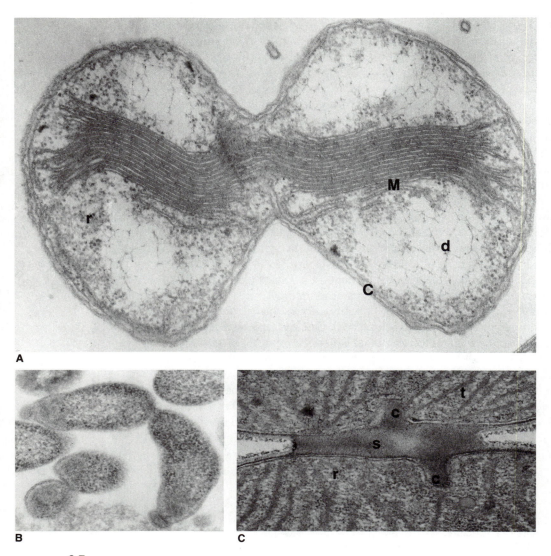

FIGURE 3-7

Cell division in prokaryotes.

(A) *Nitrocystis oceanus*, a chemoautotrophic bacterium. *M* = intracellular membranes, the presumed site of an electron-transport multienzyme system for ammonia oxidation; *C* = cell wall; *r* = ribosomes; *d* = DNA fibrils. Transmission electron micrograph, × 57,000. [Courtesy of S. Watson.] (B) *Mycoplasma gallisepticum*, a heterotrophic wall-less bacterium. Transmission electron micrograph, × 75,000. [Courtesy of J. Maniloff.] (C) Thin section of *Anacystis nidulans*, a cyanobacterium. *c* = newly forming cell wall; *s* = sheath; *r* = ribosomes; *t* = thylakoids. Transmission electron micrograph, × 75,000. [Courtesy of M. M. Allen.]

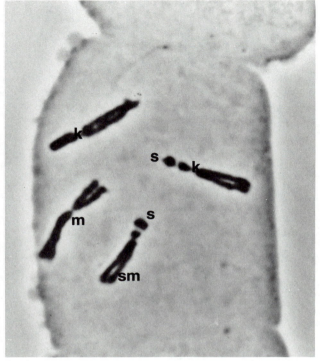

A

FIGURE 3-8

Cell division in eukaryotes. (**A** *above*; **B** and **C**, *following page*)
(**A**) Chromosomes of *Haplopappus gracilis*, a plant of the family Compositae. Diploid root cell; 2N = 4. k = kinetochore; s = satellite or heterochromatic knob. The mitotic spindle is not visible. One pair of homologues is metacentric (*m*); the other is submetacentric (*sm*). Light micrograph, × 900. [Courtesy of G. Jackson.]
(**B**) Mitotic cell division in the root top of an onion (*Allium sativa*). From left to right: interphase, prophase, early metaphase, metaphase, anaphase, and telophase. n = nucleolus; *ch* = chromatin; C = chromosomes; s = mitotic spindle; p = phragmoplast (cell plate or cell-wall precursor structure). Light micrograph of stained preparations, × 800. (**C**) Cultured mammalian tissue cell in anaphase. C = chromosomes; c = centriole; *mt* = mitotic-spindle microtubules. Transmission electron micrograph, × 8000. [Courtesy of J. R. McIntosh. In Inoué and Stephens (1975).]

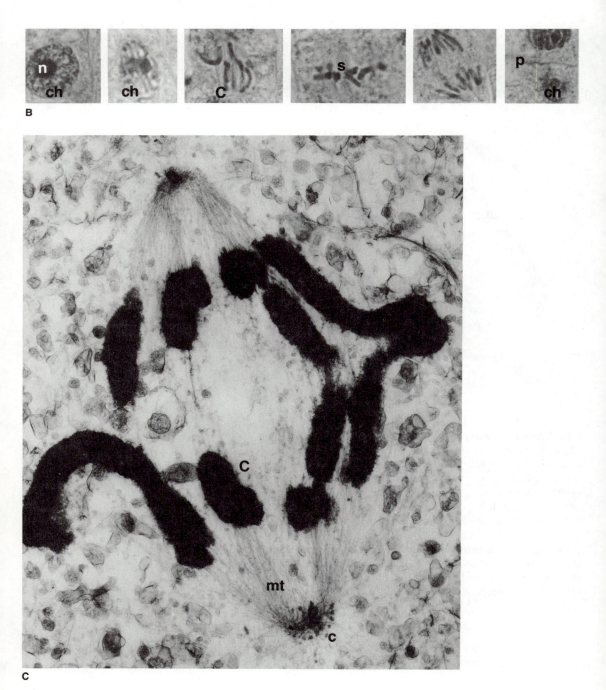

FIGURE 3-8 B, C

between nucleotide sequences of ribosomal RNA, the plastids of several types of algae and plants are more related to each other and to free-living coccoid cyanobacteria than to the cytoplasm in which the plastids reside (Figure 3-9C). Molecular phylogenies based on ribosomal RNA gene sequences and on amino acid sequences in RNA polymerases and in membrane-binding ATPases all support the prediction of the SET: organelles are more closely related to their free-living co-descendants than they are to each other or to the nucleocytoplasm in which they reside (Figure 3-9).

DISREPUTABLE THEORIES

This book, dedicated to his serene highness the Prince of Monaco, contains a lively exposition of a heresy, in regard to which the author frankly admits that if some years ago he had seen it stated at the beginning of an essay, he would have probably read no more. The heresy is that, apart from bacteria, all organisms are double, being formed by the association and "emboîtement" of two different kinds of creature. There are partners within every cell, partner-bacteria, which the author calls "symbiotes."

Nature, 1919 [Anonymous review of Portier (1918)]

Perhaps the earliest scientific idea that cell organelles came from hereditary symbiotic associations was stated in 1883 by Schimper (see Wilson, 1959). Impressed with the division behavior of the chloroplasts of green algae and plants and the similarity of these plastids to free-living cyanobacteria, investigators such as the St. Petersburg plant physiologist Famintzyn (1891) suggested that chloroplasts had a symbiotic origin. In 1909 Mereschkovsky developed concepts of the symbiotic origin of organelles; his concepts are often quoted in the Russian literature, but, for English readers, familiarity with his work has always been secondhand through Wilson (1959). Kozo-Polyansky's book (1924; Takhtajan, 1973; Khakhina, 1992), although enthusiastically supportive of a role for symbiosis in evolution, was entirely ignored outside of Russia. Altmann's 1890 theories of "bioblasts" (see Wilson, 1959) were too difficult to relate to detailed observations of the cell at a time when attention was focused on rapidly growing knowledge of the role of the nucleus and its chromosomes in heredity.

Wallin, an American biologist working at the University of Colorado Medical School, argued strongly (1927) the importance of the role of

TABLE 3-1

Multicellularity, reproduction, biochemical pathways, and cell division in the five kingdoms.

	MONERA	PROTOCTISTA	FUNGI	ANIMALIA	PLANTAE
Multicellularity	±	±	±	+	+
80S ribosomes	−	±	+	+	+
Mitosis: discrete chromosomes composed of DNA and protein	−	±	+	+	+
Microtubules, 24 nm diameter	−[a]	+	+	+	+
Chromatin attachment to microtubules via kinetochores	−	±	+	+	+
Histone chromosomal proteins	−	±	+	+	+
Intranuclear mitotic spindles (24-nm tubules)	−	±	+	−	−
[9(3) + 0] centrioles at poles in mitosis	−	±	−	+	−
[9(3) + 0] kinetosomes at axoneme base	−	±	−	+	±
Undulipodiated cells that can develop without sexual fusion (zoospores)	−	±	−	−	−
Phragmoplast (cell plate)	−	±	−	−	+
Phagocytosis, pinocytosis	−	±	−	+	+
Cellulosic cell walls	−	±	−	−	+
Chitinous cell walls	−	±	+	−	−
Golgi apparatus, dictyosomes	−	±	−	+	+
Meiosis	−	±	+	+	+
Tetrasporaceous meiotic tetrads	−	±	+	−	+
Syncytia (plasmodia, coenocytes)	−	±	+	+[b]	+[b]
Stable dikaryosis	−	±	+	−	−
α-Aminoadipic acid pathway of lysine biosynthesis	−	±	+	+	−

KEY: + probably present in all members; − probably absent in all members; ± present in some, absent in others.
[a]For cytoplasmic tubules in prokaryotes, see Chapter 9.
[b]Limited to some tissues.

TABLE 3 - 1 (continued)

	MONERA	PROTOCTISTA	FUNGI	ANIMALIA	PLANTAE
Diaminopimelic acid pathway of lysine biosynthesis	+	±	−	−	+
Mixed polyketide/shikimic acid pathway	+	±	−	−	+
Alternation of diploid and haploid generations	−	±	−	−	+
Stable endosymbiotic bacteria in cells	−	+	+[c]	+	+
Plastids	−	±	−	−	+
Mitochondria	−	±	+	+	+
Hydrogenosomes	−	±	−	−	−
Peroxisomes, glyoxisomes	−	±	±	+	+
Undulipodiated sperm (cells that develop with sexual fusion)	−	±	−	+	±
Embryos	−	−	−	+[d]	+[e]
Plasmodesmata-pit connections	−	±	+	−	+
Desmosomes, gap junctions	−	−	−	−	+
Cell survival after undulipodial transformation	−	±[f]	−	−	−
Programmed cell death in multicellular organisms	−	±	−	+	+
Nuclear transfer from cell to cell	−	±[g]	±	−	−
Multicellular sexual organs	−	±	−	+	+

[c]For bacteria in fungal cells, see Scannerini and Bonfante-Fasolo, 1991.
[d]Blastula that usually gastrulate.
[e]Retained by maternal tissue.
[f]See Handbook of Protoctista for details (Margulis et al., 1990).
[g]Goff, 1991.

symbiosis in evolution. Much as Portier (a review of his book is quoted at the beginning of this section) had done before him (1918), Wallin suggested that the little bodies inside animal and plant cells—that is, mitochondria, although they were then called chondriosomes, chondriochonts, and even plastids—can be observed to divide because they are of bacterial

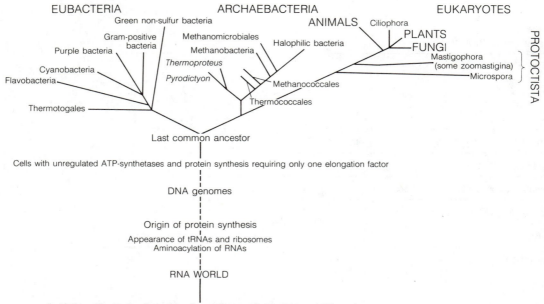

A

FIGURE 3-9 A, B

Universal trees: prokaryotic and eukaryotic organellar phylogenies based on macromolecular homology. (**A**) Protein: RNA polymerase, (**B**) ribosomal RNA, (**C**) ribosomal RNA, (**D**) ATPases. (**C** and **D**, following pages)

(**A**) Universal tree based on RNA polymerases. [Based on an original drawing by K. Delisle as modified by Lazcano (1993).]

(**B**) Small subunit ribosomal RNA sequences from eukaryotic and prokaryotic taxa were aligned using a method that considers the phylogenetic conservation of secondary structures. Similarities were computed from comparisons of 1,020 sites that could be aligned unambiguously and were converted to evolutionary distances. The line-segment lengths represent evolutionary distance between organisms. The large number of taxa included in the analysis precludes complete labeling of terminal nodes. [Based on an original drawing by K. Delisle from unpublished information courtesy of G. Hinkle and M. L. Sogin.]

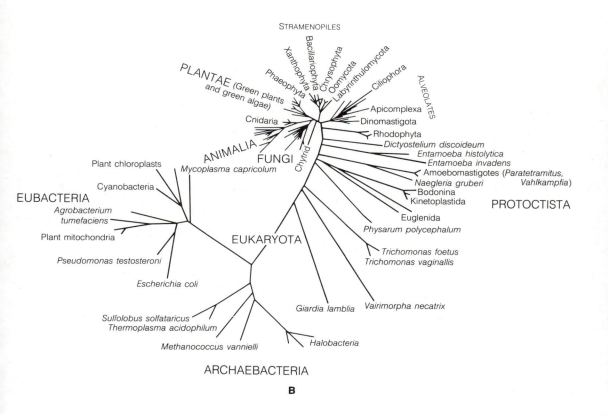

B

origin, although they now live symbiotically within cells of their hosts. He argued that, in appearance, behavior, and response to stains, mitochondria resembled bacteria greatly. The implications of such symbioses for the origin of new complex groups of organisms, including species and eventually higher taxa, did not escape him. However, his claims and fantasies extended so far beyond the data available to him that, like Portier, he was rejected summarily by biologists of the time. His downfall was the claim that he had succeeded in isolating mitochondria from animal cells and culturing them in test tubes on artificial media. His critics accused him of careless technique, which undoubtedly was true, because the growth of mitochondria outside cells is unachievable even by the sophisticated methods available today. Mitochondrial biogenesis inside cells requires many different macromolecules; a large fraction of the mitochondrial proteins, all but a dozen or so, are synthesized on cytoplasmic ribosomes under nuclear control (Gillham, 1978; Baker and Schatz, 1991). Because the requirements for mitochondrial growth and development are extraordinar-

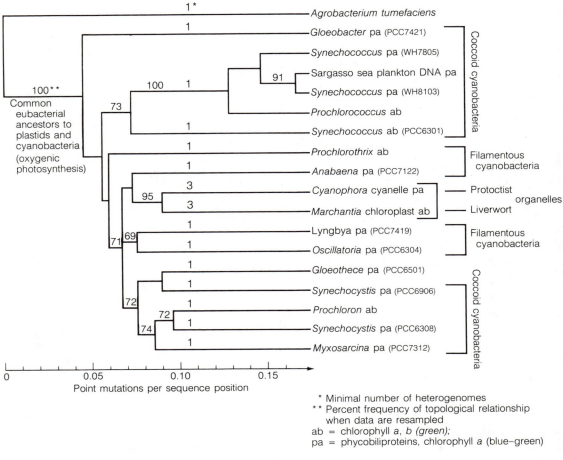

1* *Agrobacterium tumefaciens*

1 *Gloeobacter* pa (PCC7421)

Synechococcus pa (WH7805)

Sargasso sea plankton DNA pa

Synechococcus pa (WH8103)

Prochlorococcus ab

Synechococcus ab (PCC6301)

Prochlorothrix ab

Anabaena pa (PCC7122)

Cyanophora cyanelle pa

Marchantia chloroplast ab

Lyngbya pa (PCC7419)

Oscillatoria pa (PCC6304)

Gloeothece pa (PCC6501)

Synechocystis pa (PCC6906)

Prochloron ab

Synechocystis pa (PCC6308)

Myxosarcina pa (PCC7312)

100** Common eubacterial ancestors to plastids and cyanobacteria (oxygenic photosynthesis)

Coccoid cyanobacteria

Filamentous cyanobacteria

Protoctist organelles

Liverwort

Filamentous cyanobacteria

Coccoid cyanobacteria

0   0.05   0.10   0.15

Point mutations per sequence position

* Minimal number of heterogenomes
** Percent frequency of topological relationship
   when data are resampled
ab = chlorophyll *a*, *b* *(green)*;
pa = phycobiliproteins, chlorophyll *a* (blue–green)

c

ily complex, it will be some time — if ever — before mitochondria are actually grown axenically in vitro.

The shock of recognizing the similarity between some organelles and symbionts has probably been the single most important stimulus to the revitalization of symbiotic theories. The ultrastructural similarity between the chloroplasts of plants and the cells of free-living coccoid cyanobacteria is striking — especially similar are their nucleoids, the DNA-containing areas. Such morphological comparisons revived for Ris and Plaut (1962) the possibility that plastids were derived from cyanobacteria; Ris had rejected this idea earlier on grounds of insufficient evidence. In an influential review paper, Lederberg (1952) pointed out many similarities in behavior

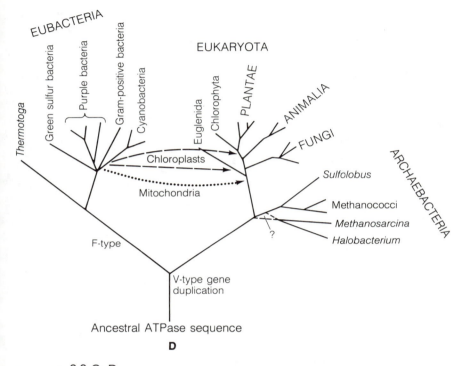

FIGURE 3-9 C, D

(C) Ribosomal RNA sequences. Chlorophyll *a*- and *b*-containing, oxygenic, phototrophic bacteria (chloroxybacteria or prochlorophytes) evolved polyphyletically from cyanobacteria. The numbers in parentheses are strain or identification numbers. [Based on an original drawing by S. Manion-Artz from information in Urbach et al. (1992).] (D) Universal tree of life as derived from the analyses of ATPase subunits. The eubacterial (F_0F_1 dissociable ATPases or F-type) and archaebacterial coupling factor ATPases/ATP synthases are homologous to the vacuolar type ATPase (V-type) that energizes the endomembrane system of eukaryotes. All of these ATPases contain two subunit types that evolved from the same ancestral gene after an ancient gene duplication. This gene duplication predates the first bifurcation leading to the extant organisms; therefore, the subunits that evolved independently since this duplication event can be used as an outgroup to root the tree of life. This figure represents a summary from several trees calculated using different algorithms for phylogenetic inference. The dashed lines and question mark indicate that the data do not yet allow one to decide with certainty whether all archaebacteria branch off in a single node or whether some of the archaebacteria branch off separately and closer to the eukaryotic nucleo-cytoplasm. [Based on an original drawing by Kathryn Delisle and Mary Jane Spring from information provided by J. P. Gogarten.]

between known endosymbionts and cell organelles, especially plastids. Karakashian (1963) was impressed with the genetic control exercised by host *Paramecium bursaria* over the green endosymbiotic chlorellas found in the cytoplasm of all normal members of that species. Karakashian and Siegel (1965) wrote a thoughtful paper in which they noted that, from a functional genetic point of view, a cyclically transmitted symbiont — that is, a symbiont that spends its entire life cycle with its host — and an organelle are virtually indistinguishable.

With the advent of molecular analysis both of eukaryotic organelles and of the protein-synthesizing machinery of bacteria, the basis for comparing organelles with free-living microbes was expanded greatly. It seemed to me (Sagan, 1967) that the assumption of a symbiotic origin of mitochondria and plastids suggested an experimental basis for protistan phylogeny. Similar ideas were stated but not developed by Goksøyr (1967).

Awareness of the importance and ubiquity of symbiotic associations and the nature of their physiology and genetics has since increased (Karakashian, 1975; Margulis, 1976; Cook, Pappas and Rudolph, 1980); Smith and Douglas, 1989; Margulis and Fester, 1991). Modern associations can serve as models for mechanisms of the origin, evolution, and functioning of cells. It is incumbent on developmental biologists to become more aware of the role of interacting symbionts in the genesis of organs: reversible ciliated squid structures (McFall-Ngai and Ruby, 1991), fruitlike galls (Pirozynski, 1991), nematode pouches harboring axenic bacterial cultures (Nealson, 1991), and many others (Schwemmler, 1991). The importance of bacterial symbioses in animal development was emphasized by Wallin (1927) and Buchner (1965). As Smith (1979) remarked, champions of symbiotic theories have not even used the current symbiosis literature to their advantage. Perhaps this book will make some amends.

Symbiotic theories have arisen and been rejected continually since the discovery of eukaryotic organelles in the nineteenth century. In the English-speaking world they have been associated with cytoplasmic genetics (Sapp, 1987), whereas farther east the Famintzyn – Mereschkovsky theory of symbiogenesis has prevailed (Khakhina, 1992; Sapp, in press).

In discussing Portier's book (1918), the skeptical anonymous reviewer quoted at the beginning of this section went on to say,

> Professor Paul Portier maintains that all organisms except bacteria have in a similar fashion [to lichens with their algal and fungal partners] a dual nature. . . . But if all cells are thus dual, why, one hastens to ask, have not the ubiquitous symbiotic, intracellular bacteria been seen before? The answer is that they have often been seen, but persistently misinterpreted. They are the components of the mitochondrial apparatus, those minute formed bodies with many an alias, which have been described in all sorts

of cells. . . . Prof. Portier is good humored enough to quote the paradox that a theory is not of value unless it can be demonstrated false. We have no hesitation in prophesying that his theory will attain that value—which is just what he would have said himself a few years ago. We are bound to admit that the author is a downright good sportsman (*Nature*, 1919).

From 1919—the year in which Lumière, inspired to anger by Portier's ideas, wrote his strongly worded *Le Myth des Symbiotes*—until the early 1970s, this kind of offhand dismissal of symbiotic theories was the rule in "polite biological society" (Wilson, 1959). Recent unprecedented advances in molecular biology have increased the value of symbiotic theories primarily because now they may be proven. Every major concept presented in this book but one, the bacterial origin of undulipodia, has been developed by others: the symbiotic origin of organelles* (Mereschkovsky, 1909, 1910; Portier, 1918; Wallin, 1927; Lederberg, 1952); the implications of symbiosis for the origins of species and higher taxa (Wallin, 1927); the correlation of the early evolution of cells with the pre-Phanerozoic fossil record (Barghoorn and Tyler, 1965; Barghoorn, 1971); and the inability of a single species to evolve independently of others (Reinheimer, 1920; Wallin, 1927). Even though little is unprecedented, the nature of the assertions has changed. That is, detailed predictions distinguishing symbiotic from direct-filiation concepts can now be made, owing to the development of ultrastructural analysis, protein and nucleic acid chemistry, especially sequence data, and modern genetic analysis. Thus, Portier's "sportsmanship" today is replaced by scholarship that, one hopes, rescues the ridiculed idea of cell symbiosis from ignominy.

---

*Kozo-Poliansky (in Khakhina, 1979) even mentioned the symbiotic origin of motility organelles. His Russian book on new concepts of organisms (1924) is still not available in any other language.

CHAPTER 4

BEFORE CELLS

THE GEOLOGICAL CONTEXT

We have no reliable chronological scale in geology
such as is afforded by the relative magnitude of
zoological change.

C. LAPWORTH, 1879

Geochronology is the science of measuring time and of dating events
based on geological information. Since a consensus on the age of the Earth
has been reached, a matured geochronology now provides a temporal
framework in which to envision the events of biological and even microbiological evolution. A brief introduction to this complex subject suffices.

How old is the Earth? The Moon and the Earth, it is assumed, formed,
with the rest of the solar system, from the solar nebula. The moon is very
old: some of its rocks are more than four billion years old, as determined by
potassium-argon, strontium, rubidium, and lead dating techniques. Thus,
studies of lunar rocks obtained by the Apollo program have greatly clarified
early solar-system history. The Earth and Moon may have accreted from
planetesimals into solid bodies about 4600 million years ago. Both owe
their form to early incessant meteorite and planetoid bombardment. Perhaps 15 percent of the Moon, by volume, is material knocked off the Earth
(Newsom and Taylor, 1989; Campbell and Taylor, 1983). The oldest
undisturbed sediments on Earth were laid down in the early waters, some
3900 million years ago. Much evidence of meteoritic bombardment on the
early Earth has been erased by subsequent weathering but meteoric impact
left its clearly visible scars on the surface of the Moon, Mercury, and Mars;

the showers of meteorites that struck our neighboring planets and the Moon must surely have hit the Earth as well. The ancient pockmarks on the Earth were erased by furiously active weathering, as well as by mountain building, subduction, and other tectonic processes. The effect of erosion on craters from recent impacts can be observed today. For example, the great meteor crater between Winslow and Flagstaff, Arizona, formed about 25,000 years ago, is now only half of its original depth.

Geological observations are organized into a framework called "time–rock divisions," the smallest detectable unit of which is the **facies**. A facies, usually seen in the field as a coherent rock layer, is a distinctive part of a sedimentary rock unit; it is characterized by lithological and biological features and is segregated areally from other parts of the unit. The names of rock units, groups of facies, from smaller to larger include, for example, **stage, series, system, group**, and **supergroup**. The corresponding names of geological time intervals from smaller to larger are **age, epoch, period, era, and eon**. Although there is much discussion in time–rock correlations about the absolute amount of time encompassed and about their nomenclature, especially for the larger sedimentary rock units and time intervals, most agree that the term "eon" describes the largest time division. Each eon (defined in some dictionaries as "a billion years") does encompass more than 500 million years. The Phanerozoic eon, the most recent and familiar, began about 600 million years ago with the Cambrian, the first period of the Paleozoic era (Table 4–1). The lowermost Cambrian is marked by the widespread appearance of trilobites and other representa-

TABLE 4 - 1

The Phanerozoic eon.

ERA	PERIOD	APPROXIMATE BEGINNING (MILLIONS OF YEARS AGO)	CHARACTERISTIC FOSSILS
Paleozoic	Cambrian	580	Trilobites, brachiopods, corals, and many other marine animals; algae
	Ordovician	500	
	Silurian	440	
	Devonian	400	First land plants (horsetails, seed ferns, cycads); first vertebrates (fish, amphibians)
	Carboniferous	345	
	Permian	290	
Mesozoic	Triassic	245	Reptiles (including dinosaurs); first mammals; conifers, ginkgos
	Jurassic	195	
	Cretaceous	138	
Cenozoic	Paleogene	66	Flowering plants; extinct and modern mammalian orders, including human primates
	Neogene	25	

tives of five or six major phyla of animals with skeletalized parts. In many museums, the Phanerozoic is the only eon represented; all of the earlier ones are grouped together as the **Precambrian**. I prefer Cloud's (1988) more logical term, the **pre-Phanerozoic**.

A continuous sedimentary series in which even an entire era is represented is extremely rare: the Paleozoic of the Grand Canyon of Arizona is one of the few intact examples. In no single region is an entire eon known to be continuously and conformably intact. It is possible that the Swaziland System, a sequence of ancient South African sediments 17 kilometers thick (see Figure 5–7), represents an entire eon (an era, at least); it formed over an estimated interval of at least 300 million to 500 million years. Because it lacks large fossils, its sediments are far more difficult to map and correlate than those of the Grand Canyon.

The chronology of an eon must be reconstructed from composite geological evidence, from sedimentary deposits of various radiometrically determined ages from several continents. Furthermore, the nature of the fossil record is quite different for each eon: Phanerozoic fossils are abundant, giving direct information about the course of evolution, whereas the pre-Phanerozoic fossil record is sparse. Extensive extrapolation and inference are needed to interpret the timing of pre-Phanerozoic events. The names and dates of the pre-Phanerozoic time–rock divisions, as well as the placement of the lower Phanerozoic boundary, currently are being debated. The geochronology of eons used in this book is, with minor modifications, that of Cloud (1974, 1978, 1988) (Table 4–2).

The **Chaotic eon** began at some indeterminately remote time; it is the period in which the Earth accreted, stabilized, and became recognizable as a planetary body. Cosmologists and astronomers, rather than geologists, concern themselves with the Chaotic events of planetary formation. The **Hadean eon** began next when the Earth became a solid mass of rock, 93 million miles away from the Sun, with the approximate diameter and mass that it has today. No Earth rocks date from either the Chaotic or Hadean eons. A more or less continuous rock and fossil record is available, however, for the next three eons: the **Archean**, the **Proterozoic**, and the **Phanerozoic** (Table 4–3). Paleontology and paleobotany have dealt almost entirely with the most recent eon, the Phanerozoic, whereas the biological evolution that is the subject of this book took place primarily during the Archean and Proterozoic.

The **Hadean eon** extended from approximately 4650 million years ago until 3900 million years ago. As its name implies, this eon represents the time of origin of the concentric structure of the Earth: core, mantle, and crust. The primary atmosphere, which had been part of the solar nebula from which the Earth-Moon system had condensed, was lost. The first

TABLE 4-2
Geochronological eons.

EON	BILLIONS OF YEARS AGO			MAJOR EVENTS
Chaotic	14 (?) – 4.6[a]		No dated Earth rocks	Origin of solar system; first major thermal events; no direct age measurements
Hadean	4.6 – 3.9	"Precambrian" (pre-Phanerozoic)		Establishment of Earth-Moon system; formation and differentiation of Earth as a planet (core, crust, atmosphere); meteorites dated from about 4 to 2.6 billion years ago; lunar rocks from Apollo 12, 14, 15 dated at 4.5 billion years, from Apollo 11 at 3.6 billion years
Archean	3.9 – 2.5			Tectonism; protocontinental rocks (Greenland); geology "immature"; earliest microbial life; age of anaerobes
Proterozoic	2.5 – 0.6		Continuous record of dated Earth rocks	Modern geological tectonic and weathering patterns; stromatolite cyanobacterial communities
Phanerozoic	0.6 – 0	Classical fossil record		Animals and plants; reef communities (see Tables 4-1 and 4-3)

[a]Cosmologists consider the universe to be less than 20 billion years old.

major outgassing took place: volatile material from the interior of the Earth was the source of the secondary atmosphere. Through volcanoes and hot springs emanated the primary outgassed product: water vapor. Thus, the Earth's extensive hydrosphere — lakes, rivers, streams, and subsurface ground water, as well as oceans — probably originated during the Hadean. Although there is no direct information concerning the gas composition of the Hadean atmosphere, most likely it was richer in hydrogen and hydrides and poorer in oxidized gases than is today's atmosphere. During the Hadean eon, transient crustal rocks probably formed but were reworked extensively under high temperature and pressure. The only rocks today, besides those from the Moon, that are old enough to be Hadean are those that fall from space as meteorites. Because there are no surviving Earth rocks from the Hadean eon, early history must be reconstructed from models of solar-system evolution. Direct studies of material returned from the lunar surface

TABLE 4-3

The eras of the Archean, Proterozoic, and Phanerozoic eons.

EON	ERA	MILLIONS OF YEARS AGO	NATURE OF FOSSIL RECORD
Archean	Early	3900	Sparse (Schopf, 1983)
	Late	3000	
Proterozoic	Aphebian	2500	⎧ Highly developed stromatolite
	Riphean (Upper)	1800	⎨ communities; microfossils in cherts
	Riphean (Lower)	1000	⎩ and shales
	Vendian	750	Sand imprints of large, well-developed soft-bodied problematica; microfossils of protocists and bacteria in cherts and shales. Metazoans?
Phanerozoic	Paleozoic	580	Widespread fossils of silicaceous and calcareous skeletons of metazoans; all animal phyla represented; first land animals and plants
	Mesozoic	245	Fossils of cycads, conifers, and reptiles
	Cenozoic	66	Fossils of flowering plants and mammals

by Apollo missions show that lunar rocks have not remelted subsequent to the Hadean or early Archean eons.

The **Archean eon** extended from about 3900 million to 2500 million years ago, although at some sites on the Earth's surface typically Archean processes terminated earlier—about 3000 million years ago. Many Archean rocks are unique. It was a time of extensive igneous activity; molten rock masses formed and flowed extensively below the solid surface of the Earth and its waters (Nisbet, 1987). Many of these igneous intrusives were not exposed to the atmosphere or water until after they had solidified. Intense heat caused much metamorphism deep in the crust. Archean rocks are characterized by very little alteration by weathering. The characteristic pattern of sedimentation has been called "immature." Sodium-rich lavas that were extruded into the atmosphere and hydrosphere sedimented close to the original sites of extrusion. Clasts—rock fragments—of volcanic origin are found in volcano- rather than in water-deposited sediments close to their source volcanoes. Archean volcanic and sedimentary rocks, transformed by heat and pressure into metavolcanics and metasediments, tend to be folded into steep, complex, huge belt-shaped structures that extend great distances. Extensive formations of these once-molten rocks have been studied in Canada, Australia, and South Africa. Such **greenstone belts,** so named for the greenish hue of their metamorphosed minerals, are known but less extensive in younger rocks. Magnesium- and iron-rich volcanic plugs are common in rocks of Archean age. The almost complete absence of granite and basaltic, potassium-rich, igneous rocks typical of later eons also marks the Archean as geologically distinct. Large, stable, open bodies of water were probably rare, although much of what became greenstone belts was probably originally deposited in shallow, not very salty water.

Peculiar sedimentary rocks called **banded iron formations** (BIFs), consisting of alternating oxidized and reduced iron-rich and iron-poor strata of magnetite, and hematite or related minerals embedded in a silica matrix, appear in the Archean. When metamorphosed, they convert into alternating layers of iron-rich and iron-poor quartzite, metamorphosed sandstone. Fewer than ten pre-Phanerozoic BIFs, notably those in Western Australia, southern Africa, the Krivoy-Rog region of Russia and the region surrounding Lake Superior in North America, are the sources of most of the world's iron ore. BIFs that are rich in the mineral hematite, for example, extend through time until the upper Proterozoic; however, they are absent from younger sediments. The existence of these iron formations suggest that reduced forms of iron, metallic and ferrous, once traveled in the surface waters. Deposition of ferrous iron at the Earth's surface is not possible in today's oxygen-rich atmosphere because iron is converted quickly to the ferric form, even in the presence of free oxygen at very low concentrations. Thus, the existence of BIFs is generally accepted as prima facie evidence for the absence of all but extremely low quantities of oxygen in the Archean atmosphere, alternating with local sources of oxygen produced at the sediment-water interface at the time of deposition.

In certain Archean sedimentary rocks, such as those of the Sheba Formation near Barberton, South Africa, there is an abundance of organic carbon, especially in the shales (Reimer, Barghoorn, and Margulis, 1979). Even coal-like and limestone deposits are known from the Archean, although they are relatively rare. Remarkable associations of gold, manganese, copper, iron sulfides, and aluminum with carbon-rich Archean sedimentary rocks are known. Deposition of photosynthetic organic carbon is related to mineral accumulation in, for example, Witwatersrand South African gold deposits (Hallbauer and van Warmelo, 1974; Mossman and Dyer, 1985).

The **Proterozoic eon** extended from about 2500 million to 600 million years ago. This eon is divided into two major intervals: the lower Proterozoic, extending from about 2500 million until 2000 million years ago, and the upper Proterozoic, extending from 2000 until 600 million years ago. By the Proterozoic eon, the large granitic continents typical of the modern Earth prevailed.

A Proterozoic geological regime was already in place in certain regions on the Earth as much as 3 billion years ago—for example, in the Zululand of southern Africa; in other regions—for example, northwestern Canada

—this regime did not begin until about 2500 million years ago. The lower Proterozoic is characterized by extensive epicratonal sediments. Cratons are the large stable surfaces of continents, and epicratonal sediments are those derived from the weathering of cratons; that is, they did not come from geothermal or ocean weathering, as did the Archean greenstone belts, but from continental masses. The lower Proterozoic epicratonal sediments differ from modern ones in that they tend to be less oxidized. However, they are associated with and intruded by igneous rocks typical of those of younger ages: potassium-rich granites and gneisses. The peculiarly Archean sodium-rich rocks by then had been replaced by modern volcanic and magmatic rocks. BIFs continued to be common. Upper Proterozoic sediments contain abundant oxidized epicratonal sediments that have been weathered from granitic continental masses. The oxidized features are interpreted as signaling a rise in the concentration of atmospheric oxygen (Cloud, 1988; Walker, 1983).

In all rocks formed since the lower Proterozoic, major igneous intrusions are composed of potassium-rich granites and gneisses. Very few BIFs bearing both ferrous and ferric iron were formed until the mid-Proterozoic. Iron, the fifth most abundant atom on Earth, is distributed primarily in oxidized iron sediments called **red beds**, which bear mainly ferric iron. Red beds are rich in dispersed iron oxides.

Stromatolites—organosedimentary layered rocks composed of limestone (calcium carbonate) or, less frequently, of carbon-rich opaline silica—became prominent in the upper Proterozoic of Western Australia, southern Africa, and Canada. Stromatolites are interpreted to be trace fossils, primarily of photosynthetic microbial communities (Awramik, Margulis, and Barghoorn, 1976; Awramik et al., 1979). Sparse in Archean rocks, stromatolites in the Proterozoic are far more abundant and well developed. The diversity of stromatolites, as well as the area they covered, expanded enormously during upper Proterozoic times. Pre-Phanerozoic stromatolites show characteristic forms and patterns that even make them useful as stratigraphic markers (Walter, 1976). The interpretation of pre-Phanerozoic stromatolites as biogenic structures has had profound implications for the reconstruction of the evolutionary history of microbial ecosystems (Monty, 1973b, c; Schopf and Klein, 1992).

Many soft-bodied fossils, about 680 million years old, some representative of metazoan phyla, have been found in sediments from several widely distant upper Proterozoic locations (Glaessner, 1968, 1971, 1976). Many unique Vendian fossils are probably not remains of animals, but of large protoctists (McMenamin and McMenamin, 1990; Seilacher, 1985; Seilacher, Reif, and Westphal, 1985). However, an abundance of difficult-to-identify microfossils has been reported in upper Proterozoic rocks. Peculiar but distinctive microfossils called **acritarchs, chitinozoa,** and **hystricho-**

spheres (the third group may be remnants of dinomastigote cysts) are common. Such fossils indicate that eukaryotes more than one billion years ago radiated as diverse protoctist lineages throughout the late Proterozoic eon (Knoll, 1990; Vidal, 1990; Bengtson, 1993).

The **Phanerozoic eon** extends from about 600 million years ago until the present. Sediments derived from large continental masses contain remains of animals, plants, and protoctists, often in great quantities. Although fossils from both Archean and Proterozoic rocks are known, the Phanerozoic eon is still considered by many to be equivalent to the "fossil record." The first era of the Phanerozoic, the Paleozoic, is distinguished by the worldwide appearance of skeletalized metazoans; it has been dubbed the "age of invertebrate animals." In early Phanerozoic rocks, the abundance and diversity of stromatolites declines. A corresponding rise occurs in the abundance of carbonate reef facies. These reefs were produced by communities of poriferans, archeocyathids, and other skeletalized metazoans, and later by colonial coelenterates, corals. By the Mesozoic, an era of the Phanerozoic eon that began about 245 million years ago, "modern" coral reefs, self-regulating barrier ecosystems, abounded in tropical oceans.

During the Phanerozoic, beginning primarily in the late Paleozoic, the continents changed their shapes and positions, coming to lie at different geographic locations at different times and enjoying changing climatic conditions. This "plate-tectonic" behavior of the Earth's crust probably began well before the Phanerozoic eon. The drifting of continents in the Phanerozoic had large consequences for the distribution of land animals, plants, and marine organisms. As the history of the Earth's surface-plate movements during the pre-Phanerozoic becomes evident, the influence of tectonic change on the course of pre-Phanerozoic evolution may become traceable.

ORGANIC COMPOUNDS

There is no special force exclusively the property of living
matter which may be called a vital force; . . . rather,
this force arises from the conflict of numerous other
[forces] and organic nature possesses no laws other than
those of inorganic nature.

J.J. BERZELIUS, 1806
(in Coleman, 1971)

Life began probably in the Hadean or early Archean eon. Most scientists would assert enthusiastically that life began spontaneously as an outcome of cosmological trends, that it was the product of the process of chemical evolution — the natural production, concentration, and interaction of organic compounds. Although scientists rarely agree about any past event, the consensus of physicists, chemists, and biologists is that life originated from prebiotically formed organic matter on the early Earth. The idea has been bolstered by the discovery that organic molecules are produced in interstellar space. In retrospect, the presence of organics between the stars might have been expected from the observation that hydrogen, carbon, nitrogen, and oxygen — the components of organic compounds and hence of living systems — are among the most abundant elements in space (Table 4-4). Another defense for this generally held concept is the detection of considerable quantities of organic matter in meteorites, objects known to have an extraterrestrial provenance (Oró, Miller, and Lazcano, 1990). Yet another is the ease with which most of the monomers of living systems can be made in laboratory simulations of conditions thought to have prevailed on the surface of the Earth during the Hadean and early Archean eons. The results of such simulations led to the concept of the spontaneous origin of the earliest cells: self-maintaining, reproducing systems enclosed in membranes.

Radioastronomers using the large antennae of radiotelescopes can detect and identify extraterrestrial sources of electromagnetic radiation. Since 1932, when Karl Jansky announced his discovery of radio emissions from space, the rate of detection of interstellar organic matter has continued to

TABLE 4-4

Chemical composition of the Sun.[a]

ELEMENT	ABUNDANCE (RELATIVE TO HYDROGEN)
Hydrogen	1.0
Helium	1.0×10^{-1}
Oxygen	6.8×10^{-4}
Carbon	3.7×10^{-4}
Neon	2.8×10^{-4}
Nitrogen	1.2×10^{-4}
Magnesium	3.4×10^{-5}
Silicon	3.2×10^{-5}
Iron	2.6×10^{-5}
Sulfur	1.6×10^{-5}
Argon	3.7×10^{-6}
Calcium	2.3×10^{-6}
Aluminum	1.9×10^{-6}
Sodium	1.9×10^{-6}
Phosphorus	$< 10^{-10}$

[a] After Ponnamperuma (1977).

TABLE 4-5

Molecules in interstellar space.[a]

MOLECULAR STRUCTURE	INORGANIC			ORGANIC		
	MOLECULE OR RADICAL		YEAR OF DISCOVERY[b]	MOLECULE OR RADICAL[c]		YEAR OF DISCOVERY[b]
Diatomic	OH	hydroxyl	1963	CH	methylidyne	1937
	H_2	hydrogen	1970	CH^+	methylidyne ion	1937
	SiO	silicon monoxide	1971	CN	cyanogen	1940
	NS	nitrogen sulfide	1975	CO	carbon monoxide	1971
	SiS	silicon monosulfide	1975	CS	carbon monosulfide	1971
	HCl	hydrochloric acid		CC		
	NH					
	PN	phosphorus nitride				
	NaCl	sodium chloride				
	AlCl					
	KCl	potassium chloride				
	AlF					
Triatomic	H_2O	water	1968	HCO^+	formyl ion	1970
	H_2S	hydrogen sulfide	1972	HCN	hydrogen cyanide	1970
	N_2H^+	protonated nitrogen	1974	HNC	hydrogen isocyanide	1971
	SO_2	sulfur dioxide	1975	OCS	carbonyl sulfide	1971
	HNO	nitroxyl ion		CCH	ethynyl	1974
				HCO	formyl	1976
				CCC		
				CCO		
4-atomic	NH_3	ammonia	1968	H_2CO	formaldehyde	1969
				HNCO	isocyanic acid	1971
				H_2CS	thioformaldehyde	1971
				C_2H_2	acetylene	1976
5-atomic	SiH_4	silane		HCOOH	formic acid	1970
				HC_3N	cyanoacetylene	1970
				H_2CNH	methanimine	1972
				H_2NCN	cyanamide	1975
				CH_4	methane	
				CCCCC		
				$H_2C{=}C{=}O$	ketene	
				$H_2C{=}C{=}C$	propadienylidene	1990
				H_2C_3	cyclopropenylidene	1984

(continued)

increase. Radicals containing up to 13 carbon atoms have been detected; organic fragments identified unambiguously by radiospectroscopy are listed in Table 4-5. However, there may be little direct relation between galactic organic matter and the origin of life on Earth. The extreme conditions of heat, pressure, impact, and general instability that are thought to have

TABLE 4-5 (continued)

MOLECULAR STRUCTURE	INORGANIC			ORGANIC		
	MOLECULE OR RADICAL	YEAR OF DISCOVERY[b]		MOLECULE OR RADICAL[c]		YEAR OF DISCOVERY[b]
6-atomic				CH_3OH	methanol	1970
				CH_3CCN	cyanomethane or methylcyanide	1971
				$HCONH_2$	formamide	1971
				$HC\equiv CCHO$	propynal	1988
				$H_2C=C=C=C$	butatrienylidene	1991
7-atomic				CH_3C_2H	methylacetylene	1971
				CH_3COH	acetaldehyde	1971
				CH_3NH_2	methylamine	1974
				H_2CCHCN	vinylcyanide (acrylonitrile)	1975
				HC_5N	cyanodiacetylene	1976
8-atomic				$HCOOCH_3$	methyl formate	1975
9-atomic				$(CH_3)_2O$	dimethyl ether	1974
				CH_3CH_2OH	ethanol	1974
				HC_7N	cyanohexatriyne	1977
11-atomic				HC_9N	cyanooctatetrayne	1978
13-atomic				$HC_{11}N$	cyanodecapentayne	

[a]After Goldsmith and Owen (1980) and Irvine et al. (1991). Because microwave spectroscopy depends on permanent dipole moment, nonpolar molecules (for example, CH_4 and N_2) have not been detected in this way; nevertheless, they probably exist in interstellar space. Some are limited to gaseous envelopes around stars. For a more complete list, see Irvine et al. (1991).
[b]If no date listed, between 1980 and 1991. See Ponnamperuma (1981), Irvine et al. (1991).
[c]If not known on Earth, no name available.

prevailed are likely to have destroyed interstellar organic compounds before they reached sufficient concentrations to form autopoietic systems. The discovery of cosmically abundant organic matter implies simply that life's chemical precursors are rather commonplace in the universe. Because they form easily in the cosmos, these carbon-hydrogen compounds also probably formed easily on the Chaotic Earth.

The crucial unsolved problem is how organic compounds organized into self-perpetuating systems. Even though the nonbiological production of organic matter is common, the origin of life itself may be a rare or unique event. No evidence exists that life has ever begun anywhere in the universe except on the surface of the Earth before or during the early Archean some 3500 million years ago.

Meteorites

Dr. J Lawrence Smith, of Louisville, KY., has made a
personal investigation of the great meteorite which fell in
Emmett County in 1879. . . . The external appearance
was that of a mass, rough and knotted like mulberry
calculi, with rounded protuberances projecting from the
surface. The larger portions were of a gray color, with a
green mineral irregularly disseminated through it. The total
weight of the portions found amounted to 307 pounds.
The stony part of this meteorite consisted essentially of
bronzite and olivine, the three essential constituents being
silica, ferrous oxides and magnesia. An analysis showed
that in composition the meteorite contained nothing that
was peculiar.

SCIENCE, 1880

Meteors frequently come to Earth as "falling stars" that enter the atmosphere and burn brightly before reaching the surface, if they reach it at all. They are a unique and regular source of extraterrestrial material amenable to laboratory study. Some, such as the Murchison meteorite that landed in September 1969 in Australia, were analyzed within days after their descent.

Meteorites are of interest because they are ancient extraterrestrial repositories of naturally occurring, nonbiological organic matter. Furthermore, they are probably the only objects formed in the solar system during the Hadean or Archean eons that can be manipulated in the laboratory. Although it is widely accepted that meteorites originate from the asteroid belt, it is perhaps equally possible that they are the remains of comets. They may even be remnants of ancient collisions between the Earth and other solid bodies.

Most meteorites collected have an aluminosilicate, stony composition. About one in twenty is conspicuously metallic — iron alloyed with nickel; depending on the proportion of metal, these are called **irons** or **stony irons**. About two percent of all meteorites collected are **carbonaceous chondrites**, a class of stony meteorites directly pertinent to questions about the origins of life. Carbonaceous chondrites actually make up a greater proportion of meteorites that enter the atmosphere, but they are recorded infrequently because they disintegrate easily. The ratio of carbonaceous chondrites to other meteorites increases in polar regions, where they are more likely to be preserved than are those falling at lower latitudes. Chondrites contain **chondrules**, spherical structures about 1 mm in diameter that are made of the minerals olivine and orthopyroxene, which probably were formed by rapid cooling.

A list of results of tests for organic substances in carbonaceous meteorites is presented in Table 4-6. Life probably evolved from carbon compounds such as these produced early in the history of the solar system. The organic constituents of meteorites, like interstellar organics, were not likely to be direct precursors of life on Earth because heat, ultraviolet light, and

TABLE 4-6

Composition of carbon-rich meteorites.[a]

FORMS	CONSTITUENTS[b]		
Total carbon (35–50 mg/g)	? Carbonate + Extractable carbon + Polymeric carbon ? Graphite − Diamond + Carbides		
Extractable organic carbon, generally 1 mg/g; racemic mixtures	+ Methane, ethane + n-Alkanes, odd and even (C_{10} . . . C_{24}) +Fatty acids (C_{14}, C_{16-20}, C_{22}, C_{26}, C_{28}) + Phenolic compounds (aromatics, alcohols, esters, acids, benzene, toluene, naphthalene) + Amino acids (glycine, ethylglycine, alanine, β-alanine, methylalanine, aminobutyric acid, amino-n-butyric acid, β-aminoisobutyric acid, valine, norvaline, isovalline, proline, pipecolic acid, aspartic acid, glutamic acid, sarcosine) + Heterocyclics (melanine, ammeline) + Sugars (mannose, glucose) + Saturated polyisoprenoids (pristane, nonpristane, phytane) + Vanadyl porphyrin − Chlorin (common derivatives of terrestrial organic matter) + Benzonitrile + Methylbenzonitrile (or indole) + Alkylpyridine (or aniline) + Alkylpyrrole ? Triazines − Hydroxypyrimidine[c] − Hydroxymethylpyrimidine[c] + Xanthine (=3 mg/g maleonitrile) + Hypoxanthine + Guanine ? Adenine		
Polymeric carbon (irregular, highly substituted aromatic polymer, structure unknown)	+ C (70%); + O (10%); + S (7%); + H (4%);	+ N (2%); + C (1%); + F; −K;	− Br; − I; − Si; + Residue, "ash" (5%)

[a] After Hayes (1967); van der Velden and Schwartz (1977).

[b] + detected; − looked for, not detected; ? questionable detection.

[c] No pyrimidines detected. Previous reports in doubt; detected pyrimidines probably formed from contaminants [Ponnamperuma (1981).]

other radiation would have destroyed these subtle chemicals before they could have concentrated sufficiently. No trace of meteoritic organic matter has been found on the surface of Mars, even though the presence of craters from ancient impacts implies that meteorites—including carbonaceous ones—struck the surface violently, early in the history of that planet (Owen et al., 1977). Only tiny amounts of carbon (from 1 to 70 parts per million) were found in the rock samples returned from the Moon, and even less (from 25 to 250 parts per billion) could be detected in the looser material —the regolith ("soil") and breccia samples. Upon pyrolysis, heating in the absence of oxygen, most of the carbon in lunar samples came off as carbon dioxide or carbon monoxide—less than 1 part per million of lunar carbon is organic. There may be no organic carbon at all in the lunar samples: the quantities measured in these studies are close to the limits of detectability of the techniques. If they did provide organic matter for early life directly, carbon compounds brought in by meteoritic impact must have been protected from destruction. Apparently on the Moon and on Mars all surface organics were destroyed throughly. Some investigators suggest that as much as 10^{23} grams of carbon was brought to the Earth on meteorites. This enormous amount is approximately equal to the total carbon stored in today's biota (Oró, Holzer, and Lazcano-Araujo, 1980).

The carbon found in meteorites testifies that complex organic compounds and abiotic structures form naturally and did so during the Hadean. Because of the extreme conditions to which meteors are subjected, most astronomers believe it extremely unlikely that life arose on them or on their parent bodies. The organic compounds detected in carbonaceous meteorites perhaps are best considered a sort of natural control for experiments in the origin of life; they represent the kinds and amounts of compounds and microstructures that were produced on the early Earth. Morphologically, the most complex organic inclusions in meteorites are rather simple, as can be seen in Figure 4-1 (Rossignol-Strick and Barghoorn, 1971).

MAKING LIFE IN THE LABORATORY

In the early decades of the [nineteenth] century . . . the issue of vitalism appeared to play a significant role in the emerging science of organic chemistry. Organic substances it seemed were formed only in organisms and were thus a product of that same vitality which maintained the living being.

W. COLEMAN, 1971

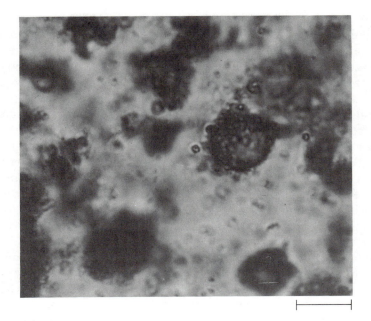

FIGURE 4-1

Microstructures of organic matter from the Orgueil meteorite. Light micrograph, bar = 5 μm. [Courtesy of E. S. Barghoorn.]

In the 1760s Spallanzani showed that decaying organic matter, if protected from mice or flies, does not give rise to infant mice or maggots. Pasteur, a century later, proved elegantly that even microbes do not arise from debris or chemicals but only from preexisting microbes. To a world that believed in an all-powerful Creator, these experiments spelled the downfall of the idea of spontaneous generation of life. The question of the origins of life reverted from the scientific to the philosophical realm, a position it still holds in the minds of many. In this century, Oparin (1924) and Haldane (1929) proposed that life originated as an interaction between organic compounds formed under the anoxic conditions that prevailed on the early Earth. The question of the origins of life became an experimental problem with Miller's (1953) pioneering discovery of the easy and rather rapid formation of amino acids from inorganic precursors. This new field of laboratory science, referred to as "primitive-Earth-model experiments," "probiotic chemistry," or "experimental chemical evolution," has generated a large literature (Ponnamperuma, 1972, 1977; Miller and Orgel, 1974; Oró, Miller, and Lazcano, 1990).

Several laboratory approaches to the problem of the origins of life exist, although some have been maligned by purists as inadequate experimental organic chemistry. One approach constructs an environment thought to have prevailed during the Hadean and provides it with inorganic reactants and energy; molecules significant to modern life are then sought among the reaction products. In another all variables—for example, amounts and types of starting materials, pH, physical environment (such as the simulation of an ocean shore with diurnal changes in temperature and light), and minerals in the medium—are specified very closely. At each addition of materials or change in conditions, attempts are made to identify the compounds produced, their yields, and the mechanisms by which they are made. Both approaches assume that compounds important in present living systems were also important precursors to life. Hydrocarbons, fatty acids, benzene and phenolic compounds, amino acids, and many other familiar organics are produced in the laboratory from carbon, hydrogen, oxygen, nitrogen, and energy; this bolsters confidence in the hypothesis that life arose from cosmic constituents under anoxic conditions. Indeed, organic compounds found in meteorites are quite similar to those produced in simulation experiments. They resemble each other more than they do the organic matter produced biotically, by cells. Definitive distinctions between abiotically and biotically produced organic compounds are not possible, although abiotic methods produce racemic mixtures and a wider variety of isomers and compounds than are produced by the present biota. This variety is independent of the energy source used. Abiotic conditions that result in the highly specific set of isomers and the compounds of high molecular weight produced by cells have not been found.

The kinds and amounts of energy thought to have been available for synthesis of organic matter during the Hadean and early Archean are shown in Table 4-7. Solar ultraviolet light, which contains the most energy, was probably more important than other sources. The Archean ultraviolet radiation flux at the Earth's surface prior to the accumulation of atmospheric oxygen may have been the major source of energy both for prebiotic organic synthesis and for early biological processes such as the induction

TABLE 4-7

Sources of energy for prebiotic synthesis.[a]

TYPE OF ENERGY	ESTIMATED FLUX[b]
Ultraviolet light	570
Electric discharge	4.0
Radioactivity	0.8
Heat (geothermal)	0.13

[a] After Ponnamperuma (1977).
[b] Calories per square centimeter per year.

of lysis and transduction by bacteriophage (Margulis, Walker, and Rambler, 1976; Rambler and Margulis, 1979).

In abiotic laboratory experiments, organic compounds of biological significance are produced mostly under mildly alkaline and totally anoxic conditions (Oró, 1970). Under acidic conditions, piperazines and other compounds not generally found in living organisms are formed. Under oxidizing conditions, organic compounds are not produced at all (Abelson, 1963). The proportion of hydrogen in the reaction mixture also affects the yield of organics: larger relative amounts lead to lower total production of amino acids but no change in the quality of the yields (Ponnamperuma, 1977). Hydrogen cyanide is a likely starting material for the synthesis of certain amino acids and nucleotides (Figure 4-2). The status of this research, much of it supported by the NASA Life Sciences Program, is reviewed by Hartman, Lawless, and Morrison (1987) and by Oró, Miller, and Lazcano (1990). The problem of the production of organic components of small molecular weight seems more or less solved, but how did aggregated organic matter make autopoietic systems? How did monomers polymerize? How were information systems bounded and metabolism maintained in aggregations of organic matter? How was reliable heritability established? The gap between the most complex mixture of organic chemicals and the simplest cell is still to be crossed both in theory and in the laboratory.

The linking of amino acid residues into polymers, including the formation of both peptide and nonpeptide linkages, occurs under conditions presumed plausible for the early Earth (Fox, 1977). Complex microstructures, some apparently with catalytic activity, have been produced spontaneously as well. But abiotic production of catalytic amino acid polymers may not be related directly to the origin of life because polypeptides have no capacity for self-maintenance nor tendency to reproduce. In principle, no single class of biochemicals can form an autopoietic system. Except for the assembly of certain small peptides such as gramicidin and the pentapeptides in the cross-linkage of bacterial cell walls, amino acids do not polymerize by direct enzymatic reaction; they polymerize on nucleoprotein ribosomes through the mediation of transfer and messenger RNAs. (Moreover, even if peptide bonds formed, they would tend to hydrolyze, so that any direct prebiotic formation of peptides is highly unlikely.) That the earliest living system made proteins in the absence of a nucleic acid system is improbable. Catalysis may have been accomplished by "enzymes" made of nucleic acid. White (1976) suggested that early systems used nucleic acid fragments with some catalytic propensity for the polymerization of more nucleic acid. This might explain the observations that (1) a large number of enzymes, 52 percent of the 1750 catalogued ones, require coenzymes for activity and that (2) most of these coenzymes either are nucleotides or are

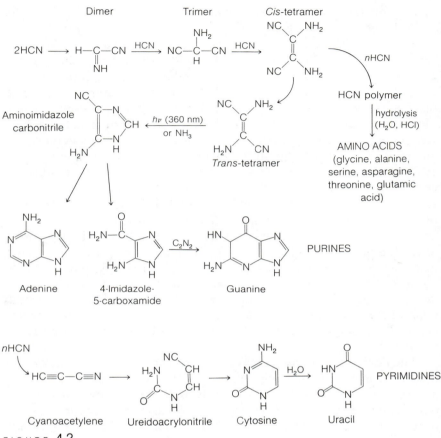

FIGURE 4-2

Possible probiotic pathways from hydrogen cyanide to nucleotide bases and amino acids. Experiments that start with methane and ammonia produce only trace quantities of adenine, whereas from 1 M HCN (hydrogen cyanide) the yield may be as much as 1 percent crystalline adenine. That HCN was a key intermediate in prebiotic synthesis is strongly indicated because it is the most abundant product in most spark-discharge experiments. Dilute HCN is concentrated easily at $-20°$ C, at which temperature water is removed from the mixture. The tetramer forms spontaneously in concentrated solutions.

clearly related to them. It seems likely, then, that the earliest cellular systems were based either entirely or primarily on polynucleotides—as imprecise and short as the early nucleic acids may have been. Nucleotides tend to form complementary-base-pairing polymers not only because they fit geometrically, but also because, in nonaqueous solution, adenine—

thymine and adenine – uracil pairs form more easily than adenine – adenine pairs, owing to electronic affinities (Rich, 1970). With discoveries by Thomas Cech and Sidney Altman and their colleagues we know that by itself RNA can both replicate and act catalytically (Lewin, 1982). RNA by itself — in certain ribonuclease enzymes — cleaves transfer RNA precursors to transfer RNAs in magnesium solution (Guerrier-Takada et al., 1983). One imagines the first autopoietic system — first life — to have been membrane-bounded, replicating-catalytic RNA; both protein and DNA synthesis came later (Lazcano et al., 1988). The establishment of any fidelity in the replication of RNA polymers was probably, from the beginning, due to interaction with amino acids, peptides, and catalytic nucleotides within membrane-bounded droplets (White, 1976).

Replication — the production of a complement itself capable of producing a complement — is a property of molecules, even in a test tube. Replication does not necessarily require life. Reproduction, however, is characteristic only of cells or organisms, that is, of autopoietic entities. Metabolism, the chemical manifestation of autopoiesis, is required to ensure continued reproduction (including molecular replication) by any material entity. No autopoietic, reproducing system (membrane-bounded, coupled polymerization of nucleic acids and proteins) ever has been formed in the laboratory from chemical precursors. But biologically active, replicating nucleic acid polymers have been synthesized abiotically: Spiegelman (1967) synthesized infective viruses of both the DNA and RNA varieties from much simpler precursors. It took only a purified enzyme (a replicase, an RNA-dependent RNA polymerase), a template nucleic acid, at least one human investigator, and much energy in the form of nucleotide precursors and dollars, under proper conditions, to make RNA viruses capable of continued replication. Analogous experiments with DNA viruses were performed by Goulian, Kornberg, and Sinsheimer (1967). Their method required two specific enzymes, namely a DNA polymerase and ligase, an enzyme that joins the ends of linear nucleic acid to form a circle. Thus, not reproducing cells, perhaps, but at least replicating infectious viruses (which are not autopoietic entities) can be made in the laboratory — by continuing human intervention (Kornberg and Baker, 1992). The investigators brought together, in reasonable concentrations, materials that tend to interact. If plausible natural phenomena could accumulate organic materials to replace those provided by the investigator in the laboratory, the problem of the origins of life could be solved, in principle, because an adequate supply of nucleic acids enclosed in membrane structures would form self-sustaining autopoietic systems. Agents tending to increase the concentrations of probiotically produced organic compounds have been suggested (Table 4-8). The enclosure of nucleic acid systems in self-made lipoprotein membranes from the

TABLE 4-8

Methods for concentrating probiotic organic compounds.

MECHANISM	REFERENCES
Coacervates and protobionts	Oparin, 1924, 1969; see Day (1984) for critical discussion; Lazcano, 1992
Apatite surface (also solves low-phosphorus problem)	Bernal, 1967; Hutchinson, 1965
"Ocean line" (thermocline)	Weyl, 1968
Microspheres (made by heating organic compounds to drive off water)	Fox and McCauley, 1968
Excitable membrane	Ponnamperuma and Gabel, 1968
Drying fringes of the sea, salt pools, environments rich in sulfur and iron	Ehrensvard, 1962
Eutectic concentration mechanisms	Orgel, 1970; Miller and Orgel, 1974
Lipid membranes	Miller and Orgel, 1974; Deamer, 1992

TABLE 4-9

Amino acids and related compounds under conditions simulating the early Earth: early examples.

DATE	INVESTIGATORS[a]	STARTING MATERIALS[b]	ENERGY SOURCE	PRODUCTS
1953	Miller	Methane, ammonia, hydrogen	Electric discharge	Glycine, alanine, β-alanine, glutamic acid
1955–1956	Abelson	Hydrogen, methane, carbon monoxide or carbon dioxide, ammonia or nitrogen, oxygen	Electric discharge	Alanine, glycine, sarcosine, serine, aspartic acid (only in the absence of oxygen)
1957	Walter and Mayer	Methane, ammonia, hydrogen sulfide	Helium spark	Ammonium thiocyanate
1957–1966	Schramm et al., Fox et al., Ponnamperuma	Tripeptides, polyphosphoric acid esters, 18 amino acids	Heat (50°–60°C)	Polypeptides, proteinoids
1959	Oró	Formaldehyde, hydrogen cyanide	Room temperature (spontaneous)	Glycine, alanine
1959–1965	Becker et al., Hanafusa et al., Ponnamperuma et al., Steinman et al.	n-Carboxyamino acid anhydrides, aminoacetonitrile sulfate, kaolin, dicyanodiamide, hydrogen chloride, aminoacetonitrile	Heat to 100°C	Aspartic acid, threonine, serine, glutamic acid, proline, glycine, alanine, valine, isoleucine, leucine, tyrosine, phenylalanine
1960	Groth and Von Weyssenhoff	Methane, ethane, ammonia	Ultraviolet (407 nm, 219.6 nm)	Glycine, alanine, aminobutyric acid
1963	Oró	Ethane, methane, ammonia		Glycine, alanine, proline, valine, leucine, asparagine

(continued)

beginning was probably a major strategy for the concentration of organics (Deamer, 1992).

Almost every monomeric component of the macromolecules of living cells has been produced in the laboratory "probiotically," that is, under appropriate chemical conditions. A variety of amino acids is produced routinely in such experiments (Table 4-9). The four that form most easily (glycine, alanine, aspartic acid, and glutamic acid) are also extremely abundant in proteins. This is probably not coincidental. Chemically rather simple, they are thought to have been the "first" amino acids. Nucleotides and other nucleic acid derivatives that have been produced under conditions simulating the primitive Earth are listed in Table 4-10; the present indispensability of adenosine triphosphate (ATP) and the other triphosphate precursors of nucleotides—guanosine, uridine or thymidine, and

TABLE 4-9 (continued)

DATE	INVESTIGATORS[a]	STARTING MATERIALS[b]	ENERGY SOURCE	PRODUCTS
1965	Grossenbacher, Knight	Methane, ammonia bubbled through ammonium hydroxide		Glycine, alanine, threonine, serine, glutamic acid, isoleucine
1965	Oró	Methane, ammonia	1300 K heat	Aspartic acid, threonine, serine, alloisoleucine, isoleucine, phenylalanine, β-alanine, glutamic acid
1966	Ponnamperuma	Methane, ammonia	Electric discharge	Polypeptides
1966	Abelson	Ammonium formate, ammonium hydroxide, sodium cyanide, ferrous sulfate	Ultraviolet	Hydrogen cyanide intermediate (glycinonitrile glycine)
1966	Ponnamperuma and Flores	Methane, ammonia, and water added to hydrogen cyanide; then hydrolysis of hydrogen cyanide		Glycine, alanine, threonine, valine, phenylalanine
1968	Steinman et al.	Ammonium thiocyanate hydrolyzed by 6N HCl; gas-phase nitrogen	Ultraviolet (260 nm for 3 hr)	Methionine

[a]The work with sulfur-containing amino acids is reported in Steinman et al. (1968); work on peptides in Keosian (1968); other work in Ponnamperuma and Gabel (1968) and Day (1984). See also Matthews (1992), Deamer (1992), Ponnamperuma (1992), and Lazcano (1992).
[b]Water was present in most experiments.

TABLE 4-10
Synthesis of nucleic acid derivatives under conditions simulating the early Earth.

DATE	INVESTIGATORS[a]	STARTING MATERIALS[b,c]	ENERGY SOURCE	PRODUCTS[c]
1960	Oró	Ammonium cyanide	Spontaneous at high concentrations	Adenine
1961	Oró	Aminoimidazole carboximide	Heat (100° – 140° C)	Guanine, xanthine
1961	Fox and Harada	Malic acid, urea	Heat	Uracil
1963	Oró	Acrylonitrile, aminopropyl malonamide semialdehydimine (found in comets) in aqueous solutions of ammonium chloride	Heat (135° C)	Uracil
1963	Ponnamperuma et al.	Ribose, adenine, and phosphate in dilute aqueous solutions	Ultraviolet	Adenosine
1963	Ponnamperuma and Sagan	Adenine, adenosine, AMP, ADP, ethyl phosphate	Ultraviolet	Adenosine, AMP, ADP, ATP
1964	Ponnamperuma and Kirk	Deoxyribose, adenine, hydrogen cyanide	Ultraviolet	Deoxyadenosine
1964	Miller and Parris	Hydroxyapatite and cyanate salts (reaction takes place on surface of mineral)		Pyrophosphate
1965	Ponnamperuma	Hydrogen cyanide (dilute)	Ultraviolet	Adenine, guanine, urea
1968	Ferris et al.	Cyanoacetylene, cyanate, cyanogen	Electric discharge (cyclical temperature changes to freeze water lead to eutectic concentration)	Pyrimidines, cytosine, uracil
1974 to present	Orgel, others (see Day, 1984)	Bases, ribose, sea salt	Heat cycles (simulated volcano, hot springs)	Inosine, guanosine, adenosine, xanthosine

[a]See Oró et al. (1990) and Lazcano (1992) and references therein for details.
[b]Water was present in most experiments.
[c]AMP, ADP, ATP = adenosine mono-, di-, and triphosphate.

cytidine triphosphates (GTP, UTP or TTP, and CTP) — probably is due in part to the ease with which they were made in probiotic chemical processes. Figure 4-2 shows suggested pathways for the probiotic synthesis of purines and pyrimidines. Various carbohydrates have also been produced abiotically (Table 4-11). Some authentic energy sources, such as ultraviolet light and heat, are used in these experiments; other sources, such as high-energy particles produced in an accelerator, are unlikely to have been

TABLE 4-11

Synthesis of carbohydrates under conditions simulating the early Earth: Early examples.

DATE	INVESTIGATORS[a]	STARTING MATERIALS[b]	ENERGY SOURCE	PRODUCTS[c]
1861	Butlerow and Leow	Formaldehyde in presence of alkali		Mixture of sugars, optically inactive hexoses
1933	Orthner and Gerisch	Glyceraldehyde, dihydroxyacetone		Erythrose, pentoses
1962	Oró and Cox	Glyceraldehyde, acetaldehyde (catalyzed by divalent metal ions or ammonia)		Deoxyribose
1964	Steinman et al.	Glucose, dicyanodiamide, phosphoric acid	Ultraviolet, spontaneous	Disaccharides, glucose-6-phosphate
1965–1966	Ponnamperuma	Formaldehyde, hydrogen cyanide	Ultraviolet, X-rays, and γ-rays from linear accelerator	Ribose, deoxyribose
		Glyceraldehyde and glycoaldehyde		Ribose
1967	Ponnamperuma and Gabel	Aqueous solution of formaldehyde refluxed over kaolinite (hydrothermal spring simulation)		Triose, tetrose, pentose (including ribose and deoxyribose), hexose

[a]See Oró et al. (1990) for review of current literature.
[b]Water was present in most experiments.
[c]Sugars are unstable in basic aqueous solution. At last 0.01 M formaldehyde is needed as a precursor. The time in which the sugars are destroyed is not much longer than the time required for synthesis (Miller and Orgel, 1974; Day, 1984).

available. Considering the number and variety of experiments that have achieved fairly large yields of organic compounds, the precise source of energy does not seem to be critical.

Because compounds such as porphyrins and isoprenoids are not required for reproduction, their prebiotic production (Hodgeson and Ponnamperuma, 1968), like that of non-peptide-bonded amino acid residues, was probably gratuitous. Many organic compounds not universal in living cells and not required for reproduction were probably first made abiotically and then catabolized as food or retained by early cells because they conferred some selective advantage on them. Such dispensable compounds have become secondary metabolites; the primary ones are compounds necessary for reproduction (amino acid and nucleic acid derivatives and some lipids). Whether or not investigators can identify all of the compounds produced in their laboratory experiments, bacteria recognize them!

If not protected from attack by airborne bacteria and fungi, even the most complex polymers made in these experiments are recognized as food and eaten. Experiments in "chemical evolution" therefore must be performed under sterile conditions.

The question of how the first autopoietic systems (probably now-extinct "RNA bacteria") originated from aggregates of organic compounds cannot be answered. Yet progress continues to be made. Subsequent chapters of this book assume that minimally complex prokaryotes, living at first by the fermentative and light-driven metabolism described in Chapter 5, did originate in some way by nonbiological processes from abiotic organic matter early in the history of this planet. Some of these tiny bacteria augmented their short-chain, inaccurate genetic storage in RNA molecules by production of complementary, longer and reparable DNA molecules and thus became the ancestors of all modern organisms.

EVOLUTION
BEFORE OXYGEN

CRITERIA FOR RELATEDNESS

Bacteria do not form species; instead of genetic isolation, which exists in plants and is even more complete in animals, any bacterium may obtain genes from other types of strains. Gene exchange is a generalized phenomenon involving mostly plasmids and prophages. . . . Since their origin, the eukaryotes have been immersed in the all-pervading, dispersed, global bacterial organism.

S. SONEA, 1991

Assuming that prokaryotes evolved extensively by direct filiation, what was their early evolutionary history? How did they diversify metabolically? By what criteria may they be related to each other? What evidence for anaerobic bacteria is there in the early fossil record?

Living or autopoietic systems are organizationally determined: all are self-makers. These membrane-wrapped, chemically reacting entities incessantly synthesize nucleic acids and proteins from smaller molecules. Their fluid, interactive boundaries are far from rigid; metabolism, or the incorporation of matter into self—using light or chemical energy—leads to growth. Growth results in reproduction, of which most autopoietic entities are capable. Reproduction by budding, by fission, by mitosis, or by other means tends to occur with incredible fidelity. Because accurate reproduction, which is the same as a low spontaneous mutation rate, is a characteristic of all cells, the means that ensure it must be both ancient and continuously selected. The triplet genetic code is universal: in all organisms the same rules govern the translation of the sequence of three-nucleotide units—codons—in the linear structure of DNA into the sequence of RNA nucleotides, protein amino acids, or the beginning or end of a protein. Some stretches of a DNA molecule, however, do not code for sequences of protein amino acids or RNA nucleotides, but merely separate the informational sequences of DNA. Even this nontranscribed DNA probably has been subject to natural selection: in eukaryotes, it is complexed with histone proteins, forming a part of the chromatin, which comprises chromosomes.

The genetic code may have arisen from a less faithful translating system. In earlier codes, only the first two bases of the triplet are thought to have determined the amino acid sequence of proteins (Crick, 1968). This is inferred from the observation that the third base today is often redundant: for example, CGU, CGC, CGA, and CGG sequences in messenger RNA all code for arginine. The existence of an even earlier, simpler code is indicated by the fact that the middle base often determines the simplest and most common amino acids: most codons with the middle base G code for glycine, arginine, or serine; most codons with the middle base C code for serine, proline, or threonine; and most codons with the middle base U code for leucine or valine. Stabilization of the genetic code must have been one of the earliest events in the evolution of cells.

Once the original DNA-based protein-synthesizing machinery worked well in some bacteria, these outreproduced the others and became ancestors of all later cells. Only those stereoisomers of the amino acids already fixed by use in the genetic system would continue to be used unmodified. At first, different isomers of carbohydrates such as D- or L-ribose could be taken up by the system. Eventually, however, either by selective absorption from the medium across the lipid-protein membrane or by selective utilization of the "correct" isomers from racemic mixtures, only the familiar L-amino acids and D-sugars came to be used for reproduction. Because all cells now distinguish stereoisomers of amino acids and sugars with remarkable precision, this ability must have evolved very early. At some point isomerases, enzymes for making useless isomers usable, appeared.

Although the use of visible light to maintain ionic homeostasis may have been an early microbial trait, the earliest microbes probably reproduced most rapidly in environments that supplied them abiotically with the nutrients and minerals of which they were composed. Considering compounds that are produced easily in laboratory simulations and are universal cell components or precursors of cell components, one can surmise that the list of abiotic foodstuffs included, at least, ribose, glyceraldehyde, purines and their precursors, orotic acid and other pyrimidine precursors, and a large variety of protein and nonprotein amino acids. The phosphate salts of sodium, calcium, and potassium were probably plentiful as products of the weathering of Archean rocks. In addition, a multitude of as yet undiscovered or unidentified substances, including some tarlike long polymers, were available as food. What was the minimal cell, the simplest self-maintaining system, like? A possibility based on the principles developed by Morowitz (Morowitz and Wallace, 1973) is shown in Figure 5-1. Table 5-1 lists some of the compounds and organelles whose presence can be inferred because they have the basic functions required for reproduction. Morowitz's minimal cell can be compared with certain extant anaerobic bacteria. Even the smallest known — *Bacteroides, Veillonella,* and some mycoplasmas — are larger in diameter by a factor of ten and more complex than Morowitz's calculated minimum, probably in part because he failed to take into account all of the enzymes and protein factors necessary for maintenance and reproduction. Nevertheless, his analysis is a useful guide to the minimal primary metabolism required for autopoiesis.

How may we reconstruct the trends in evolution subsequent to the origin of minimal anaerobic heterotrophic bacteria? What criteria are to be used to trace the evolutionary relationships between prokaryotes? Some inferences from the better-established fields of animal and plant evolution may be pertinent. In the evolution of complex traits such as lungs, eyes, or flowers, each newly accepted mutation confers some small advantage upon the affected organism. These complex traits depend on interacting metabolism and therefore on the presence of many enzymes and a corresponding quantity of DNA to code for them. For animals and plants, the evolution of traits requiring a number of genes can be traced through time. Such multigenic traits are called **semes**. The same principle, the origin of new semes by the accumulation of selectively advantageous mutations, must have operated in the evolution of bacteria. Bacteria that made the metabolites used for reproduction were selected for because they were more adept at obtaining environmental metabolites (food) than their competitors were. Once a seme evolves — for example, a metabolic pathway yielding some needed biochemical — selection generally acts to retain it. Semes are altered slightly or profoundly in the subsequent evolution of descendant populations. In eukaryotes, populations became new species by the accumulation of seme changes. (For the development of the seme concept, see Hanson (1977) and Table 5-2.)

The concept that natural selection acts on isolated traits — enzymes, RNA molecules, cell walls, pigments, or any single property of organisms — is absurd, as Simpson (1967) and others have pointed out. Rather, natural

CELL COMPONENTS	FUNCTION	CHEMICAL COMPOUNDS IMPLIED
DNA	Replication	Deoxyribose, guanine, adenine, thymine, cytosine, phosphoric acid
RNA (transfer, messenger, ribosomal, etc.)	Protein synthesis	Ribose, guanine, adenine, uracil, cytosine, phosphoric acid, amino acid— acylating enzymes
Enzymes	Catalysis	Proteins, peptides, amino acids, nucleotide coenzymes
Membrane	Maintenance of internal cellular environment	Ether-linked or ester-linked lipids; proteins controlling Mg^{2+}, Na^+, K^+, H^+, etc.
Ground cytoplasm	Solvent system, food	All of the above in water; metabolites (acetate, lactate, etc.)

F I G U R E   5-1

Biochemical components of a universal minimal cell (smallest autopoietic entity).

TABLE 5-1

Minimal biochemical functions needed for reproduction.

FUNCTION	REQUIRED COMPONENTS, REACTIONS, AND SYSTEM	IMPLIED COMPOUNDS AND ORGANELLES
DNA synthesis	Synthesis of nucleosides as well as their phosphorylated derivatives; template and primer DNA; DNA polymerases	Nucleotide triphosphates (TTP, GTP, ATP, CTP), about 20 amino acids, genophore (nucleoid)
Messenger RNA synthesis	Messenger RNA, RNA polymerase	Nucleotide triphosphates, proteins
Nucleotide-triplet-coded protein synthesis; ribosome synthesis	Transfer RNAs for each amino acid; amino acyl transferases; amino acids; hydrogen, potassium, sodium, chlorine, calcium, phosphate, and magnesium ion regulation	About 4 ribosomal RNAs and 50 proteins organized into ribosomes
Membrane proteins for ion transport
Membrane synthesis	Various lipids	Acetyl coenzyme A, biotin, saturated and monounsaturated fatty acids, glycerol, cell membrane
Food degradation for energy to maintain low entropy	Metabolites	Lactate, acetate, many other organic acids and sugars

TABLE 5-2

Semes and their evolutionary fates.[a]

FATE	NAME	EXAMPLES
Appearance of seme	Neoseme	Locomotion, leaves, hair, feathers
Change of seme[b]		
Alteration of trait	Episeme	Vertebrates: lungs from swim bladders
Arthropods: differentiation of supernumerary body parts
Plants: flower petals modified from leaves
Repeated trait	Polyseme	Annelids: increase in number of body segments
Vertebrates: increase in number of vertebrae and somites
Arthropods: increase in number of legs
Decrease in trait	Hyposeme	Mammals: reduction in size of appendix, of horse toes, of primate snout
Increase in trait	Hyperseme	Mammals: increase in size and branching number of antlers, increase in body size of horse
Loss of trait	Loss	Eyes of cave animals

[a] After Hanson (1977). Semes of eukaryotes, since they are not plasmid- or viral-borne, are never lost without leaving vestiges of their former existence.

[b] Semes identified as altered forms of earlier semes are called "aposemes."

selection acts on specific populations of organisms in particular environments at definite times by permitting reproduction of some members of the population and preventing reproduction of others. In bacteria, too, evolution acts to preserve or alter semes in particular populations of organisms at definite times and places. Just as in the construction of animal and plant phylogenies, in determining evolutionary relationships among bacteria one asks when and under what conditions particular semes evolved. With bacteria, the "where" question is far less pertinent because of the environmentally available nucleic acid—the tendency of the genes for semes, as plasmids and carried by phages, to travel from organism to organism through the aquatic medium, that is, to be acquired "horizontally."

The semes of related organisms may be compared: the more semes two organisms have in common, the more recently they diverged from a common ancestor. However, before semes can be compared, they must be identified. For example, living and fossil vertebrates show morphological patterns that differ by steps related to adaptive value. Distinctive features such as the pineal gland, hooves, the retina, nails, antlers, hair, and feathers are identifiable as semes. But the morphology of bacteria is so limited that it is impossible to base the identification of semes on it. Possible exceptions are endospores, hormogonia, heterocysts, stalks, and aerial spore-bearing structures. How, then, can semes—primarily metabolic pathways whose end products have some selective advantage—be identified?

Semes are not equivalent to the "phenotypic" traits used in practice to identify isolates of bacteria. Metabolic pathways must be known in detail before it can be determined whether they are semes. Two groups of bacteria may produce an identical metabolic end product—and thus share the same trait. However, the genetic bases of the metabolic pathways may reflect very few or tens of thousands of differences in DNA nucleotide sequence. For example, both *Zymomonas* and *Escherichia* oxidize glucose, releasing carbon dioxide and water. However, the two genera differ in every enzyme of glucose catabolism. Thus, glucose fermentation cannot be a single seme: the same trait with the same selective advantage has arisen in two bacterial genera by **metabolic convergence**. Many bacteria share traits that appear to be semes but probably evolved convergently, such as coccus morphology or the production of gaseous nitrogen by nitrate reduction.

Single mutations, such as the loss of the capacity to make a functioning enzyme, may be due to minute genetic changes but nevertheless have profound effects on an organism. In bacteria, which have only a single set of genes and hence are formally equivalent to haploid eukaryotes, most mutations are expressed as soon as they arise; however, the existence of gene control means that not all genetic potential in bacteria is expressed at all times. The estimation of the evolutionary distances between bacteria may be erroneous if not semes but repressible traits or those determined by single mutations are used for evolutionary analysis. This analysis based on semes, which is implicit in DeLey's (1968) use of metabolic pathways of bacteria for taxonomic purposes, is ignored by Woesean ribosomal RNA sequencers (see below), who designate their results "phylogenetic" and discount all other genetics and physiology.

Bacterial genome organization which involves large and small replicons, is unlike that of animals and plants (with their familiar karyotypes). As emphasized by the Sonea quote at the beginning of this section, bacteria exchange genes freely. They engage in a kind of reversible sex that precludes the formation of stable species. Correlated with the genus of bacteria (for example, *Bacillus, Serratia, Anacystis,* or *Clostridium,*) is the genophore, also called the large replicon, that bears the genes for the cell wall, metabolic pathways, and other features that ensure continuity of the bacterial cell. Yet, as Sonea insists, a single clone of bacteria is an incomplete and incompetent "individual" in nature. Sonea (1991) writes, "a bacterium is not a unicellular organism; it is an incomplete cell . . . belonging to different chimeras according to circumstances." The common recognition of species by the bacteriologist depends on his maintenance of constant environmental conditions such that the small replicons (plasmids and prophages) are retained. On return to nature, plasmids are gained and lost within short periods of time. "Species" that change up to 15 percent of their genetic constitution on a daily basis can no longer be called species.

The evolution of the generic groups of the bacterial world can be reconstructed by partial phylogenies of ribosomal RNA because the large replicon has maintained genes for the generic features on the same genophore as those for the RNA genes; however, "speciation of bacteria" is an oxymoron, since all bacteria are linked through their promiscuous genetic system. The dispersed bacterial body follows the contours of the habitable globe. Thus, our discussion of the evolutionary expansion of bacteria from fermenters to chemoautotrophs becomes an integrated list of the global bacterial system as it responded, by genetic refinement, to the global catastrophes, many of which it brought upon itself. The consequence of its DNA-ordained expansionist tendencies was environmental pressures on other parts of its complex, dispersed being. A single example here suffices: the production of methane by methanogenesis from atmospheric hydrogen and carbon dioxide in some bacteria led to great selection pressures for methylotrophy in others. Some of the bacteria choking on methane evolved the capacity to survive this "waste product"; they depleted the suffocating methane by using it as an energy source. Growing on the oxidation of methane, ancient methylotrophs thrived then as they do today by converting methane "waste" into atmospheric CO_2 and water.

No single bacterial lineages (unlike those of, for example, cercopithecan monkeys or leguminous plants) can be outlined and depicted in two dimensions; rather, the coevolving global biosphere as an entity can be sketched roughly as we trace the consequences of its own evolution.

Woese (with his colleagues Pace, Zillig, Stackebrandt, Wheelis, and others, including those who tend not to agree with Woese on all issues: Doolittle, Gogarten, Hori, Lake, and Sogin) developed ingenious methods; many are reported in the *Journal of Molecular Evolution*. These involve isolating, amplifying, and sequencing the ribosomal RNA genes from a great many microbial sources. The sequence data are then aligned, collated, and represented as a dichotomous tree in which those organisms with the most nucleotide sequences in common are closest to each other. The usefulness of the Woesean discrete ribosomal RNA sequences for organizing knowledge of bacterial genera is not precluded by my analysis, which incorporates metabolic semes and geological information consistent with Sonea's view of promiscuous gene flow. Rather, I dispute Woese and his colleagues on the validity of their claims of evolutionary significance. This leaves us with two (Woesean) phenomena: (1) the most powerful methods invented for distinguishing the differences between extant microbial genera and (2) the questionable application of the dichotomous trees to the past history of life on Earth. Woese and his colleagues measure differences; indeed, as a definitive diagnostic maneuver (for example, for pathogens or biopolymer producers), these techniques are spectacular. However, the relationship of these trees to time and co-evolution of bacteria of the biosphere remains to be demonstrated. The difficulty of the Woesean analysis is compounded when applied to any eukaryote, all of which contain multiple genomes and are products of symbiotic events (as we discuss on pages 4–8 and 209–214).

Attempts are made here to identify homologous semes. If the DNA sequences in all past and present bacterial genera and their relation to bacterial phenotype were known, phylogenies could be constructed directly, without the use of strong inference. Unfortunately, this feat is beyond the capability of any present method and is likely in the near future to remain an unattainable goal—certainly for fossils or for the genera that cannot even be grown in axenic cultures. In the absence of complete DNA sequence data, DNA sequences corresponding to the ribosomal RNA genes are used. These and other partial phylogenies can be ranked by their degree of approximation to direct information about the genome. Table 5-3 lists criteria for relatedness ranked roughly according to how reliably each tests the homology of transcribed DNA sequences. These techniques for

TABLE 5-3
Criteria for relatedness between prokaryotes.[a]

CRITERION	APPLICABLE TECHNIQUES
Homology of DNA sequences	Total DNA nucleotide sequencing,[b] sequencing and comparison of ribosomal genes (5S, 16S, 28S)
Homology of metabolic pathways and identity of the enzymes	Biochemical methods, including radioisotope tracers, electrophoresis, chromatography
Identity of ultrastructural morphology	Transmission and scanning electron microscopy (fixed material)
Identity of gross morphology and life cycle	Light microscopy and cytology (live material)
Homology of single genes and their RNA or protein products	DNA, RNA and DNA, DNA homologies, amino acid sequencing
Same molecular structure of a single enzyme, pigment, or other metabolite	Physical chemistry (X-ray crystallography and others), spectroscopy, enzymology, electrophoresis, chromatography, immunocytochemistry (epitope analysis)
Common phenotypic traits (ability to grow on same carbohydrate, production of same end product, isolation from same type of enrichment medium, same method of spore production, similar patterns of motility, etc.)	Standard microbiological identification techniques (culturing, staining, etc.)

[a] Applies to any organisms with single genomes and to their associated systems for protein synthesis. Criteria are listed roughly in order of importance.
[b] Not yet completed for the genome of any cell.

comparing semes of bacteria were developed by extrapolation from those used for animals and plants (Hanson, 1977).

If a given seme is invariant among populations or is present in one population and absent in a second, that seme is less useful in determining evolutionary relationships between taxa than is one that varies. Many semes in plants and animals, for example, evolved in bacteria and were retained without modification: protein synthesis and the triplet code, glucose fermentation to pyruvic acid, aerobic respiration, and oxygenic photosynthesis. Some semes are restricted to only certain eukaryotes: megasporophylls, leaf stomata, feathers, beaks, amniote embryos, cartilage, and jaws are a few examples. Other semes, such as ossification in bone development, the dorsal hollow nervous system of chordates, and the presence of vascular bundles in tracheophytes, evolved in remotely ancestral populations of eukaryotes and were retained by all mammals or all flowering plants without much change. Such restricted and unchanging semes are inadequate for constructing phylogenies.

To be most useful for evolutionary studies, a seme must have undergone detectable changes in the populations of organisms under study. Identifying a population in which a seme has evolved depends on identifying groups of related organisms that demonstrate a range of small variations in the seme, variations that may be correlated with definite selective environmental factors. Essentially, this is the method of comparing "characters in common," used so unselfconsciously in the construction of animal and plant phylogenies (Grant, 1971; Simpson, 1953). Some eukaryotic semes analyzed by this method are the triploblastic (three-layered) body of metazoans, the four appendages with five digits of tetrapod vertebrates, mammalian and reptilian embryogeny, the beak and feet forms of Galapagos Island finches, the placental uterus and mammary glands of mammals, the upright posture of manlike apes, the speech of humans, the operculae and setae on the sporangiophores and sporophytes of fungi and plants, the antheridia and archegonia of bryophytes, the composite flowers of angiosperms, and cyanogenic glucoside metabolism in bracken fern. Such semes change in the ways noted in Table 5-2; thus, their study is pivotal in tracing the course of evolution.

The task of identifying comparable semes and tracing their changes in populations of bacteria is made more complex by the fact that so many bacterial semes are plasmid-borne and mobile—for example, arsenic resistance and manganese oxidation. Even today the absolute requirement of the cell to rid itself of toxic heavy metals and their derivatives (mercury, organomercurials, silver, arsenic and its oxides, borate, cadmium, cobalt, chromium, copper, nickel, lead, tin, zinc, and others) has led to a "Gaian" distribution of these substances—for example, as lead, silver, and copper sulfide deposits (see pages 363–367). In some cases bacterial cells (using gene products that are plasmid-borne) chemically convert the offending toxin to a waftable chemical derivative, in other cases bacteria "deport" the ions without change using adenosine triphosphatases (ATPases)—now known, through a battery of sequence homologies, to be related, to each other (Silver et al., 1989).

The phyto- or zoogeographic analysis employed for the far less mobile plants and animals is not possible for the ever-changing members of the Kingdom Prokaryotae. Thus, bacterial seme accumulation is best understood as sequential acquisition by a dispersed global bacterial entity responding to its own excesses; that is, to its tendencies toward exponential growth coupled with metabolic viruosities and their inevitable waste products.

ANAEROBIC INNOVATIONS

Nihil ideo quoniam natum est in corpore, ut uti possemus,
sed quod natum est id procreat usum.*

T. LUCRETIUS, First Century B.C.

The Earth did not go the way of Venus, when the sun
began to increase its luminosity, because atmospheric
CO_2 may have been reduced by the activity of oxygenic
photosynthesis and the consequent carbon fixation,
especially in the form of stromatolitic carbonates.

HSU, 1992

The universal biochemical functions of primary metabolism that define the minimal cell require, by Morowitz's estimate, approximately 50 genes (Morowitz and Wallace, 1973). The term **gene** here refers to the hundreds of nucleotide pairs that together code for a single product, either RNA or protein. Indeed, the earliest anaerobic heterotrophic bacteria presumably contained little more than such primary metabolites as are listed in Table 5-1. Probably in all organisms, most gene products, measured as amino acid sequences in proteins, are necessary primarily for autopoiesis; that is, for "housekeeping" functions of self-maintenance (Loomis, 1988). The discontinuity between life and nonlife ought not to be underestimated (Fleischaker, 1990).

The iron-sulfur proteins, ferredoxins, lack coenzymes and are found in all organisms, including fermenting anaerobic bacteria. Their biosynthesis must be an ancient capability, representing ancient semes. From the homology of amino acid sequences, it can be inferred that the small ferredoxin proteins are among the most ancient of gene products and that rubredoxins and cytochromes were derived from them directly by gene duplications and subsequent differentiation of function. Another early seme (probably several semes) is fermentation, the metabolism of small carbon compounds in which ketones and aldehydes are converted to organic acids and other small compounds, yielding energy in the form of ATP. Fermentation may be defined as the enzymatic transformation of organic compounds in which other organic compounds serve as electron

---

*For nothing is born in the body in order that we may be able to use it, but rather, having been born, it begets a use.

acceptors. That the intermediary metabolism of all organisms involves such transformations is an argument for their antiquity. The appearance of fermentation semes was followed by the evolution of nearly universal anaerobic glycolytic pathways, schemes for the catabolism of sugars. The anabolism and catabolism of five-carbon compounds required for nucleic acid replication, deoxyribose and ribose, probably preceded that of larger molecules, such as glucose, fructose, and lactose.

While some early anaerobic bacteria evolved the heterotrophic feat of anaerobic glycolysis, others—at least three major groups of bacteria—evolved the capacity to fix atmospheric carbon dioxide into various reduced organic compounds. Fixation by the acetate pathway, used by clostridia, is probably the most ancient, because it is found in a great group of successful endospore-forming bacteria that (1) are not sensitive to visible light (S. J. Giovannoni, 1979, personal communication) and that (2) lack porphyrin biosynthesis and other semes which they would not have lost entirely if they had ever acquired them. Fixation of carbon dioxide by phosphoenol-pyruvate carboxylase probably evolved next, because it occurs in anaerobic photosynthetic bacteria (although it is better known in *Escherichia coli*). Fixation of carbon dioxide by ribulose bisphosphocarboxylase, a pathway in many aerobic organisms and typical of many photosynthesizers and chemoautotrophs, must have evolved after the two anaerobic pathways. The dissimilarity of carbon dioxide fixation in several groups suggest that the capability evolved convergently in different ancestors (Figure 5-2).

Because carboxylases, the enzymes that catalyze carbon dioxide fixation today, also react with molecular hydrogen, N. G. Carr (1977 personal communication) suggests that oxidation of hydrogen to water may have been their first selectively advantageous function; only as hydrogen was depleted by escape into space and by reaction with other gases, including oxygen, did the carboxylation functions of these enzymes emerge. Either simultaneously or closely thereafter, the fixation of atmospheric nitrogen evolved. Carr also suggests that enzyme complexes required for nitrogen fixation evolved first to oxidize ethylene to acetylene under earlier, stronger reducing conditions. Fixation of atmospheric nitrogen invariably requires the nitrogenase complex, an enzyme system containing ferredoxins. Nitrogenases react strongly and specifically with acetylene, an energy-rich compound probably synthesized prebiotically in quantity from cyanide. Thus, the nitrogenases and carboxylases may represent bacterial preadaptations, semes that were retained because of new selective advantages in altered environments.

As abiotically produced organic matter became depleted, competition for the organic prerequisites for reproduction ensued. As the carboxylation and nitrogen-fixing functions were achieved, a new, abundant, and direct

*Clostridia (heterotrophs)

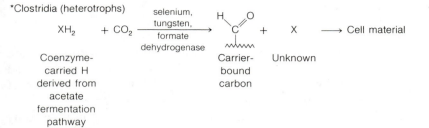

$$XH_2 + CO_2 \xrightarrow[\text{formate dehydrogenase}]{\text{selenium, tungsten,}} \overset{H}{\underset{\text{Carrier-bound carbon}}{C}}\overset{O}{\diagdown} + \underset{\text{Unknown}}{X} \longrightarrow \text{Cell material}$$

Coenzyme-
carried H
derived from
acetate
fermentation
pathway

Escherichia coli (heterotroph)

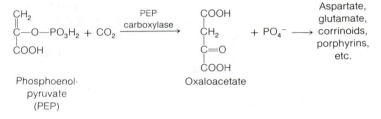

$$\begin{matrix} CH_2 \\ \| \\ C-O-PO_3H_2 \\ | \\ COOH \end{matrix} + CO_2 \xrightarrow[\text{carboxylase}]{\text{PEP}} \begin{matrix} COOH \\ | \\ CH_2 \\ | \\ C=O \\ | \\ COOH \end{matrix} + PO_4^- \longrightarrow \begin{matrix} \text{Aspartate,} \\ \text{glutamate,} \\ \text{corrinoids,} \\ \text{porphyrins,} \\ \text{etc.} \end{matrix}$$

Phosphoenol-
pyruvate
(PEP)

Oxaloacetate

†Methanogenic bacteria (chemoautotrophs)

$$\underset{\substack{\text{Unknown} \\ \text{carrier}}}{XH} + CO_2 \xrightarrow{?} XCOOH \longrightarrow XCHO \longrightarrow XCH_2OH \longrightarrow XCH_3 \longrightarrow CH_4$$

$$\text{Methylcobalamin} + \underset{\text{Coenzyme M}}{HSO_3(CH_2)_2SH}$$

*Phototrophic bacteria (some), all algae and plants

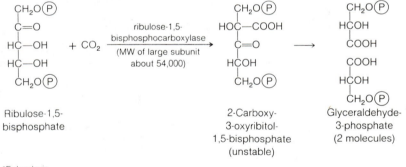

$$\begin{matrix} CH_2O\,\text{(P)} \\ | \\ C=O \\ | \\ HC-OH \\ | \\ HC-OH \\ | \\ CH_2O\,\text{(P)} \end{matrix} + CO_2 \xrightarrow[\substack{\text{(MW of large subunit} \\ \text{about 54,000)}}]{\substack{\text{ribulose-1,5-} \\ \text{bisphosphocarboxylase}}} \begin{matrix} CH_2O\,\text{(P)} \\ | \\ HOC-COOH \\ | \\ C=O \\ | \\ HCOH \\ | \\ CH_2O\,\text{(P)} \end{matrix} \longrightarrow \begin{matrix} CH_2O\,\text{(P)} \\ | \\ HCOH \\ | \\ COOH \\ \\ COOH \\ | \\ HCOH \\ | \\ CH_2O\,\text{(P)} \end{matrix}$$

Ribulose-1,5-
bisphosphate

2-Carboxy-
3-oxyribitol-
1,5-bisphosphate
(unstable)

Glyceraldehyde-
3-phosphate
(2 molecules)

*Eubacteria
†Archaebacteria

FIGURE 5-2

Distribution of some pathways of carbon dioxide fixation. The structure of methylcobalamin is identical to that of cyanocobalamin (vitamin B_{12}) except that bound to the cobalt atom in the former is a methyl group; in the latter, it is a cyanide group (see Figure 5-3). Some phototrophic bacteria (e.g., *Chlorobium*) use a reverse citric acid cycle to incorporate CO_2 (see Figure 5-4).

source of carbon and nitrogen for organic syntheses became available — the atmosphere. Obviously, the ability to take up atmospheric carbon and nitrogen would be of great selective advantage. A tentative sequence of seme acquisition by anaerobic bacteria is presented in Table 5-4.

Recognizing that the appearance of a bacterial seme was probably followed by its rapid recombinatory spread through the entire biosphere, this microbial phylogeny becomes a sketch of the appearance of new semes consistent with the evidence for deposition of biogenic organic compounds in sediment and for atmospheric changes resulting from bacterial innovation (see Figure 6-3). Of course, like any phylogeny, it is only tentative. The suggested sequence is based on metabolic products of bacteria before they died. Changes wrought by community interactions, including decomposition, weathering, and diagenesis, further complicate direct verification. Tracing the evolution of microbial ecosystems from the evidence of their

TABLE 5 - 4

Sequence of seme acquisition by anaerobic bacteria.

SEME	GROUP IN WHICH SEME APPEARED	APPROXIMATE TIME OF APPEARANCE ($\times 10^9$ YEARS)	SEDIMENTARY EVIDENCE[a]
None	None	4.5–3.5	Abiogenic organic matter (no optical activity, wide varieties of isomers)
Synthesis of optically active compounds: membrane-bounded nucleic acid replication	First anaerobic prokaryotes	3.5–2.0	Diagenetically altered products of nucleic acids and proteins; organic phosphorus, lipid derivatives, mono- or bisaturated fatty acids; carbon and sulfur isotope fractionation; microfossils
Carbon dioxide fixation (using molecular or organic hydrogen), glycolytic pathways, nitrogen fixation, methanogenesis	Anaerobic fermenting bacteria	3.5–2.0	Reduced carbon compounds, deposited sulfide, intermediates in carbon dioxide reduction pathways, intermediates in glycolytic pathways, molybdoferredoxins, methane, coenzyme M derivatives
Synthesis of carotenoids, bacterial chlorophylls, and cytochromes	Anoxygenic bacterial photosynthesizers	3.5–2.0	Porphyrin and isoprene derivatives, organic carbon, sulfate, sulfur
Synthesis of cytochromes and catalase	Anaerobic sulfate and nitrate respirers	3.0–1.5	Porphyrin and isoprene derivatives

[a]Expected correlated sedimentary evidence; because of microbial community interactions and diagenesis, this list is inevitably oversimplified.

sedimentary remains is a central problem of geochemistry. Unfortunately, it is still impossible to draw definite conclusions from geochemical evidence; in fact, trends in the metabolic evolution of anaerobic prokaryotes and their communities are inferred better from studies of living microbes.

I assume that no seme — that is, no metabolic novelty requiring a large number of interacting genes — ever evolved and became extinct without leaving a trace of its former existence in its descendants. This assumption is justified in the evolutionary history of plants and animals. Semes whose products are always of great selective advantage are generally the outcome of interaction among many genes. After they first appear, semes may be altered profoundly, but detailed study reveals clues to their earlier forms. Higher taxa, such as phyla and classes, group organisms that retain ancient semes in common. Once a higher taxon appears, a seme may recede in importance by extinction of its members and become largely replaced, but as a rule it does not become entirely extinct. Thus, living organisms and their communities store in their morphological and metabolic semes immense quantities of information about their past.

Heat-shock or stress proteins and their highly homologous relatives provide an example of an extremely ancient seme, probably necessary for all surviving autopoietic systems. Heat-shock proteins (whose sizes range from 60 to 90 kd) protect other proteins from irreparable denaturation and shuttle all kinds of easily victimized proteins through membranes. Naked DNA survives freezing, boiling, and other insults that would devastate proteins; death is far more often due to protein denaturation than it is to gene loss (nucleic acid hydrolysis). Heat-shock proteins, then, produced obligatorily in prodigious quantity, maintain the cell system by many required activities not understood in detail (Patrusky, 1990).

Some heat-shock protein genes (for example, *dna* genes) code for primases, making possible primer RNA required to initiate DNA synthesis. One imagines massive production of these proteins inspiring DNA synthesis in response to desiccation, heat or other severe stress. At least some products of *dna* genes are essential to life and growth because they are required for distribution of the large replicon, that is, they are necessary for the distribution of membrane-associated nucleoid DNA to offspring cells (Grompe et al., 1991).

Extensive information on the origins of sex has been compiled (Margulis and Sagan, 1986). Although bacteria lack meiotic sex — two parents do not donate equal quantities of genetic material to an offspring — they participate in DNA recombination on the molecular level. Bacterial recombinants typically contain unequal numbers of genes from each "parent." When in the history of life did molecular DNA recombination evolve? The repair of ultraviolet-induced damage to DNA and DNA recombination require some of the same enzymes; probably, the evolution of enzymatic repair of ultraviolet-induced damage preadapted bacteria for DNA recombination. Transfer of DNA among different bacteria is mediated by bacteriophage, and this process may be induced by exposure to ultraviolet. Thus, an important component of the sexual process, the enzymatic construction of DNA-recombinant molecules, evolved early, long before meiosis—perhaps in obligately anaerobic prokaryotes.

Prior to the accumulation of atmospheric oxygen, microbes must have been threatened by solar ultraviolet reaching the Earth's surface. Mechanisms for resisting ultraviolet damage and for repairing DNA—for example, photoreactivation—must have arisen very early; many are known, even in obligate anaerobes (Rambler and Margulis, 1980). At least some of these mechanisms have been retained because their chemistry became intimately involved with the sexual processes that were later selected for in their own right. However, the chance of finding direct evidence in the sedimentary fossil record for the evolution of microbial mechanisms of protection against ultraviolet seems remote indeed.

In living bacteria, different anaerobes are closely related (in the same genus or family) to aerobes, suggesting that bacterial metabolism evolved before or during the transition to the oxygen-rich atmosphere and, hence, before the appearance of any animals or plants, all of which are obligate aerobes. All cells require energy sources and specific electron donors and recipients in addition to essential elemental nutrients. As energy-storing prebiotically produced compounds were depleted, a great adaptive radiation ensued: new ways of using energy evolved, leading to photolithoautotrophy.

Photoheterotrophic metabolism in obligate anaerobes (that is, heliobacteria and halobacteria) suggests that visible light had a role in metabolism before strict photolithoautotrophy appeared. Photoheterotrophs can use light to generate ATP and to create osmotic gradients, but they require complex organic carbon compounds as food. Some photoheterotrophs can also fix carbon dioxide. Thus, both carboxylation of atmospheric carbon dioxide and photoproduction of ATP probably evolved before these pathways were combined in photolithoautotrophic bacteria. Methanogenic archaebacteria obtain energetic and material requirements from hydrogen and carbon dioxide. Some use cobalt-conjugated compounds (for example, methylcobalamin) in the methanogenic fixation of carbon dioxide (see Figure 5-2). Probably an ancient group, they adapted to living on Archean volcanic gases. The existence of carbon dioxide fixation in various nonphotosynthetic anaerobes with limited or no ability to synthesize tetrapyrroles and carotenoids (for example, *Desulfovibrio*) implies that phototrophic systems were preceded by other kinds of metabolism and their effects:

isotopic fractionation of carbon and sulfur; the production of sulfide from sulfate, thiosulfate, and hydrogen gas; methane production; and the deposition of organic compounds derived from biogenically reduced carbon dioxide.

Another extremely ancient microbial seme is isoprenoid biosynthesis. This pathway leads from acetate through mevalonic acid and isopentenyl pyrophosphate (see Figure 6-1) to such isoprene derivatives as squalene, ubiquinone (coenzyme Q), vitamins A and K, and the phytyl chain on chlorophylls. Isoprene derivatives, present in all photosynthetic microbes including strict anaerobes, may have evolved as protectors from photo-oxidation (Krinsky, 1966), retained because of their expanded versatility. Vitamin A is the precursor of rhodopsin, the visual pigment of vertebrates as well as the purple membrane pigment of halophilic bacteria. Some isoprenoids came to be used by halobacteria in reactions that used visible light to pump sodium out of the cells against a gradient (Lanyi, 1980).

The antiquity of isoprenoid biosynthesis can be argued from its presence in archaebacteria, distinct prokaryotes, of which many are obligate anaerobes. Archaebacteria include many different methanogens, heat- and acid-tolerant microaerophils such as *Sulfolobus* and *Thermoplasma*, and the aerobic, and therefore more recent, halobacteria *Halococcus* and *Halobacterium*. Archaebacteria are distinguishable from other prokaryotes, eukaryotic cytoplasm, and organelles by several criteria: unique nucleotide 16S ribosomal RNA sequences, the presence of phytyl ethers rather than fatty acid esters in the lipoprotein cell membranes (Kates et al., 1966; Tornabene et al., 1979), the absence of glycopeptides in the cell wall, and unusual patterns of post-transcriptional modification in the transfer RNAs.

Nitrogen fixation is an anaerobic, energy-requiring process restricted to prokaryotes. It probably evolved very early. Nitrogenases, present in clostridia, most likely preceded the origin of the most important new metabolism ever to have appeared—photolithoautotrophy. Bacterial photosynthesis, which freed life from reliance on probiotic organics, certainly evolved in anaerobic microbes, some of which could fix nitrogen. The differences—in 16S RNA, Gram-negativity, responses to oxygen, spore formation, phototrophic lamellar structure, and other features—among groups of bacterial synthesizers suggest early diversification—perhaps even convergent evolution—of photoautotrophy in the purple sulfur bacteria (Chromatiaceae), the green sulfur bacteria (Chlorobiaceae), the purple nonsulfur bacteria (Rhodospirillaceae), the green nonsulfur bacteria (Chloroflexaceae), and the heliobacteria. The fixation of carbon dioxide by a reductive Krebs cycle and the corresponding lack of a Calvin-Benson cycle in green sulfur bacteria also suggest the polyphyly of photosynthesis (Almassy and Dickerson, 1978) and its spread through the evolving biosphere by horizontal gene flow. Bacterial photosyntheses resulted in the deposition of organic carbon, elemental sulfur, and perhaps even sulfates. In the Archean, before the transition to the oxygen-rich atmosphere, there were complete anaerobic ecological sulfur cycles: iron oxides were reduced from ferric to ferrous iron, and sulfates were reduced by desulfovibrios to hydrogen sulfide. Phototrophic bacteria completed the cycle by using hydrogen sulfide gas as a hydrogen donor, producing sulfate or sulfur and depositing organic carbon.

The reduction of large quantities of iron hydroxide and iron oxide to ferrous iron compounds in the anoxic Archean probably preceded even sulfate reduction. Newly isolated bacteria such as "GS-15" (Lovley, 1991) or *Shewanella* (Nealson and Myers, 1992) not only vigorously reduce large quantities of iron compounds under conditions of anaerobic growth but, by the same respiratory mechanism, simultaneously oxidize acetate entirely to atmospheric carbon dioxide and water. Hence, prior to the time when large quantities of oceanic sulfate provided a terminal electron acceptor for sulfide-producing bacteria, anaerobic, respiring iron reducers most likely left evidence for their geochemical activities in the Brazilian Cajiras Formation of the Archean and later in the banded iron formations of the Proterozoic. Microbial sedimentary activities were involved directly in the formation of banded iron in at least three ways: (1) The dissimilatory reduction of iron oxides and hydroxides by bacterial metabolism, a process in which iron oxides are "breathed," serving as electron acceptors, resulted in a sediment rich in reduced (and valuable) iron compounds. (2) The precipitation of iron oxides by many types of iron-oxidizing bacteria and (3) the production of the oxidant (gaseous O_2) by communities of cyanobacteria were critically related to the enormity of early Proterozoic iron deposits.

The discovery of the class of microbes that use elemental sulfur as a terminal electron acceptor in the formation of hydrogen sulfide — the **desulfomonads** — adds another heterotrophic dimension to this cycle. Elemental sulfur was probably used as an electron donor also in anaerobic photosynthesis, as it is today by *Chromatium buderi* and other microbes (Trüper and Jannasch, 1968).

Closely contemporary with or even prior to the rise of isoprenoid synthesis must have been the evolution (as one or more semes) of pathways for making corrin derivatives. Corrinoids, the biosynthesis of which stops just short of porphyrins, seem to be distributed universally. Judging from their present function and the ubiquity of the first steps of their biosynthesis, corrinoids may have appeared during the evolution of DNA, in the reduction of ribose to deoxyribose sugars.

The first corrinoids may have been like today's cobalt corrinoid coenzyme, a form of vitamin B_{12} Corrinoid synthesis probably preceded porphyrin synthesis; the earliest porphyrins may have resembled iron-chelated porphyrin coenzymes of the antioxidant catalase. Porphyrin derivatives, metal-chelated tetrapyrroles, are found today in all organisms except some anaerobic bacteria. In sulfate reducers they serve as carriers of electrons from sulfide to sulfate. Thus, porphyrin derivatives, such as the iron-containing coenzymes of catalase, peroxidase, and cytochromes, the heme of sulfate-reducing proteins and hemoglobins, and the magnesium-chelated chlorophylls, were synthesized by microbial phototrophs prior to any oxygen release. The absence of porphyrins in clostridia and their presence in desulfovibrios imply that the porphyrin-synthetic pathways are semes. Figure 5-3 shows the structure of some biologically important porphyrins. Porphyrin itself is not found in living organisms (Roberts and Caserio, 1964). Although they can be produced in minute quantities under conditions simulating those on the primitive Earth (Hodgeson and Ponnamperuma, 1968), fossil porphyrin derivatives in sediments are probably biogenic. All life synthesizes tetrapyrroles via the five-carbon universal precursor aminolevulinic acid (ALA). The relation between isoprenoid and one major mode of porphyrin biosynthesis — with ALA via glycine and succinyl coenzyme A (succinyl CoA) — is shown in Figure 5-4. In a second major mode of ALA synthesis (in *Methanobacterium*, cyanobacteria, algae, and plants), also shown in Figure 5-4, ALA is formed from the intact carbon skeleton of glutamic acid by at least three enzyme-catalyzed reactions and glutamyl–transfer RNA (Scheer, 1991).

Selective pressures for the retention of porphyrin syntheses were probably various. Corrinoid derivatives used first in the reduction of ribose to deoxyribose and in protein synthesis were then selected for as methyl-transfer agents both in biosyntheses and in gas ventilation — toxic elements may be neutralized and eliminated by the formation and release of their methylated derivatives, such as methyl arsenate. The slightly more complex porphyrins may first have been retained because they reduced mutagenic oxidizing agents in the environment. Whatever these compounds were originally selected for, the genetic capacity to form them has been of great and lasting value. All photosynthetic bacteria must synthesize chlorophyll, itself a porphyrin derivative, and they all must have evolved from porphyrin-synthesizing ancestors. Eventually, respiring heterotrophic bacteria used porphyrin proteins as electron carriers in the anaerobic respiration of food materials: hydrogen atoms from two- and three-carbon food molecules such as acetate, lactate, and pyruvate were transferred by conjugated proteins, such as cytochromes, to the final hydrogen acceptor, nitrate; molecular nitrogen and nitrous oxide were produced as end products. In methanogens, carbon dioxide is the terminal electron acceptor, and methane is produced.

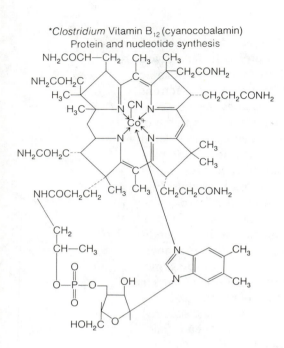

Clostridium Vitamin B₁₂ (cyanocobalamin)
Protein and nucleotide synthesis

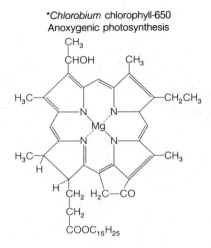

Chlorobium chlorophyll-650
Anoxygenic photosynthesis

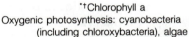

*†Chlorophyll a
Oxygenic photosynthesis: cyanobacteria
(including chloroxybacteria), algae

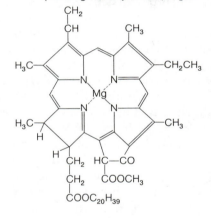

*Bacteriochlorophyll *a*
Anoxygenic photosynthesis, purple bacteria

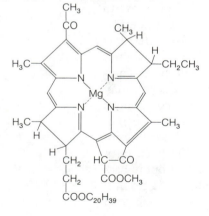

*†Heme
Respiration, many

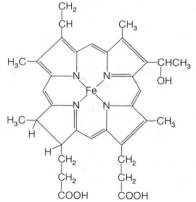

FIGURE 5-3

Porphyrins synthesized by prokaryotes and metabolic pathways in which they function. * = present in eubacteria; † = present in eubacteria and eukaryotes.

Porphyrin derivatives retained as electron carriers preadapted cells for phototrophy, because most porphyrins absorb visible light. The earliest photosynthetic microbes were obligate anaerobes, as many phototrophic bacteria still are. Because they use the least complex process, photosystem I only, photosynthesis is believed to have evolved in anaerobic bacteria. In these bacteria, photosystem I generates ATP by cyclic phosphorylation; in cyanobacteria and plants it also produces reducing power in the form of reduced nicotinamide-adenine dinucleotide phosphate ($NADPH_2$) (Figure 5-5). The conversion of light to ATP is the lowest common denominator, occurring in all photosynthesizers. In oxygenic photosynthesizers, such as cyanobacteria, photosystem II always supplements but never replaces pho-

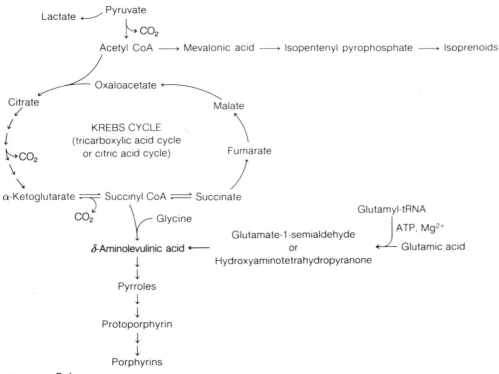

FIGURE 5-4

Relation of porphyrin and isoprene synthesis to the primary metabolism of Krebs cycle intermediates and protein synthesis. [After Jordan (1990).]

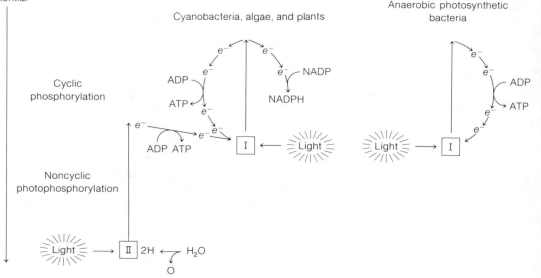

FIGURE 5-5

Photosystems I and II. The boxed letters I and II represent chlorophyll reaction centers of the two photosystems. The *e*'s represent an unspecified number of electron carriers, such as ferredoxin, ubiquinone, and cytochrome *c*. No photosynthetic organism contains only photosystem II.

tosystem I. Anaerobic photosynthesis, which requires only photosystem I, preceded the more complex oxygenic form, which requires both photosystems. Anaerobic photosynthetic sulfur bacteria are ancestral to oxygenic cyanobacteria.

Oscillatoria limnetica, a filamentous cyanobacterium, is an obligate autotroph. Yet it releases oxygen only facultatively (Cohen, Padan, and Shilo, 1975). In photosynthesis, this organism can use, instead of water, either hydrogen sulfide or sodium sulfide as an electron donor, depositing sulfur. It probably uses hydrogen gas as a hydrogen donor during photosynthesis in the light and in the dark produces hydrogen and hydrogen sulfide along its other metabolic pathways. In anaerobic environments, the photosynthetic sulfur bacteria can outcompete facultative hydrogen- and sulfide-using cyanobacteria, which thrive in intermittently aerated zones. Thus, Jack-of-all-trades species such as *O. limnetica* were selected against in all but very special hot, saline, intermittently aerated environments, to which they are apparently restricted today.

Knowing about *O. limnetica* led Stanier and his colleagues to stress other cyanobacteria with high sulfide concentrations under anaerobic conditions; about half the laboratory cultures tested were able to photosynthesize anoxically and could grow using sulfide as the electron donor. Such physiological studies support the morphologically based idea that cyanobacteria descended from photosynthetic sulfur bacteria.

Until the mid-1970s, photosynthetic oxygen release by prokaryotes was thought to be limited to cyanobacteria. These bacteria synthesize several phycobiliproteins, such as phycocyanin, allophycocyanin, phycoerythrin, and chlorophyll *a*, but never chlorophyll *b*. However, oxygenic photosynthetic prokaryotes lacking blue-green pigments were discovered (Lewin, 1976, 1977). *Prochloron* and *Prochlorothrix* are prokaryotes with a photosynthetic physiology and pigmentation like that of plants. They contain chlorophylls *a* and *b* in the approximate proportion 1:4 but lack the typically cyanobacterial phycobiliproteins and the cellular structures on which they are borne, the phycobilisomes. Thus, in pigmentation but not in cell structure, these organisms resemble green algae (Figure 5-6). Structurally and in ribosomal RNA sequence analyses, even though they are cyanobacteria, they constitute a link between the photosynthetic prokaryotes and the chloroplasts of green algae, other protists, and plants. In my opinion, they should be called **chloroxybacteria**, a name that indicates prokaryote affiliations but does not imply that these organisms are direct ancestors of green algae and plants.

The discovery of *Prochloron* and *Prochlorothrix*, and their tiny spherical relative, an unidentified marine bacterium (Chisholm et al., 1988) greatly increases the probability that photosynthetic pigment systems containing chlorophyll *b* evolved in prokaryotes before eukaryotes acquired them symbiotically. "Green-plant" oxygen-releasing photosynthesis did not evolve in green plants at all, but in oxygen-releasing photosynthetic bacteria, both green and blue-green.

Several authors have constructed prokaryotic phylogenies. Broda's (1975) is based on many traits; Almassy and Dickerson (1978) based theirs primarily on the amino acid sequences of proteins; Woese and Fox (1977) based theirs primarily on nucleotide sequences of ribosomal nucleic acids. In broad outline, they are consistent with those presented here. Refinement and perhaps even profound alterations will follow as new data become available. However, it is certain that the main metabolic pathways of life evolved in the prokaryotes—most likely during the Archean Eon. The evidence for these spectacular metabolic achievements in the Archean fossil record, although sparse, is available, and its interpretation is productively debated (Schopf, 1983).

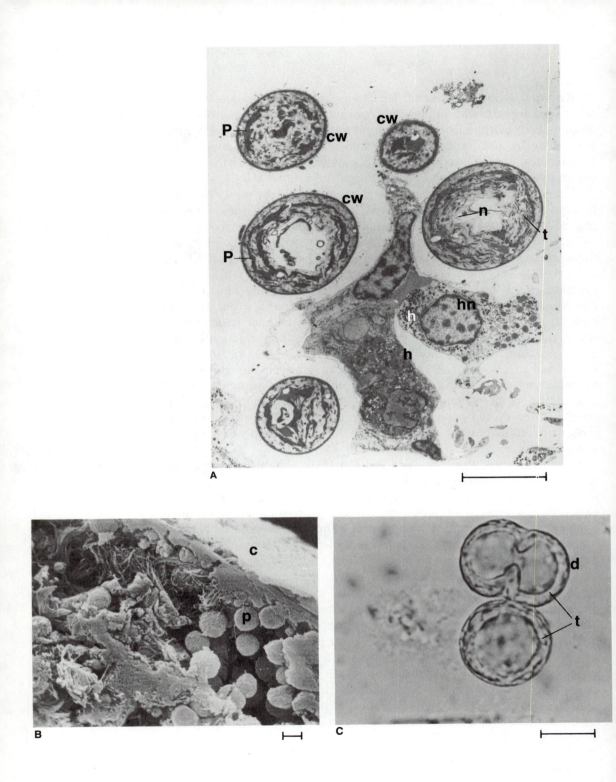

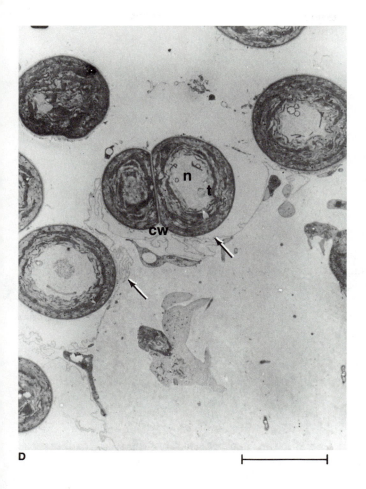

FIGURE 5-6

Prochloron. Parts **A** and **D** are photographs of *Prochloron* species found as an extracellular symbiont in a didemnid, *Diplosoma virens*, taken by Ralph A. Lewin in water 1.5 meters deep off the coast of Palau. **A**–**D**, bars ≐ 5.0 μm. (**A**) *Prochloron* sp. (*p*) cells in the common cloacal cavity of the tunicate host. *h* = host cell; *hn* = host nucleus; *t* = thylakoids; *cw* = *Prochloron* cell wall; *n* = nucleoid. (**B**) Cloacal wall (*c*) of a didemnid host showing population of *Prochloron* sp. (*p*). Scanning electron micrograph. (**C**) *Prochloron* sp. in division (*d*). The arrangement of the thylakoids (*t*) at the periphery gives the appearance of a thick wall. Light micrograph of artificially silicified *Prochloron*. [From Francis, Barghoorn, and Margulis (1978).] (**D**) *Prochloron* cells in division in the cloaca, showing the adhesion between the host tissue (arrows) and the green photosynthetic prokaryotes. [Courtesy of K. Lee.]

The Archean record

There seems to be no evidence at all to favor the view
that the metabolism of the stromatolites and oncolites
found in the carbonate rocks of the Bulawayan
(approximately 2.7 billion years), Ventersdorp
(approximately 2.3 billion years), and Transvaal (greater
than 2.0 billion years) systems, and in nearly all later
carbonate sequences of Precambrian age, was any
different from that of their Paleozoic and modern analogues.

C. F. DAVIDSON, 1965

Evidence for Archean life comes primarily from three sources: microfossils, stromatolites associated with well-dated rocks, and chemical fossils. Careful studies of the deposition environments, as well as of the fossils themselves, have been used to trace the early history of life (Cloud, 1988). Well-preserved, extensive Archean and Proterozoic sedimentary outcrops are known on many continents; important sites are listed in Table 5-5.

Microfossils, microscopic remains of organisms or their parts, are studied in petrographic thin section with the light microscope. The best preserved are in cherts, rocks composed primarily of silica. These microcrystalline quartz rocks are cut into more or less transparent slices no more than 100 μm thick. The fossils and the mineral matrix in which they are preserved are studied with white light and polarization optics. Most of the best microfossils are embedded in black cherts retaining significant quantities of carbon, usually less than one percent by weight, although some are found in shales as well. Microfossils in shales must be released by maceration with hydrochloric and hydrofluoric acids because the shales are opaque. If the concentration of microfossils is low or the observation of thin sections is hindered by obscuring minerals, the cherts, too, are dissolved by brief treatment with hydrofluoric acid. The released organic material is collected for study. When organic microfossils are studied morphologically, a part of the same sample can be analyzed by gas chromatography and mass spectroscopy. The analysis and identification of organic compounds released from sedimentary rocks is a major activity of organic geochemistry. The discovery of abundant and varied organic compounds in "rock juice" from ancient sediments has led to the concept of chemical fossils (Schopf, 1983).

In principle, could probiotic organic matter be distinguished from biotic? Not easily. The earliest organic compounds required for replication, those produced originally by solar radiation, must have included easily formed monomers such as amino acids. Also made probiotically were many organic isomers of low molecular weight, an assortment of compounds more diverse than those in living systems. Tarlike polymers similar to the unidentifiable materials often produced in "origin of life" experiments were probably formed in abundance; however, it is difficult to distinguish them from altered biogenic polymers. Hydrocarbons such as the aromatics identified in carbonaceous meteorites were probably also formed probiotically on Earth. In the absence of life, the isomers of amino acids and sugars were probably not formed in optically active proportions (Fox, 1965).

Organic compounds produced by biosynthesis, as in photosynthesis or chemoautotrophy, are enriched in ^{12}C relative to ^{13}C when compared with abiotically synthesized organics (Barghoorn et al., 1977). The reduction of sulfate to sulfide by microbes leads to an enrichment of ^{32}S relative to ^{34}S. The interpretation of Archean isotopic effects is detailed by Schopf (1983). Probiotic organic matter in sedimentary rocks would be expected not to contain amino acid polymers uniformly peptide-bonded like the proteins of living cells; rather, mixtures of linkages like those in polyanhydrous copolymers would be likely (Ponnamperuma, 1977). However, because meteoritic organic carbon may show similar trends, optical activity, ^{12}C or ^{32}S enrichment, and lack of normal peptides do not rigorously distinguish abiotic (meteoritic) or inferred prebiotic from biotic organic matter. Thus, these criteria for biogenesis must be assessed carefully in each case before any Archean sedimentary carbon can be identified as probiotic or biotic (Strother, 1992).

Sedimented organic matter is vulnerable to many hazards. Matter deposited by one kind of organism becomes food for another or is mixed with another organism or its waste products. Even if buried and shielded from microbial decay, organic materials may dissolve or volatilize or be affected by physical and chemical changes in the sediments — that is, by diagenetic alteration. Changes in organic matter wrought by degradation and diagenesis are greater than evolutionary changes with time. Diagenetic alteration must be determined before inferences can be drawn about ancient living communities. The order of deposition in pre-Phanerozoic sediment (summarized in Table 6-11) is only surmised, inferred from the order in which the major microbial pathways may have evolved. This order is an oversimplified reflection of the fate of biogenic matter during 3000 million years.

Once the earliest Archean microbes had evolved, highly specific, optically active isomers of carbon compounds were biosynthesized consistently. Amino acid polymerizations would be linked obligatorily with nucleic acid polymerizations, because these polymers must be synthesized simultaneously for reproduction to occur. Acetate, lactate, pyruvate, alanine, glycine, ribose, deoxyribose, and other metabolizable small molecules would have been concentrated in cells. Ancient rocks have yielded various soluble compounds — amino acids, prophyrins, and isoprenoids — although because of porosity it is unlikely that soluble carbon compounds are syngene-

TABLE 5-5
Evidence for pre-Phanerozoic life.

APPROXIMATE AGE ($\times 10^9$ YEARS)	FORMATION AND LOCATION	ORGANISMS	NATURE OF EVIDENCE[a]	REFERENCES
>3.7	Isua Formation, West Greenland	None, highly metamorphosed sediments	Extensive $^{12}C/^{13}C$ enrichment of graphite	Schidlowski et al., 1979
>3.4	Sheba Formation (Fig Tree Group), Barberton, South Africa; Onverwacht	Unidentified forms	Microfossils: bacteria in division, carbon-rich sediments	Knoll and Barghoorn, 1977; Reimer et al., 1979; Schopf, 1976; Walsh and Lowe, 1985
>3.4	Warrawoona North Pole, W. Australia	Unidentified forms	Microfossils, stromatolites, $^{12}C/^{13}C$ enrichment, pristane, phytane, extensive reduced carbon	Dunlop et al., 1978; Awramik, 1980
3.0–2.7	Bulawayan limestones, Zimbabwe	Cyanobacteria, other bacteria	Stromatolites, microfossils, $^{12}C/^{13}C$ enrichment of organic reduced carbon compared to carbonate in the same rock	Walter, 1993; Hoering, 1967; Nagy et al., 1977; Schidlowski et al. 1975
		Bacterial spores, metal-oxidizing bacteria	Extensive stromatolites, microfossils, organic geochemicals	Schopf and Barghoorn, 1967; Walter, 1976; Nagy and Zumberge, 1976; Nagy and Nagy, 1976
2.5	Soudan Iron Formation, northeastern Minnesota and western Ontario	Microstructures, unidentified	$^{12}C/^{13}C$ and $^{34}S/^{32}S$ enrichments, biogenic iron and sulfur deposits	Cloud et al., 1965; Goodwin et al., 1976
	Michepocoten, Woman River, Ontario		Stromatolites	H. G. Thode, 1977 (personal communication)
2.0	Gunflint Iron Formation (shales, limestones), Ontario	Eoastrion, Kakabekia (Metallogenium-like microfossils; see Figure 5–9G)[b]	Stromatolites, diverse microfossils, some found with oxides of iron and pyrite	Barghoorn and Tyler, 1965; La Berge, 1967; Awramik and Barghoorn, 1977; Schopf et al., 1965
	Nabberu Formation, W. Australia	Siderocapsa-like, Metallogenium-like microfossils; Sphaerotilus-like and Siderococcus-like bacteria (iron bacteria)		Licari and Cloud, 1968; Schopf and Barghoorn, 1967; Cloud, 1988
		Cyanobacteria	$^{12}C/^{13}C$ enrichment, pristane, phytane	

2.0–0.7	Belt series, Wyoming and Montana; Nonesuch shale, northern Michigan; cosmopolitan	Sporelike cysts, multichambered protoctist-like forms, cyanobacterial sheaths, cyanobacteria, protoctista	Stromatolites, ^{12}C/^{13}C enrichment, diverse organic compounds, phytane, pristane, vanadyl porphyrins, optically active organic compounds	Schopf, 1975; Bengtson, 1993[d]
1.8	Many worldwide (younger BIFs are found)[c]	As in Gunflint Iron Formation (*Eoastrion, Kakabekia*)	End of BIFs; replacement by red beds	La Berge, 1967; Cloud, 1988
1.4	Beck Springs, Alberta, Canada	Cyanobacteria	Microfossils in stromatolites	Cloud, 1988; Vidal, 1993
0.75–0.55	Namibia, Australia, Newfoundland, Wales, etc.	Vendobionta (plasmodia)	Sandstone body fossils, carbonaceous films[e]	Seilacher, 1993; Fedonkin, 1993; Hofmann, 1993
0.75	Ediacara, South Australia; over 20 other sites worldwide	Several major animal phyla	Well-developed metazoan fossils	Glaessner, 1976; McMenamin and McMenamin, 1990; Bengtson, 1993
50 million years[f]	Green River Shale, North America		Wide diversity of organic compounds: many odd carbon isoprenoids; branched-chain, ring, and straight-chain alkanes; carotane; phytane; pristane; steroid derivatives (sitostane, triterpenes, steranes, cholestane)	Eglinton and Calvin, 1967; Calvin, 1969a, b; Eglinton and Murphy, 1969

[a] For a review of the extensive literature on pre-Phanerozoic microfossils, see Schopf (1983). About 200 well-developed microfossil chert microbiota are known, and the number is increasing. For a comprehensive discussion of stromatolites, which are present in the Archean but become overwhelmingly dominant in the Proterozoic, see Walter (1976). There are hundreds of known Proterozoic stromatolite locales. For a bibliography of the entire pre-Phanerozoic, see Awramik and Barghoorn (1978).

[b] For discussion of the remarkable *Kakabekia*-like living organism, a complex prokaryotic heterotroph 5 to 10 μm in diameter on a centrally attached stalk 5 to 15 μm in length, found growing in ammonia-rich environments, see Siegel et al. (1967).

[c] BIF = banded iron formation.

[d] Superbly preserved Proterozoic fossils belonging to at least five new families (e.g., Grypaniaceae) are now known (Bengtson, 1993).

[e] Chintinozoa, *Charnia, Chuaria*, and other large pre-Phanerozoic fossils both from shales and from cherts are coming to light; many are eukaryotic (Knoll, 1979; Schopf, 1992).

[f] Though not pre-Phanerozoic, this is included for comparison. Fossils from all five kingdoms are known from the Phanerozoic.

tic with the rocks. Much ancient organic matter has been preserved in the form of kerogen, an unextractable and highly complex organic polymeric material abundant in shales and other carbon-rich sedimentary rocks. Kerogen has the virtue of being autochthonous—free from contamination by younger organic material (Nagy et al., 1977). It probably harbors secrets of the early evolution both of organisms and of their ecosystems. Kerogen has

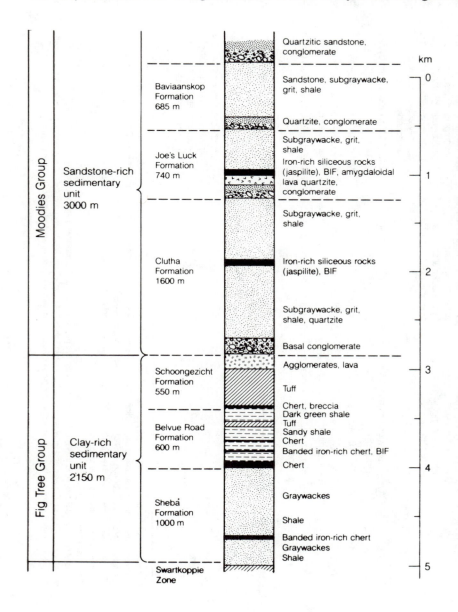

been degraded for study by the rather drastic method of ozonanalysis (Nagy and Zumberge, 1976); however, it is difficult to assess the effect of the analytical procedure on the original organic matter.

Inclusions that are probably fossils of bacteria have been seen in the oldest known unmetamorphosed sediments, from the Swaziland System in South Africa (Figure 5-7). These extensive deposits were laid down from

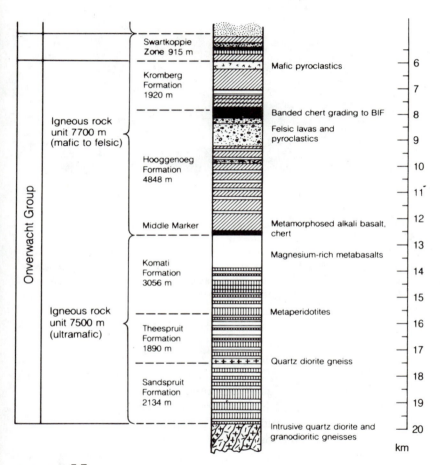

FIGURE 5-7

Composite section of the rocks of the Swaziland System, Barberton Mountain Land, South Africa. BIF = banded iron formation. [After Cloud (1974). For fossils see Walsh and Lowe (1985), Walsh (1992) and Figure 5-8.]

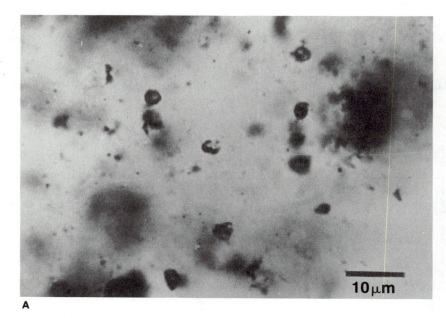

A

FIGURE 5-8

Modern bacteria and microfossils compared.
(A – F, K) Light micrographs of microfossils in thin sections of cherts more than 3400 million years old from the Swaziland System in South Africa; scale as shown in **A. (G – J)** Light micrographs of modern coccoid cyanobacteria; scale as shown in **A. (L)** Electron micrographs of putative fossil bacteria two billion years old. [All photographs courtesy of E. S. Barghoorn. **A – K** from Knoll and Barghoorn (1977). **L** from Schopf et al. (1965).]

about 3400 million to 3100 million years ago. In Fig Tree cherts, Knoll and Barghoorn (1977) discovered microfossils apparently of bacteria in the act of cell division (Figure 5-8). Equally impressive 3400-million-year-old microfossils have been identified in the Warrawoona Group of northwestern Australia (Awramik, Schopf, and Walter, 1983). Some cherts, at the lower boundary of the Sheba Formation, are rich in reduced carbon. The quantity of carbon in these sediments is quite impressive; apparently it was laid down under conditions and at rates comparable to the deposition of carbon in far younger rocks. These observations suggest that diversified ecosystems based primarily on microbial photosynthesis functioned during the Archean about 3400 million years ago (Reimer, 1979). Entities thought to be fossils of bacteria, including cyanobacteria, have now been found in carbon-rich cherts throughout the pre-Phanerozoic geological column. Many are magnificently preserved (Schopf, 1967, 1972, 1983). The simplest are

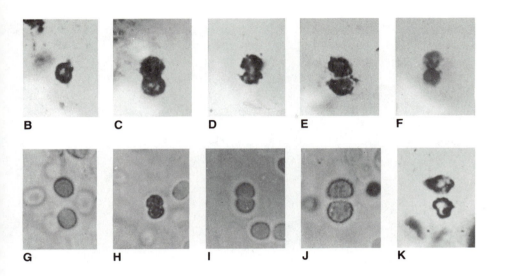

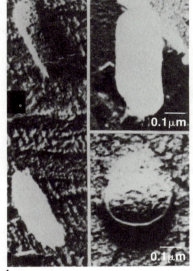

A

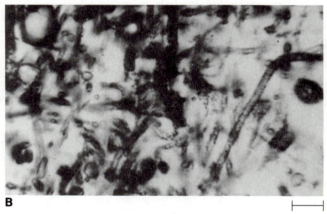

B

FIGURE 5-9

Microbiota from cherts of the Gunflint Iron Formation.
(A) Rocks of the Gunflint Iron Formation near Schreiber, Ontario. (B) Petrographic thin section through Gunflint chert showing abundance of microfossils. Light micrograph, bar = 10 μm.

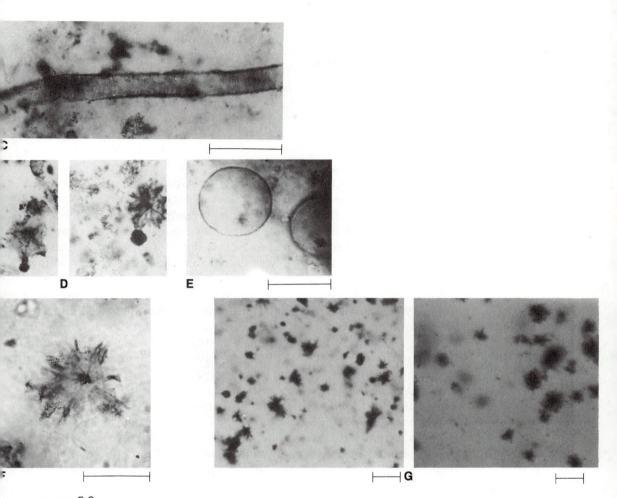

FIGURE 5-9

(C) *Gunflintia grandis* Barghoorn, microfossils from Schreiber, Ontario, facies of the Gunflint chert. Light micrograph. (D) Two views of *Kakabekia umbellata* Barghoorn, a microfossil very abundant in certain facies of the Gunflint chert. These fossils show a range of star shapes; some are indistinguishable from the modern putative manganese- and iron-oxidizing *Metallogenium* bacteria. Light micrograph. (E) *Leptoteichus golubicii* Barghoorn, a microorganism abundant in the Gunflint chert, named for our colleague Stjepko Golubic (Awramik and Barghoorn, 1977). Light micrograph. (F) *Eoastrion*, a microfossil from the Frustration Bay facies of the Gunflint Iron Formation; remarkably like *Metallogenium*. Light micrograph. (G) Left: *Eoastrion* in thin section prepared by E. S. Barghoorn. Right: precipitates caused by live enigmatic manganese-oxidizing organisms (*Metallogenium*) supplied by A. Zavarzin and grown by E. Gong in bactoagar with 0.000001% manganese acetate. Nomarski differential phase-contrast light micrographs [A courtesy of P. E. Cloud, Jr.; B–F courtesy of E. S. Barghoorn.], bars = 10 μm.

the spheres of the Fig Tree Group of rock formations, which are among the most ancient. Preceding them and below them in the section are the upper Onverwacht filaments, from the Hooggenoeg and Kromberg Formations (Walsh and Lowe, 1985). Schopf (1976) has argued that these spheres are not fossil bacteria, but remains of probiotic organic entities like the microspheres that can be produced spontaneously in the laboratory by mixing certain organic compounds. Perhaps probiotic and biotic entities coexisted in these earliest communities. In any case, the detailed studies of the Swaziland System (Reimer, Barghoorn, and Margulis, 1979), together with carbon-isotope fractionation studies of the even older (about 3800 million years old) Isua Formation of Greenland (Schidlowski et al., 1979), suggest that the origin of life occurred long before Swaziland times.

Although the Archean fossil record is still limited—perhaps because of a geological regime not favorable to the establishment and preservation of extensive microbial mats—the Proterozoic record is plentiful. By the time the Gunflint sediments of southern Ontario and adjacent northern Minnesota were deposited, about 2000 million years ago, many types of photosynthetic prokaryotes were widely distributed (Figure 5-9; Barghoorn and Tyler, 1965; Knoll, 1979). The extensive beds of pre-Phanerozoic stromatolites were probably formed just as such beds are today: by the trapping, binding, and precipitation of sediment by subtidal and intertidal communities of cyanobacteria and other bacteria—primarily, filamentous sheathed forms (Monty, 1973a; Walter, 1976). Some Proterozoic microfossils are remarkably similar to modern microbes in size, morphology, patterns of degradation, distribution, and other attributes (Schopf, 1992; Awramik and Barghoorn, 1977; Knoll and Barghoorn, 1977). For example, the coccoid cyanobacterium *Eosynochococcus* from the 2300-million-year-old rocks of Belcher Island (Hudson Bay, Canada) is so similar to the modern coccoid *Gloeothece* that only an expert can distinguish them (Golubic and Campbell, 1979).

Oxygen-releasing photosynthesis had spread widely to littoral, submerged, and terrestrial habitats by the early Proterozoic, if not before. Photosynthetic prokaryotes, not plants, produced the first oxygenic atmosphere. Direct geological evidence, in the form of oxidized iron and other minerals, supports the interpretation that oxygen was abundant in the atmosphere by 2000 million years ago (Cloud, 1974). The presence of banded iron in the Swaziland sequence—and maybe even in the Isua rocks—means that oxygenic photosynthesis may have evolved far earlier. Banded iron formations (BIFs) (Chapter 4) are characteristic of late Archean and early Proterozoic sediments (Cloud, 1988). Such sediments become less frequent in younger rocks and by late Proterozoic times they disappear, signaling the conversion to an oxygen-rich atmosphere.

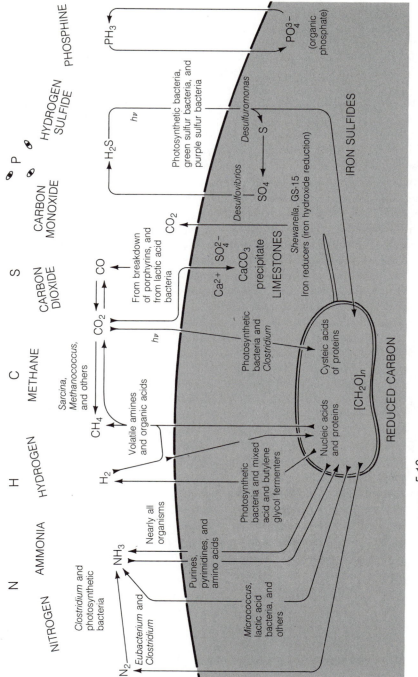

FIGURE 5-10

Biogeochemical cycles in an anaerobic world. The dark curved line crossing the diagram represents a horizon—above is the atmosphere; below is water and sediment. Because phosphorus has no volatile phase, except perhaps phosphine (PH_3), it may have been spread through the atmosphere as combined phosphates by propagules such as actinospores (of actinobacteria), endospores of clostridia, and others. (Dévai et al., 1988.) hv = light.

During the Archean age of the anaerobes, before molecular oxygen accumulated sufficiently to be released into the atmosphere by photosynthetic prokaryotes, it is probable that all of the major prokaryotic metabolic and enzymatic systems had evolved: nucleic acid–protein-based autopoiesis and DNA repair; fermentation; glycolysis; biosynthesis of ester- and ether-linked lipids; methanogenesis; cell wall and spore wall formation (requiring diaminopimelic acid); alkanoate reserve deposition (polyhydroxybutyrate, valerate, etc.); tricarboxylic acid pathways both for synthesis and for ATP generation; the various pathways of carbon dioxide fixation; nitrogen fixation; anaerobic photosynthesis, including the biosynthesis of isoprene and porphyrin derivatives and associated proteins; the oxidation of hydrogen sulfide to sulfur; the deposition of elemental sulfur, sulfate, and sulfide; iron and manganese oxidation and reduction; and so forth. Elemental cycles in an anaerobic world are diagrammed in Figure 5-10. This was the late Archean scene on the eve of the revolution induced by overabundance of the waste product of oxygen-releasing phototrophs.

CHAPTER 6

ATMOSPHERIC OXYGEN FROM PHOTOSYNTHESIS

AEROBIOSIS IN MICROBES

Oxygen is toxic! We, whose lives depend upon a considerable supply of oxygen do not easily comprehend its toxicity. Our apparent comfort at the ambient level of oxygen is due to elaborate defenses against its very considerable toxicity.

I. FRIDOVITCH, 1977

The planetary consequences of microbial release of waste oxygen were profound and have persisted until the present. The course of subsequent evolution was permanently deflected.

Conspicuous life forms of today require oxygen to live. All plants and animals are obligate aerobes; they fail to complete their life cycles if the ambient oxygen concentration decreases below the range to which they are accustomed. Species that thrive in the absence of oxygen are rare in protoctists and fungi; they are common among the monera. The five classes of responses to oxygen are tabulated on the kingdom level in Table 6-1. Many moneran genera are strictly anaerobic; others contain both anaerobic and aerobic strains; still others contain only obligately aerobic isolates, including microaerophils, organisms adapted to less than the current atmospheric concentration of oxygen. For information concerning any specific moneran genus, consult *Bergey's Manual of Systematic Bacteriology* (Holt, 1984–1989) or the *The Prokaryotes* (Balows et al., 1991).

TABLE 6 - 1
Responses to molecular oxygen.

RESPONSE	DEFINITION	KINGDOM	DISTRIBUTION
Obligate anaerobe	Requires absence of O_2 at all times	Prokaryotae Proctoctista Animalia Fungi Plantae	Many Few[a] None None None
Facultative anaerobe = facultative aerobe	Alters metabolism to use O_2 (aerobic) or other electron acceptor (e.g., nitrate)	Prokaryotae Proctoctista Animalia Fungi Plantae	Many None None Some yeasts None
Anaerobe: aerotolerant	Grows in absence of oxygen equally as well as in presence; neither uses nor is inhibited by free O_2	Prokaryotae Proctoctista Animalia Fungi Plantae	Many None None None None
Aerobe: microaerophil	Requires O_2 for growth but in concentrations less than ambient	Prokaryotae Proctoctista Animalia Fungi Plantae	Many Some None Some None
Obligate aerobe: intolerant of anoxic conditions	Requires O_2 for growth in concentrations comparable to present 7 to 10 ppm in H_2O, 20% in air	Prokaryotae Proctoctista Animalia Fungi Plantae	Many Most All Nearly all All
Obligate aerobe: tolerant of anoxic conditions	Requires ambient O_2 for growth but tolerates less or no O_2 for extended periods	Prokaryotae Proctoctista Animalia Fungi Plantae	Many Few[b] Few[c] Many[d] Few[e]

[a] Protoctista: some zoomastigina, chytrids, and ciliates.
[b] Protoctista: encysting amoebomastigotes, ciliates, etc.
[c] Animalia: sediment-burying animals.
[d] Fungi: in spore form only.
[e] Plants: seeds and other dormant portions.

The great range of responses to oxygen found in bacteria, even in members of the same genus, is not seen in eukaryotes; all those that have mitochondria are strict aerobes. In the few eukaryotes that are tolerant to anoxic* conditions, the mitochondria shrink (sometimes until they are invisible) and become nonfunctional under anoxic conditions. These orga-

---

*"Aerobiosis" refers to a metabolic mode; for example, an organism is an aerobe. The terms "anoxic" and "oxic" refer to the environment; for example, an anoxic habitat is one with low levels or no free oxygen.

nisms retain the capacity to redifferentiate their mitochondria. Certain symbiotic or mud-dwelling protoctists lack mitochondria and thrive (Cleveland and Grimstone, 1964; Bloodgood and Fitzharris, 1977) in environments depleted of oxygen (Daniels, Breyer, and Kudo, 1966; Fenchel and Finlay, 1991). Some sand-dwelling ciliates also physiologically survive anoxic conditions. In certain eukaryotic anaerobes, ones that clearly evolved from aerobic ancestors, such as yeast, anaerobiosis is a derived, more recent condition. Although some animals and plants can survive oxygen depletion temporarily, the completion of an animal or plant life cycle in totally anoxic conditions has never been documented.

The variety of prokaryotic responses to atmospheric oxygen is understood most easily as a result of a rising oxygen concentration that applied significant selective pressure at the time of major prokaryotic adaptive radiations. That tolerance, and eventually utilization, of oxygen evolved convergently in prokaryotes is deduced from the different responses to oxygen found within groups of close relatives, such as the enterobacteria or actinobacteria. Indeed, a direct relation between DNA structure and the physiology of growth is known in the bacterium *Salmonella*: a DNA topoisomerase, which leads to relaxation of the DNA nucleoid, is required for expression of genes for aerobic growth, whereas a gyrase, associated with supercoiling of DNA, is necessary for expression of genes needed for anaerobic growth (Yamamoto and Droffner, 1985). Such enzymatically mediated, rapid changes in DNA leading to switches from anaerobic to aerobic metabolism are clearly a prokaryotic virtuosity. In the few eukaryotes capable of anaerobiosis, the entire genome and its associated mitochondrial protein-synthesizing system are altered.

Although most cyanobacteria respire oxygen, some species of *Phormidium* are poisoned by the oxygen that they themselves produce. The existence of many obligate aerobes in Kingdom Procaryotae indicates that prokaryotes continued to evolve and diversify during and after transition to oxic environmental conditions.

Only the bacterial release of oxygen during photosynthesis can explain its planet-wide accumulation in the atmosphere over two billion years ago (Fischer, 1965; Cloud, 1968a, 1974, 1978; Walker, 1980). The hazardous oxygenic environment to which the biota was forced to respond was produced by the biota itself. In the mid-Proterozoic eon, that biota was prokaryotic.

Why is oxygen toxic? How is this toxicity circumvented? Apparently, many mechanisms are involved in both the toxicity of and the protection from oxygen (Haugaard, 1968; Fridovitch, 1977). Oxygen reacts with organics to form short-lived superoxide (O_2^-) and hydroxyl (OH^-) radicals and singlet oxygen. Oxygen poisons by means of hydrogen peroxide, which oxidizes ubiquitous metabolites such as lipoic acid and acetyl coenzyme A

TABLE 6-2
Bioluminescence.

ORGANISMS	LUCIFERIN[a]
Monera *Photobacterium* *Vibrio* *Xenorhabdus*	 Reaction in *Vibrio* requires FMNH$_2$.
Protoctista Dinomastigota *Gonyaulax* (and others) Actinopoda *Aulosphaera* *Cytocladus*	Unknown; molecular weight less than 1000 Unknown
Animalia Cnidaria (coelenterates) Anthozoa *Renilla* *Stylatula* *Ptilosareus* *Cavernularia* *Acanthoptilum* Hydrozoa *Aequorea* *Obelia* Ctenophora *Mnemiopsis* *Beroe* Mollusca *Watasenia* (more than 75 genera of cephalopods) Arthropoda 5 species of decapods (shrimp) *Gnathophausia* (opossum shrimp) Chordata *Neoscopelus, Diaphus* (fish)	
Arthropoda *Cypridina* (ostracod) Chordata *Apogon* *Parapriacanthus* *Porichthys* *Pyrosoma* (tunicate)	

POSSIBLE FUNCTION

Maintenance of symbiosis with fish
Habitat attraction? (fish gut)
Attraction of vertebrates to glowing insect carcass; dispersal agents for nematodes and their
associated bacteria

?

?

Many; for example, camouflage, communication, mate attraction, defense by baffling or startling

Many; examples include mate attraction, interspecific communication, camouflage, defense,
illumination

(continued)

TABLE 6 - 2 (*continued*)

ORGANISMS	LUCIFERIN[a]
Hemichordata *Balanoglossus* *Ptychodera*	Unknown
Arthropoda Lantern flies (Fulgoridae) Glowworm flies (Mycetophilidae) Beetles (Elateroidea, Cantharoidea, more than 1000 species) *Photinus* (firefly)	(firefly luciferin structure)
Annelida *Diplocardia* *Octochaetus* Several genera of earthworms	(luciferin structure)
Mollusca *Latia* and other gastropods	(Latia luciferin structure)
Fungi Ascomycota *Xylaria* Basidiomycota (more than 40 species) *Armillariella* *Pleurotus* *Panellus*	Not known; reaction in basidiomycota requires NADPH or NADH
Plantae None	

[a] $FMNH_2$ = reduced flavin mononucleotide; NADH and NADPH = reduced nicotinamide adenine dinucleotide and its phosphate.

(acetyl CoA), and also by inhibiting iron- and sulfhydryl-containing flavoproteins and by reacting with carbon-carbon double bonds in polyunsaturated fatty acids (Fridovitch, 1977). Several classes of compounds evolved originally as protective agents against rising oxygen concentrations: luciferins, which emit light in enzyme-mediated reactions and react with oxygen so that toxic radicals are not produced; superoxide dismutases (iron- manganese-, or copper-containing enzymes that destroy the superoxide radical); and some isoprenoids and porphyrin derivatives.

Bioluminescence, a scientific curiosity for more than half a century (Harvey, 1952), may be related to oxygen detoxification (McElroy and Seliger, 1962). Bacteria, dinomastigotes, fireflies, fungi, coelenterates,

POSSIBLE FUNCTION
Mate attraction
?
?
Attraction of nematode or bacterial dispersal agents?

pyrosomes (tunicates), and fish radiate cold light (Table 6-2). The luminescent system of the bacterium *Photobacterium fischeri* requires oxygen and is sensitive to an oxygen pressure of 0.0005 mm Hg. When peroxide is formed, it reacts immediately with luciferin aldehyde to form the corresponding organic acid and water. No toxic radicals are produced, and light is emitted. Since the tasks of detoxification and, eventually, utilization of oxygen were taken over by other oxygen-handling metabolic agents (peroxidases, catalases, epoxides, flavins, isoprenoids, and so on), luminescence was lost by the ancestors of most extant organisms.

Bioluminescent systems in diverse organisms are interpreted as ancient, independently evolved oxygen-detoxification mechanisms that have undergone selection pressures leading to new functions. Unrelated substrate–enzyme (luciferin–luciferase) systems have been identified. Whereas the use of bioluminescence by fireflies for finding mates is a recent evolutionary development, some eukaryotes use luminescent systems that evolved long ago in prokaryotes. Although the extent to which eukaryotic luminescence is due to prokaryotic symbionts is not yet known, many bacteria–animal symbiotic luminescent systems have been discovered, such as those of tunicates and deep-sea fishes (Harvey, 1952; Herring, 1978). The fish *Photoblepharon* has chambers under its eyes in which it cultures luminescent photobacteria, for example. These systems are certainly not homologous, whether or not they are related to oxygen detoxification.

Isoprenoid derivatives, distributed widely in prokaryotes (Table 6-3), are synthesized from compounds that arise from acetate. Molecules of acetyl CoA are linked successively to form an intermediate compound, mevalonic acid. The alcohol group of mevalonic acid is pyrophosphorylated; a decarboxylation follows, producing the universal biological isoprene unit, isopentenyl pyrophosphate (Figure 6-1). Isoprene itself is not an intermediate in the biosynthesis of isoprene derivatives, whereas isopentenyl pyrophosphate is the starting point for the synthesis of hundreds of isoprenoid compounds. In anaerobic bacteria, some major end products of isoprenoid biosynthesis are (ubiquinone) coenzyme Q, vitamin K, and several types of saturated and monounsaturated fatty acids and related compounds. Phytol, a C_{20} isoprenoid alcohol bound to chlorophyll, is found in photosynthetic organisms, as are the carotenoids, diverse orange, yellow, and red C_{40} isoprenoid ring compounds.

TABLE 6-3

Isoprenoid distribution in prokaryotes.

COMPOUND	DISTRIBUTION[a]
Vaccenic acid: $CH_3(CH_2)_5CH{=}CH(CH_2)_9COOH$	Most bacteria
Oleic acid: $CH_3(CH_2)_7CH{=}CH(CH_2)_7COOH$	Most bacteria
Vitamin K: $C_{10}H_4O_2CH_3(CH_2{-}CH{=}\overset{\underset{\displaystyle CH_3}{\mid}}{C}{-}CH_2)_nH$	See Table 6-5
Coenzyme Q_n: $C_9H_9O_4(CH_2{-}CH{=}\overset{}{C}{-}CH_2)_nH$	See Table 6-5
$\underset{CH_3}{\mid}$	
Phytol: $C_{20}H_{39}OH$	Part of chlorophyll in all photosynthetic bacteria; archaebacteria[b]
Squalene and phytyl ethers	Arachaebacteria[b]
Carotenoids (for example, β-carotene, $C_{40}H_{56}$)	All photosynthetic bacteria; many heterotrophs

[a]Miller (1961); Krinsky (1966); Tornabene et al. (1979).
[b]Methanogens, halobacters, thermoacidophils, etc.

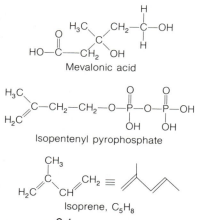

FIGURE 6-1

Isoprenoid formation. Mevalonic acid is the precursor of isopentenyl pyrophosphate, which is the universal building block in isoprenoid biosyntheses. Isoprene itself is not used in such biosyntheses.

The cyclization of the end rings of carotenoids generally requires atmospheric oxygen. How carotenoids are synthesized in obligately anaerobic photosynthesizers is a mystery; perhaps light is a source of energy for their biosynthesis. It has been observed that isoprenoids protect cells from light-induced oxidations (Table 6-4). Some heterotrophic anaerobes (for example, *Spirochaeta aurantia*) are white when grown anaerobically but synthesize carotenoids in the presence of oxygen. However, many species of purple nonsulfur bacteria — facultative photoautotrophs — are brightly colored only in the absence of oxygen. When stressed with oxygen, Rhodospirillaceae are forced to repress carotenoid synthesis and grow heterotrophically; in the dark, too, they turn white because they are unable to synthesize carotenoids. Not only is the physiology of isoprenoid biosynthesis in individual organisms sensitive to oxygen but so is its distribution among organisms. A relationship between aerobic metabolism and the capacity for synthesis of vitamin K and quinones is shown in Table 6-5. Some isoprenoid derivatives are complex isomers that are highly stable and restricted to very few organisms; these make excellent biochemical and evolutionary markers.

Since steroids are synthesized from the isoprenoid squalene, they are complex isoprenoids. These lipid-soluble, fused-ring compounds are components of all eukaryotic cell membranes; steroids must either be synthesized by the cells or supplied to them. Steroid synthesis is limited in prokaryotes to some aerobic bacteria and cyanobacteria. (Some mycoplasmas contain steroids, but there is no evidence that they synthesize

TABLE 6-4
The protective role of isoprenoid derivatives.

COMPOUND	ORGANISM

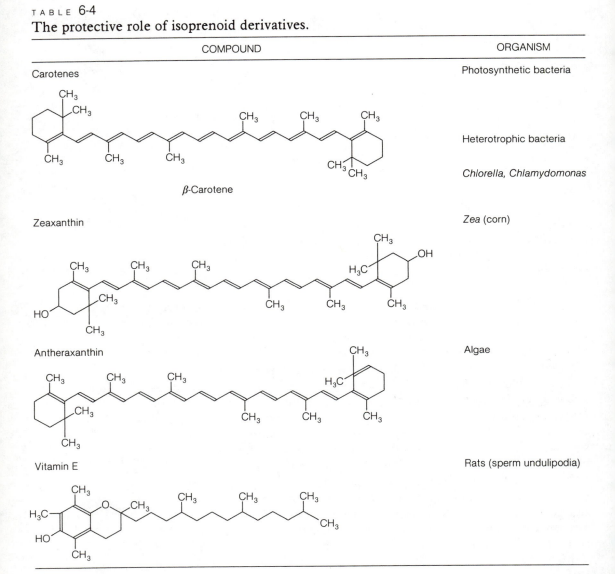

Carotenes — Photosynthetic bacteria

β-Carotene — Heterotrophic bacteria

Chlorella, Chlamydomonas

Zeaxanthin — Zea (corn)

Antheraxanthin — Algae

Vitamin E — Rats (sperm undulipodia)

them. The steroid requirement of mycoplasmas is probably due to their symbiotrophic life in steroid-rich environments.) Hopanoids, lipid-soluble hydrocarbon ring compounds, are far more abundant in prokaryotes than are steroids (Ourisson, Albrecht, and Rohmer, 1984). Steroids arose as a late development in the evolution of metabolic pathways; most diversifica-

ROLE	COMMENTS	REFERENCE
Protection against photo-oxidation	Mutants lacking carotenoids survive dark-aerobic and light-anaerobic conditions; they die under light-aerobic conditions.	Krinsky, 1966
Protection against photo-oxidation	Carotenoids afford protection against visible light only, not against ultra-violet or X-rays.	Krinsky, 1966
Protection against photo-oxidation		Krinsky, 1966
Protection against photo-oxidation	In albino mutants, chlorophyll and catalase are destroyed in the presence of light and oxygen, but not in the presence of light and a nitrogen atmosphere.	Krinsky, 1966
Protection of chlorophyll in the presence of light and oxygen		Bamje and Krinsky, 1965
Protection against oxidation	Number of centrioles increases if vitamin E is deficient and oxygen pressure is high.	Hess and Menzel, 1968

tion of steroid metabolism is correlated with eukaryotic innovation. Because the squalene cyclization requires atmospheric oxygen, steroids, too, may have originated in response to minute quantities of this gas and were selected subsequently for the remarkable flexibility they confer on membranes.

TABLE 6-5

Isoprenoids and aerobic metabolism.[a]

ORGANISMS	CAN SYNTHESIZE UBIQUINONE (COENZYME Q)	CAN SYNTHESIZE VITAMIN K	RESPONSE TO OXYGEN	CONDITION OF MITOCHONDRIA
Prokaryotes				
All obligate anaerobes investigated (for example, *Clostridium*)	−	−	Cannot tolerate	
Rhodospirillum	+	−	}	
Hydrogenomonas	+	−	} Facultative anaerobes[b]	
Escherichia coli	+	−	}	
Bacillus mesentericus	−	+	}	
Streptomyces	−	+	} Obligate aerobes	
Mycobacterium	−	+	}	
Eukaryotes				
Saccharomyces cerevisiae (yeast)				
Grown anaerobically	−	− }	Facultative anaerobe	Dedifferentiated
Grown aerobically	+	− }		Functioning
Neurospora crassa (ascomycote)	+	−	Obligate aerobe	Functioning
Mucor corymbias (zygomycote)	+	−	Obligate aerobe	Functioning

[a] After Miller (1961).
[b] Some *Rhodospirillum* species are obligate anaerobes; some can carry out oxidative metabolism in the dark under oxic or micro-oxic conditions.

The biosynthesis of both porphyrins and isoprenoids is connected intimately to autopoietic metabolism. Isoprenoids require the synthesis of acetyl CoA, and porphyrin biosynthesis involves glutamate and its transfer RNA or glycine and succinate, suggesting very early evolution of both classes of compounds.

Porphyrin synthesis from either glutamate or succinate proceeds, apparently, by the same steps, no matter what the end product is. The common steps on a pathway are the most ancient (Bernal, 1967); in porphyrin synthesis, these are the steps from succinate through δ-aminolevulinic acid or from glutamate to levulinic acid. Table 6-6 lists some porphyrin-related compounds synthesized by prokaryotes; some structural formulas are shown in Figure 5-3. The distribution of the ability of microbes to synthesize porphyrins, comprehensible as an evolutionary response to aerobiosis, casts doubt on the supposition that porphyrins occurred by chemical evolution prior to the origin of life (Hodgeson and Ponnamperuma, 1968; Calvin, 1969a).

Just as luciferins, isoprenoids, and porphyrins can be related to evolutionary innovations of prokaryotes to the increasing concentration of oxy-

TABLE 6 - 6

Corrinoid and porphyrin biosyntheses in prokaryotes.

DERIVATIVE	PORPHYRIN OR CORRINOID[a]	CHELATED METAL ION	ORGANISMS AND PROBABLE SELECTIVE ADVANTAGE
None synthesized	—	—	Only a very few: some clostridia, lactic acid bacteria, some rickettsias, and spirochetes. All these organisms are oxygen-intolerant.
Vitamin B_{12}	C	Cobalt	Synthesis limited to propionic acid bacteria, streptomycetes, and clostridia. Eukaryotes cannot synthesize B_{12}, yet all seem to require it. Used in methylation of RNA, biosynthesis of purine and pyrimidine, reduction of ribose to deoxyribose, and synthesis of protein.
Catalase coenzyme	P	Iron	Most prokaryotes are catalase-positive; catalase-negative organisms include some lactic acid bacteria, some clostridia, and *Ruminococcus*, which are extremely oxygen-intolerant. The enzyme protects against oxidation, breaking down inorganic peroxide (H_2O_2) into oxygen and water.
Peroxidase coenzyme	P	Iron	Synthesized by many, if not all, facultative and obligate aerobes. Enzyme breaks down organic peroxide into water and oxygen.
Chlorophylls	P	Magnesium	Synthesized by all photosynthetic bacteria. Chlorophyll is the main light-absorbing pigment in photosynthesis.
Hemes (coenzymes of cytochrome proteins)	P	Iron	Synthesized by many microbes—most, if not all, facultative and obligate aerobes. Used in oxidation of two- and three-carbon food molecules; respiration; photophosphorylation of ATP. Cytochromes are present in all photosynthetic and respiring organisms.

[a]P = porphyrin; C = corrinoid.

gen, other classes of compounds can be understood as indicators of later steps toward the use of free oxygen in metabolic pathways. The coming of the oxic world opened up many new metabolic options. Polyunsaturated fatty acids, universal in eukaryotes, appeared first in cyanobacteria that were growing aerobically. This is suggested by the observation that whenever polyunsaturates are found in cyanobacteria, they have been produced during autotrophic growth, when oxygen is available to the cell as a waste product of photosynthesis (Table 6-7; Holton, Blecker, and Stevens, 1968). Tetracycline synthesis is apparently limited to aerobic actinobacteria, such as *Streptomyces* (Miller, 1961). That the synthesis of steroids, polyunsaturates, and tetracycline is restricted to aerobes implies that acquisition of such synthetic capabilities was linked to the availability of molecular oxygen.

Eukaryotes with mitochondria either synthesize or ingest the steroid compounds they require. About seven percent of the dry weight of the eukaryotic endomembrane system is steroid; all eukaryotes, it seems, de-

TABLE 6-7

The production of unsaturated fatty acids.

| ORGANISMS | SATURATED AND MONOUNSATURATED (MOSTLY C_{16} AND C_{18} ACIDS)[a] | POLYUNSATURATED[a] | |
		2 DOUBLE BONDS (MOSTLY LINOLEIC)	3 DOUBLE BONDS (MOSTLY LINOLENIC)
Heterotrophic bacteria (many)	+ (v)	−	−
Photosynthetic bacteria (many)	+ (v)	−	−
Beggiatoa (sulfur oxidizer)	+ (o)	−	−
Cyanobacteria[b]			
Synechococcus	+ (o)	−	−
Haplosiphon	+ (o)	−	−
Chlorogloea			
Autotrophic growth	+ (o)	+	+
Heterotrophic growth	+ (o)	−	−
Nostoc	+ (o)	+	+
Oscillatoria	+ (o)	+	+
Algae (eukaryotes)	+ (o)	+	+

[a]− = less than 1% of total fatty acids; + = greater than 1% of total fatty acids; o = oleic acid; v = vaccenic acid.
[b]Data from Holton et al. (1968).

pend upon polyunsaturated fatty acids and steroids. For example, although yeasts are capable of meiosis and fertilization in the absence of oxygen, they dedifferentiate their mitochondria and do not synthesize steroids and poly-unsaturated fatty acids. Lacking mitochondria and the products of their aerobic respiration, fermenting yeasts will not reproduce unless they are supplied with these metabolites (Table 6-8). Two inferences can be made from these observations: First, polyunsaturated fatty acids, steroids, tetracyclines, and other metabolites requiring atmospheric oxygen for their synthesis were made by metabolic pathways that evolved only after photosynthetically produced oxygen became available. Second, eukaryotes evolved only during or after transition to an oxygen-rich environment.

TABLE 6-8

Requirements of the yeast *Saccharomyces cerevisiae* for steroids and polyunsaturated fatty acids.[a]

GROWTH CONDITIONS	SUBSTANCES REQUIRED TO SUPPLEMENT BASIC MEDIUM
Aerobic	None
Anaerobic	Ergosterol and linoleic acid (or other steroid or polyunsaturated fatty acid)

[a]From Proudlock et al. (1968).

Some photosynthetic sulfur bacteria must have accumulated mutations that permitted the use of water instead of hydrogen sulfide as the hydrogen donor in the reduction of carbon dioxide. The evolution of photosystem II (see Figure 5-5) led to the origin of blue-green and green oxygen-releasing prokaryotes. The oxygen excreted was at first simply waste.

Unlike aerobic algae, cyanobacteria tolerate conditions ranging from anaerobic to fully aerobic; often they live, with anaerobic bacteria, in oxygen-depleted environments. In tidal flats and pools near the Gulf of Mexico, in Baja California, and in Abu Dhabi, the cyanobacteria are generally found in layers just above anoxygenic phototrophs, where the concentration of oxygen is low (Horodyski, 1977; Cohen and Rosenberg, 1989). The physiology of cyanobacteria can be understood best if it is assumed that aerobiosis evolved independently within the group by developments analogous to those in actinobacteria and archaebacteria.

In cyanobacteria, aerobic respiration of carbohydrates occurs in the dark, whereas photosynthesis in the light is an anaerobic process, as it is in other photosynthetic bacteria (Echlin and Morris, 1965; Stanier and Cohen-Bazire, 1977). That is, whereas eukaryotic photosynthesizers, plants and algae, respire regularly in the light, cyanobacterial respiration and photosynthesis are often mutually exclusive. Some cyanobacteria (for example, *Anacystis*) are sensitive to peroxide (Marler and Van Baalen, 1965). These physiological differences are comprehensible from an evolutionary perspective: like other bacteria, cyanobacteria evolved as the concentration of ambient oxygen increased, whereas eukaryotes, which compartmentalize photosynthesis (in plastids) and respiration (in mitochondria), originated in an already oxic world.

Some cyanobacteria bear a striking morphological resemblance to certain heterotrophs (Table 6-9) that lack chlorophylls but synthesize carotenoids and, often, oxidize sulfur to sulfate (Fox and Lewin, 1963). Some, like their cyanobacterial analogues, move by gliding without any apparent motility organelles. The mechanism of this gliding is mysterious; it does require contact with a surface (Walsby, 1968; To and Margulis, 1978). An evolutionary explanation of these aerobic analogues of cyanobacteria is that each evolved from the cyanobacteria that it resembles. Comparisons of the structures of cytochrome proteins isolated from these organisms support this interpretation (Almassy and Dickerson, 1978). Like cyanobacteria, some analogues use the ribulose diphosphate carboxylase pathway to fix carbon dioxide, consistent with their evolution from cyanobacteria by loss of photosynthetic ability. Cyanobacteria, and even their morphological analogues have been reported in the Proterozoic fossil record (Figure 6-2).

Chemoautotrophic microbes such as *Beggiatoa* and *Thiothrix* would have evolved when photoautotrophy was replaced by the ability to derive

TABLE 6-9

Possible heterotrophic descendants of some cyanobacteria.[a]

CYANOBACTERIUM	HETEROTROPH[b]	DESCRIPTION OF HETEROTROPH
Lyngbya[c]	Herpetosiphon	Gliding organism; produces orange pigment; hydrolyzes starch, gelatin, casein, and tributyrin, but not cellulose; sheathed filament.
Oscillatoria[b]	Vitreoscilla	Gliding straight filaments.
	Beggiatoa	Straight filaments; oxidizes hydrogen sulfide; deposits sulfate.
Rivularia[b]	Leucothrix	Filaments arranged in rosettes.
	Thiothrix	Filaments arranged in rosettes; oxidizes hydrogen sulfide.
Spirulina[d]	Saprospira	Gliding aerobic marine and freshwater organisms; produce carotenoids; helical.
	Thiospirillopsis	Gliding organisms; oxidize hydrogen sulfide; helical.

[a] Although they vary in color from white to orange, these organisms are capable of carotenoid biosynthesis.
[b] See Balows et al. (1991) for extensive descriptions of morphologically complex bacteria.
[c] Holt and Lewin (1968).
[d] Lewin (1965b).

energy from the oxidation of sulfide to sulfate. By this reckoning, sulfide oxidation evolved convergently in prokaryotes, as did other forms of chemoautotrophy and mixotrophy—the ability to obtain some energy from direct oxidation of inorganic substances coupled with a limited requirement for certain organic nutrients. Eventually, the ultimate biosynthetic talent appeared: chemoautolithotrophic bacteria that live on atmospheric oxygen and carbon dioxide, a few salts, a reduced inorganic energy source, and water (for example, *Nitrocystis*; see Figure 3-7). Even in the dark, these bacteria synthesize all their organic nutrients. A scheme of the order of appearance of metabolic innovations based on the concept of the polyphyletic response of prokaryotes to the central role of atmospheric oxygen as a selective agent is diagrammed in Figure 6-3.

Organic geochemistry

Truly, the history of oxygen is the history of life!

TRAUBE (1826–1894)
(in Coleman, 1971, p. 135)

A goal of organic geochemistry is to reconstruct the course of early evolution by analysis of organic compounds in well-dated sediments. This ap-

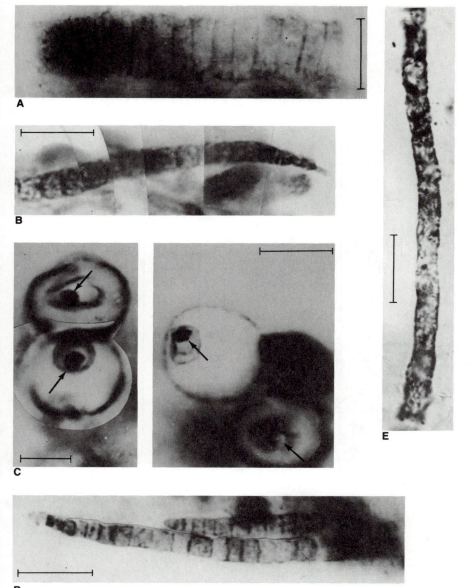

FIGURE 6-2

Microfossils from thin sections of Proterozoic cherts. Light micrographs; bar = 10 μm.
(A) *Lyngbya*-like filament from the Bitter Springs chert, Central Australia. (B)
Rivularia-like filament from the Bitter Springs chert. (C) Spheroidal microfossils
from the Bitter springs, chert, originally interpreted to be remains of eukaryotic
algae (arrows at "nuclei"), but see Golubic and Barghoorn (1977); Francis,
Barghoorn, and Margulis (1978); and Francis, Margulis, and Barghoorn (1978). (D)
Fossil filament from the Skillogalee dolomite, Port Augusta, South Australia. (E)
Oscillatoria-like filament from the Bitter Springs chert. [Courtesy of J. W. Schopf.]

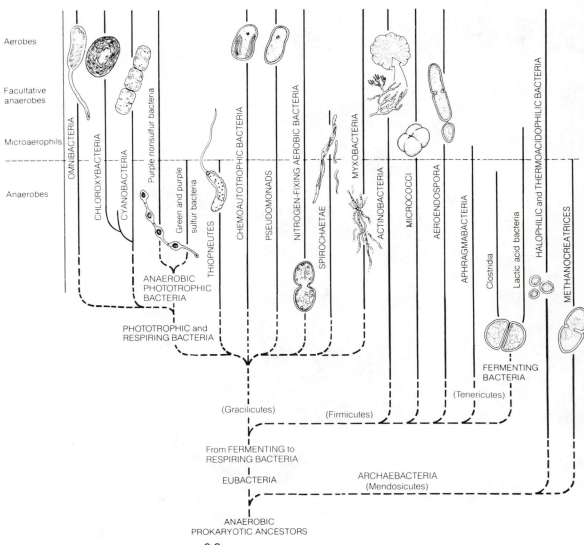

FIGURE 6-3

A prokaryote phylogeny (see Margulis and Schwartz, 1988, for detailed descriptions).

proach has its promise and its limitations. To be of use in reconstructing the evolutionary past, organic compounds must be syngenetic with the rocks in which they are embedded. They must have been deposited as biochemical components or excretions of organisms. Such biochemicals must be forced to yield information about the population of organisms that made them; this is seldom possible. Fortunately for evolutionists, life is extremely conservative both chemically and morphologically. Nearly all symbionts and fossil organisms for example, have free-living and modern morphological analogues.

Some biogenic compounds and classes of compounds are very rare; others are distributed widely (Florkin and Mason, 1960–1964). The mere presence of a compound inside an organism is not evidence that the organism synthesized it. Evolutionary inferences based only on distributions of specific compounds can be erroneous. Mycoplasmas require steroids, for example, yet none make them: steroids are supplied by the host or by the fluid medium. Some caterpillars become red from eating carrots. Accumulated undigested carotenoids, which are not synthesized by the larva itself, can be seen through the translucent skin. The spectacular pink coloration of flamingos is due to dietary carotenoid. The presence of carotenoids in quantity does not mean that caterpillars and flamingos have genes coding for the synthesis of these bright orange pigments. Many other examples of compounds in cells and tissues that lack the ability to synthesize them are known. Many biosyntheses that have been attributed to host cells are actually performed by inconspicuous bacterial symbionts (Margulis, 1976; Ragan and Chapman, 1978). The presence of even very specific carotenoids in sediments is not evidence for caterpillars, carrots, or flamingos.

If biosynthetic ability rather than spoken language were the measure of evolutionary advancement, metazoans would be considered much less advanced than the chemolithoautotrophic bacteria that grow on simple gases in salt water. Autolithotrophic bacteria evolved long before humans, although they are metabolically far more independent. Like all metazoans, humans have complicated nutritional requirements; we depend on plants and microbes to supply those that we are unable to synthesize. In animals, as in most eukaryotes, there is no a priori relationship between metabolic complexity and time of evolutionary appearance.

In using relics of recognizable biochemical pathways in the sediment as clues to evolutionary history, one must beware of pitfalls. Certain metabolic pathways and products are universal, and others are present in most organisms. Neither ubiquitous pathways nor selected enzymes or metabolites from them are useful as evolutionary markers. The distribution of most compounds has little or no phylogenetic significance. For example, cellulose, once thought to be a universal polymer in plant and algal cell walls, is absent in many green and red algae, whereas it is synthesized and excreted by certain bacteria, and it is present in several protoctists, such as cellular slime molds, as well as in many fungi. Furthermore, chitin, thought to be characteristic of animal exoskeletons, is the primary wall constituent in many fungi (Alexopoulos, 1962). Nearly every compound synthesized by animals, plants, or fungi is a metabolite in some prokaryote as well (Miller, 1961).

Variations in specific biochemical capabilities may reflect evolutionary diversification. For example, animals, plants, and fungi synthesize steroids in significant quantities. The cyclized sterol precursor is different in each kingdom: it is cholesterol in animals, cycloartenol in plants, and, generally, ergosterol in fungi. These compounds, which are precursors of a plethora of steroids, are rarely, if ever, synthesized by bacteria. The steroids of methane-oxidizing bacteria and cyanobacteria are produced in lesser quantity than those of eukaryotes and are structurally different. Specific steroids, then, may be useful indicators of chemical phylogeny.

Unfortunately, such indicator compounds are not found unchanged in sediment from the time of deposition. The organic compounds extracted from shales and other sedimentary rocks are chemical fossils; like other fossils, they have been altered since they were parts of living things. To determine just how they were altered is never easy. Until the chemical details of decay and diagenesis are understood, identification of the sources of organic compounds found in rocks must be tentative. Most evolutionary history must be based on comparisons of living forms. The fossil record can corroborate inferences and establish broad chronologies. Postmortem changes in organisms, the possibility of contamination in laboratory and field, the intractability of the kerogen fraction, the paucity of preservable and unique biochemicals correlated with specific taxa, and other factors limit attempts to reconstruct fossil communities from organic geochemical data. Table 6-10, which lists several classes of compounds potentially useful as phylogenetic markers, should be read with these caveats in mind. Postmortem chemical studies of living species by high-resolution gas chromatography and mass spectroscopy must be combined with chemical, sedimentological, and mineralogical studies in order to analyze and compare past and present microbial ecosystems. Complete studies are rare; the work on the microbial communities of Solar Pond (Krumbein, Cohen, and Shilo, 1977; Cohen and Rosenberg, 1989) and Laguna Figueroa (Horodyski, 1977) is an exemplary beginning.

By the Proterozoic eon, communities of prokaryotes had diversified so extensively that the ecological consequences of their presence (that is their effects on the atmosphere and on sediments) would be notable with few exceptions. Prokaryotes can synthesize any compounds found in other living things except for some unique alkaloids (such as some notorious hallucinogenic compounds), lignin and other polyphenolics, exotic nerve toxins, lichenic acids, and flavonoids. Furthermore, many biosyntheses and metabolic pathways are found only in prokaryotes (Table 6-10; see also

TABLE 6-10

Metabolic products as phylogenetic markers.

TAXA[a]	METABOLIC PRODUCTS
Prokaryotes	
All	Nucleotides, amino acids, DNA, RNA, ATP, etc. (see Figure 5-1)
Oxygen-tolerant	Porphyrin derivatives, flavoproteins, isoprene derivatives
Photosynthesizers (Moneran phyla 6, 7, and 8)	Chlorophyll, carotenoids
Actinobacteria (Moneran phylum 15)	Tetracycline derivatives
All phyla except Aphragmabacteria and some Archaebacteria	Diaminopimelic acid, muramic acid, N-acetyl glucosamine
Spore-formers (moneran phyla 2, 4, and 11)	Dipicolinate
Eukaryotes	
All with mitochondria	Steroids, steranes, tubulin, polyunsaturated fatty acids, F_0F_1-ATPase (mitochondrial ATPase)
Undulipodiated eukaryotes	Dynein ATPase
Angiospermophyta (plant phylum 9)	Flavonoids, higher-order terpenoids, polyphenols (lignans), betalains

[a]Phyla numbers are taken from Margulis and Schwartz, 1988.

Table 2-6). By the Proterozoic eon, therefore, most biochemical evolution was over. Microbial communities dominated the soil, air, and waters much as they do today. The elements that are cycled by the biota through the fluids of the Earth today were already cycling in much the same way. Carbon, hydrogen, sulfur, nitrogen, and oxygen were cycled, as well as many nutrient elements required in smaller amounts—for example, calcium, iron, manganese, zinc, copper, potassium, and chlorine. Observed from space even 2000 million years ago, the Earth would have seemed an improbable planet, especially if compared with Mars and Venus. Not only would its atmosphere have been conspicuously depleted in carbon dioxide, but it would have had far too large a concentration of oxygen to allow for the presence of free nitrogen, methane, hydrogen, and hydrocarbons. As known from fossil evidence, for the last two billion years the Earth's atmosphere has been altered greatly by an expanding oxygenic biota. Table 6-11 lists the tentative order of seme acquisition in aerobic organisms with implied consequences for the original deposition of organic materials in Proterozoic sediments.

TABLE 6-11

A postulated sequence of deposition of organic compounds in Proterozoic and later sediments.

PROTEROZOIC AND PHANEROZOIC PATHWAYS	EARLIEST GROUPS WITH PATHWAYS	APPROXIMATE TIME OF APPEARANCE ($\times 10^9$ YEARS AGO)	SEDIMENTARY EVIDENCE[a]
Lipid, including hopanoid synthesis	Many heterotrophic bacteria	3.4–1.5	Phytane,[b] pristane, hopanoids
Oxygen-eliminatiing photosynthesis using chlorophylls *a* and *b*	Blue-green and green oxygenic phototrophs	3.4–1.5	Some oxidized sediments, some polyunsaturated fatty acids, and all chlorophyll derivatives
Oxidation of iron, sulfur, and nitrogen	Aerobic prokaryotes	3.0–2.0	Iron and manganese oxides, nitrates, diaminopimelic acid, pectin, cellulose, glycogen, muramic acid, tetracycline, aminoglycoside antibiotics, and all other prokaryote products
Synthesis of steroids, sterols, polyunsaturated hydrocarbons, microtubular proteins, cellulose, and chitin	Eukaryotic cells	1.0–0.7	Steranes and polyunsaturated hydrocarbon derivatives
Synthesis of quinoids	Fungi	0.7–0.4	Derivatives of radicinin, citrinin, fulvic acid, penicillins, and other fungal metabolites
Synthesis of flavonoids, lignin, lignans, alkaloids, pectins, and hemicelluloses	Vascular plants	0.5–0.3	Flavonoid derivatives, alkaloids, and polyterpenes
Synthesis of depsides, depsidones, dibenzofurans, and "lichenic acids"	Lichens	0.5–0.3	Depside derivatives

[a]Expected correlated sedimentary evidence; because of microbial community interactions and diagenesis, this list is inevitably oversimplified.
[b]Ourisson et al., 1984.

THE PROTEROZOIC RECORD AND THE EARLIEST EUKARYOTES

These . . . cherts are widely scattered throughout the Precambrian. I'm showing you now a thin slice of sliceous sinter from Yellowstone. As you know, thermophilic blue-green algae [cyanobacteria] are notoriously abundant and widely speciated in these thermal waters, growing at temperatures ranging up to perhaps 75°C. . . . The silicate, which may run up to 800 parts per million in these extremely supersaturated solutions, precipitates. It

essentially forms an embedding matrix for the . . .
[cyanobacteria] colonies. The blue-green algae . . .
continue to grow at the upper surface as the silicate
encases and imprisons the column below. This forms the
curious thimble-like arrangement and the development of
pillars. The extant phenomenon at Yellowstone, I think,
has a remarkable counterpart in an existing situation to
what we see in [certain] 2 billion-year-old Gunflint rocks [of
North America].

E. S. BARGHOORN, 1970

Although rates and quantities are difficult to estimate, that biogenic oxygen accumulated on a worldwide scale during the pre-Phanerozoic is considered uncontestable (Cloud, 1974, 1988). Davidson (1965) and Dimroth and Kimberly (1976) have argued that the atmosphere contained significant quantities of free oxygen even in Archean times. The abundance, diversity, and complexity of stromatolites is the most direct evidence that cyanobacterial communities dominated the Proterozoic landscape (Walter, 1976; see Table 5-5). The magnificent profusion of well-preserved cyanobacteria in Proterozoic cherts (Schopf, 1975, 1983) also suggests that oxygen-releasing photosynthesis by prokaryotes existed at least two billion years ago. Measured $^{12}C/^{13}C$ ratios in ancient sediments and traces of isoprenoids that may have derived from lipids by decarboxylation are also interpreted as evidence of early oxygenic photosynthesis (Schidlowski et al., 1979).

Geologically, the middle and late Proterozoic orogenies, emergent continents, sedimentation profiles, neutral to alkaline hydrosphere comprising oceans and lakes, and weathering styles were virtually modern. Extensive Proterozoic sediments have been studied near the eastern arm of the Great Slave Lake in Canada. There, near the northwestern margin of the Laurentian Shield, a succession of unmetamorphosed sedimentary and volcanic rocks more than 50,000 feet thick is exposed. Age brackets have been established by potassium–argon biotite dating of igneous intrusives. The sequence is from 2.5 billion to 1.7 billion years old. In discussing these sediments, Paul Hoffman of the Canadian Geological Survey concluded (unpublished manuscript) that

1. Red beds predominate in the formations deposited between 2.17 and 1.7 billion years ago and occur sparingly in the earlier formations (2.55 to 2.17 billion years ago).
2. Crystal casts, stable replacements of other materials in cavities originally filled by the evaporite minerals halite and gypsum, are abundant in some formations deposited between 2.17 and 1.85 billion years ago.

3. Ironstones found in some formations deposited between 2.17 and 1.85 billion years ago consist of calcite-cemented, cross-bedded, oolitic, and granular hematite sandstones, rather than the banded siliceous iron formations commonly considered typical of the Precambrian era.

4. Glauconite occurs as pellets in cross-bedded, white, deltaic sandstone in one formation deposited between 2.17 and 1.85 billion years ago.

5. Stromatolites are abundant in many formations deposited between 2.55 and 1.7 billion years ago. In one formation they constitute a major "reef" complex, 1500 feet thick, which separated a shallow-water carbonate platform from a deep-water graywacke-shale-filled basin. Stromatolites in the marine formations are of the "sediment-trapping" type, whereas those in the nonmarine formations resulted from in situ precipitation of calcium carbonate (Figure 6-4).

6. Shelf-type Archean rocks, deposited more than 2.5 billion years ago, including 10,000 feet of cross-bedded orthoquartzite and dolomite, indicate that the absence of such rocks from the Archean of most areas is a function of their low preservation potential rather than their nondeposition.

7. *Skolithus*-like tubes are found in the glauconitic sandstone units described above and are similar to lower Paleozoic structures attributed to burrowing metazoans.

Hoffman said further that the geosynclinal model of these rocks "is surprisingly similar to that of the Appalachians. No major difference in tectonics from Phanerozoic geosynclines is apparent from the sedimentary record." These and other studies, indicating a lack of environmental discontinuity at the Phanerozoic border, suggest that the "Cambrian discontinuity" in the fossil record was due primarily to the evolution of preservation potential in biological communities interacting with their environment.

The Cambrian discontinuity—the sudden and widespread appearance of skeletalized metazoans in the (Lower Cambrian)—is interpreted primarily as an obvious manifestation of more obscure biological and plate-tectonic events (Lowenstam and Margulis, 1980a; McMenamin and McMenamin, 1990). This discontinuity is not inconsistent with the earlier accumulation of atmospheric oxygen as a result of prokaryotic photosynthetic activity. It has been suggested that the Cambrian "explosion" was the result of a rise in oxygen concentration above a minimum (about one percent of the present atmospheric concentration) required to support metazoan life (Berkner and Marshall, 1965; Fischer, 1965; Cloud, 1978). Although animal life certainly requires at least that much oxygen, this alone is not sufficient to explain the Cambrian discontinuity Lowenstam and Margulis, 1980a,b; McMenamin and McMenamin, 1990). It seems doubtful that either ultraviolet irradiation or the scarcity of oxygen ever

A

B

FIGURE 6-4

Ancient and modern carbonate stromatolites.
(A) Proterozoic carbonate stromatolites from the Great Slave Lake, Northwest Territories, Canada. The geology hammer (lower left) is about one foot long. A Pleistocene glaciation groove is at the lower right. (B) These two-billion-year-old stromatolite heads from the Great Slave Lake were cut by Pleistocene glaciation in this bedding plane exposure. The internal laminated structure of these pre-Phanerozoic stromatolites is very similar to that of recent ones. (C) Recent carbonate stromatolites in a warm, hypersaline environment: Hamelin Pool, Shark Bay, Western Australia. Sediment particles are trapped and bound by the microbes, primarily filamentous photosynthetic prokaryotes with well-developed sheaths. They are cemented by the precipitation of inorganic aragonite. These modern lithified intertidal stromatolites contain thriving communities of microorganisms, including ones that bore into carbonate and decrease the rate of growth of the heads. [Courtesy of Paul Hoffman.]

C

directly restricted the evolution of metazoa (Margulis, Walker, and Rambler, 1976).

Animals and vascular plants did not radiate until the prerequisite genetic systems had evolved in their protist ancestors; systems that culminated in mitosis preceded the evolution of metazoan skeletons. Aerobic eukaryotes arose only after significant quantities of atmospheric oxygen had accumulated and Mendelian genetic systems had evolved within aerobic heterotrophic protists, themselves the result of symbiotic events that led to the origin of cell organelles. If photosynthesis and respiration evolved in prokaryotes and were only then acquired, symbiotically, by eukaryotes, then fossil evidence for oxygen-intolerant microbes, oxygen-releasing cyanobacteria, and a large assortment of other oxygen-tolerant and oxygen-requiring microbes should well antedate evidence for animals and plants. No "missing" higher taxa of photosynthesizers connecting cyanobacteria or *Prochloron* with nucleated algae ever existed; thus, these should not be found either in the fossil record or among living forms. The peculiar metabolic virtuosities of fungi and plants — for example, the production of lignin, certain triterpenes, steroids, flavonoids, and indole alkaloids — should appear in the organic geochemical record only during the Phanerozoic.

There are conspicuous differences between the Archean and the Proterozoic regimes; from about 1500 million years ago, however, the sedimentary record has a modern aspect, although there is no direct evidence for either skeletalized or soft-bodied metazoans until later (Cloud, 1968b; McMenamin and McMenamin, 1990). Is there direct fossil evidence that the eukaryotic level of organization had been reached by protoctists, the presumed ancestors of plants and fungi? If so, when?

Fossil "spot cells," "meiotic tetrads," and other microfossils interpreted as pre-Phanerozoic eukaryotes (Schopf and Blacic, 1971) are probably misinterpretations (Francis, Barghoorn, and Margulis, 1978; Francis, Margulis, and Barghoorn, 1978; Golubic and Barghoorn, 1977). However, provocative evidence for the appearance of eukaryotes before the Ediacaran metazoans (about 680 million years ago) does exist. Acritarchs — probably protoctist cysts or other propagules — are common in clastic facies after about 1400 million years ago. *Chuaria*, distinctive globular fossils as much as 1 mm in diameter, date back to at least 1000 million years ago (Hofmann, 1993). Macroscopic organic sheets resembling the chlorophyte *Ulva* have been found in the late Proterozoic Dal Group of the Northwest Territory in Canada. Chitinozoans, probably protoctists, have been discovered in the Grand Canyon (Schopf, Ford, and Breed, 1973), and other eukaryotic fossils in phosphate nodules from the late Proterozoic of Sweden (Knoll, 1979; Vidal, 1990).

Eukaryotes that originated more than a billion years ago gave rise to several groups of organisms, including the enigmatic ones just mentioned

(Vidal, 1984). Although no acritarch or chitinozoan provides unequivocal documentation of the eukaryotic level of organization, they are suggestive. The indirect actions of communities of microbes that alter their environment document that the modern era was under way prior to the appearance of the hypertrophied animals and plants that we hold so dear. The absence of skeletalized animals is the main distinction between the upper Proterozoic and the lower Phanerozoic.

The effects of microbial communities on the atmosphere and sediments are understood only poorly. Some will be mentioned here, mainly as a plea for further investigation. Suitably equipped scientific teams — which must include, at least, sedimentologists, geochemists, microbiologists, and atmospheric chemists — must cooperate before we will be able to decipher the effects of the biota on the surface of the planet and their changes with time.

Of the many thousands of minerals known, nearly sixty are biogenic, the major ones of which are shown in Table 6-12 (Lowenstam and Wiener, 1989). In rocks unaltered by high temperatures and pressures, the presence of these minerals is acceptable evidence of the organisms that synthesize them (Table 6-12). Microbial metabolism in evaporite flats and sabkhas may even have played a crucial role in the genesis of economically important mineral ores (Renfro, 1974).

Bacteria reduce manganese and calcium sulfates to sulfide; in the sea, this often results in the formation of pyrite (FeS_2) and metallic manganese. Sodium chloride brines are purified of sulfate by microbes. In fresh water, biogenic sulfides can be precipitated in the form of hydrotroilite ($FeS \cdot nH_2O$). Iron can be reduced from the ferric to the ferrous state and even precipitated as metallic iron (Ferguson-Wood, 1967). Ferrous iron can be oxidized to iron oxides, which are precipitated in morphologically conspicuous ways by a number of "iron bacteria." Even in today's oxic world there are anoxic iron-rich habitats in which bacteria thrive — under entirely oxygen-free conditions, carbon compounds can serve as reductants, and reduced iron can be formed in quantity by bacteria in sediments (Myers and Nealson, 1990; Lovley, 1991). Todorokite, magnetite, calcite, aragonite, and barite are at least partly biogenic minerals that are present as clasts in sediments. Until recently they were assumed to be produced by physical processes above and dismissed as fossils, not only from the Proterozoic, but from the Phanerozoic eon as well (Lowenstam and Wiener, 1989).

Pre-Phanerozoic rocks sometimes show a sedimentary sequence that corresponds to different depths of an ancient body of water: from deeper to shallower, iron sulfide, iron carbonate, and iron oxides (Fe_2O_3) were deposited. That the sulfide is a microbial product can be deduced from sulfur-isotope analysis, knowledge of the environment of deposition, and, in some cases, direct observation of adjacent stromatolites and embedded microfos-

TABLE 6-12

The distribution of bioinorganic minerals[a]

MINERAL[b]	MONERA	PROTOCTISTA														FUNGI		ANIMALIA											PLANTAE	
	Monera	Sarcodina	Dinomastigota	Prymnesiophyta	Bacillariophyta	Phaeophyta	Rhodophyta	Chlorophyta	Conjugaphyta	Rhizopoda	Heliozoa	Actinopoda	Granuloreticulosa	Myxomycota	Ciliophora	Basidiomycota	Deuteromycota	Porifera	Cnidaria	Platyhelminthes	Ectoprocta	Brachiopoda	Annelida	Mollusca	Arthropoda	Sipuncula	Echinodermata	Chordata	Bryophyta	Tracheophyta
Calcium carbonates																														
Calcite	+	+	+	+			+	+					+	+				+	+	+	+	+	+	+	+	+	+	+	+	+
Aragonite	+			?		+	+	+					+					+	+				+	+	+	+		+		+
Vaterite	+						+																	+	+			+		+
Monohydrocalcite	+																							+				+		
Amorph. hydr. carbonate														+						+				+	+			+		
Phosphates																														
Dahllite	+														+		+											+		?
Francolite																						+		+				+		
$Ca_3Mg_3(PO_4)_4$																							+							
Brushite																						+		+						
Amorph. dahllite precursor																					+			+				+		
Amorph. brushite precursor																								+						
Amorph. whitlockite precursor																				+			+	+	+					
Amorph. hydr. ferric phosphate																				+			+	+	+		+			
Halides																														
Fluorite																								+	+		+			

Amorph. fluorite precursor				+			+			
Oxalates										
Whewellite		+	+	+		+	+	+	+	
Weddellite		+ ?	+ ?	+		+	+ +	+ +	+ +	
Sulfates										
Gypsum				+				?	+	
Celestite	+									
Barite	+	+	+ +	+						
Silica										
Opal	+	? + + +	+	+	+ + + +	+	+ + +	+ +	+ +	
Iron oxides										
Magnetite	+				+ +		+ +	+		
Maghemite	?									
Goethite					+		+			
Lepidocrocite				+	+		+			
Ferrihydrite	+	+ +		+	+ +	+	+ +	+	+ +	
Amorph. ferrihydrates		+			+		+	+		
Manganese oxides										
Burnessite	+									
Iron sulfides										
Pyrite	+									
Hydrotroilite	+									

[a]Modified from Lowenstam (1980).
[b]Amorph. = amorphous; hydr. = hydrated.

A

FIGURE 6-5

Modern and fossil microbial mats.
(A) Modern *Microcoleus cthonoplastes* mat from North Pond, Laguna Figueroa, Baja
California Norte, Mexico. **(B)** Ancient laminated chert from the Fig Tree Formation
of southern Africa.

sils (Goodwin, Monster, and Thode, 1976). Studies of the stratigraphy,
sedimentation, deposition, environment, and isotope chemistry of iron and
sulfur minerals suggest that large, open, stable bodies of water dominated
by diverse communities of microorganisms prevailed over much of Ontario
in early Proterozoic times.

Precipitation of calcium carbonate today is biogenic and probably has
been since before the Proterozoic eon. Calcium carbonate is precipitated
from sea water by the photosynthetic removal of carbon dioxide or by other
biogenic increases in alkalinity. The extent to which calcium carbonate
would precipitate in the absence of life would be useful for comparing
Earth with Mars and Venus, which have retained much more carbon
dioxide in their atmospheres.

B

The biota exchanges nitrogen with the air. Organic nitrogen in sediments such as peat and coal is released into the atmosphere by microbial attack. Nitrates produced either abiotically by lightning or biotically from ammonia are removed from the atmosphere as soluble salts and fixed into organic nitrogen by photosynthetic organisms.

Hydrogen and methane are readily formed in large quantity by phototrophic, fermenting, and methanogenic bacteria. Oxygen and acids are also produced. Microbial gas production permits the transport of soluble reduced minerals in anoxic locales (Renfro, 1974). The accumulation of copper, aluminum, manganese, nickel, vanadium, gold, and uranium ores may be credited in part to biogenic "reducing power." Many of these ores are found with undisturbed carbon-rich sediments. The highly concen-

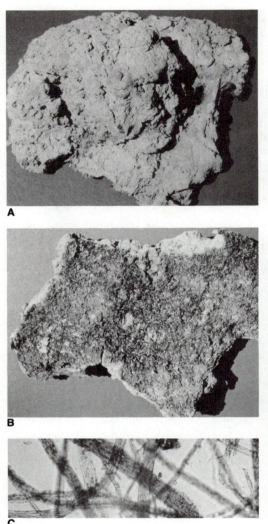

FIGURE 6-6

Trapping and binding by filamentous microbes. Live mats of this sort may have been precursors of the lithified stromatolites found in the fossil record.
(A) Unconsolidated marl forming a live microbial mat; from southern Florida. Sample is about 15 cm in diameter. (B) Underside of the mat in A. Dark patches are the bacteria, especially the filamentous cyanobacteria that, by trapping and binding sediments, give the mat its integrity. (C) Filamentous cyanobacteria that trap and bind the carbonate-rich sediments. (Light micrograph, ×200).

trated manganese-iron and nickel-rich deep-sea nodules were formed partly by bacteria; both freshwater and marine laminated nodules are probably stromatolites (Monty, 1973a). The degree to which silica can be depleted in surface waters by diatoms and radiolarians is impressive: the quantity of silica in the photic zone may be reduced several orders of magnitude to below the level of chemical-analytical detection (less than one part per million). Even the concentration of particulate matter suspended deep in the ocean is influenced by biological processes. For example, iron, manganese, vanadium, cobalt, chromium, nickel, copper, silver, antimony, gold, and mercury are found in concentrations greater than would be expected

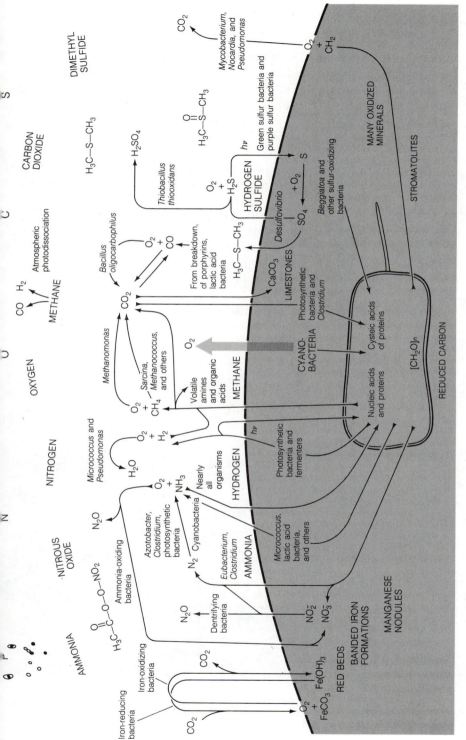

FIGURE 6-7

Biogeochemical cycles in an aerobic world. The heavy, dark, curved line represents the liquid or solid surface of the Earth. Under oxic conditions phosphorus cannot move through the atmosphere in gaseous form but is distributed in the form of organic phosphates in spores and other particles.

The accumulation of oxygen in the atmosphere suppressed the cycling of elements in reduced compounds such as hydrogen sulfide and ammonia. However, it led to the photoproduction of new oxidized volatiles such as paracetylnitrate, $H_3C(CO)O_2NO_2$, which may be of geochemical significance. Also, dimethyl sulfide, $(H_3C)_2S$, dimethylsulfoxide, $(H_3C)_2SO$, and other sulfur-containing gases probably replaced hydrogen sulfide as important volatile components of the sulfur cycle.

under abiotic conditions of weathering and physical settling processes. The biota strongly alters the predicted concentrations of these elements (Buat-Menard and Chesselet, 1979; Whitfield, 1988). In short, the differences between the sterile regoliths of Mars, the Moon, and Venus and the soil and sediments of the Earth are profound. Many, perhaps most, of the differences are the consequences of microbial activity. The details deserve better scrutiny.

The similarity between some ancient sedimentary deposits and modern environments nearly entirely dominated by bacteria is astounding. A chert from ancient African rocks is compared with a living microbial mat from Baja California Norte, Mexico, in Figure 6-5. Interaction of microbes with sediment is easy to find but, often not very conspicuous, it tends to be overlooked. For example, a low-lying, dark, crusty material covers and consolidates the loose desert soil at many places in the southwestern United States and northern Baja California. A marl sediment consolidated by cyano- and other bacteria is shown in Figure 6-6. The first extensive land biota to stabilize the ancient soil must have included such desert prokary-

FIGURE 6-8

Panorama dominated by prokaryotes: the salinas, and tidal channels at Laguna Figueroa, Baja California Norte, Mexico.

otes, represented today by *Microcoleus vaginatus* in drier upland environments and *Microcoleus cthonoplastes* in wetter intertidal ones (Campbell, 1979).

A summary of interactions of bacteria with sediments and the atmosphere is sketched in Figure 6-7. Well developed in the Proterozoic, these biogeochemical cycles have persisted through the Phanerozoic, where they can still be observed. Microbial gas and sedimentary interaction still dominate the biosphere, but observers tend to be distracted by the larger eukaryotes that have supplemented microbes. In crucial but inconspicuous regions—mud flats, swamps, hypersaline lagoons, estuaries, and salt marshes—one can still observe the landscapes of Proterozoic times (Figure 6-8). In the murky Proterozoic waters of such a world, subtle comings and goings gave rise to the protoctist ancestors of our animal cells (Bengtson, 1993).

CHAPTER 7

SYMBIOGENESIS

SYMBIOSIS AS PARASEXUALITY

I called this process symbiogenesis, which means: the
origin of organisms through combination and unification
of two or many beings, entering into symbiosis.

K.S. MERESCHKOVSKY, 1920
(in Khakhina, 1979)

The term **symbiosis** entered the biology literature when defined by Anton
De Bary in the nineteenth century as the "living together of differently
named organisms" (Smith and Douglas, 1989). Symbiosis here more
sharply defines De Bary's concept; it refers to the physical association of
members of different species of organisms (the **bionts**) for significant portions of the life history. The partners (bionts) come from many different
taxa and vary greatly in the details of their relationships. The extent to
which symbioses are amenable to analysis by modern science is evident
from the experiments reported by Smith and Douglas (1989) and Nardon et
al. (1989). The recognition of the near ubiquity of the symbiotic state, the
persistence through time of most symbioses, and the profound consequences for the partners led to a reexamination of the evolutionary implications of symbiosis (Margulis, 1976). I attempt here to replace the problematic and restricted cost–benefit, economic vocabulary in current use
with biochemical and biological analysis appropriate to the subject matter,
escaping from the stultifying concern with economic terms such as mutualism or parasitism (Table 7-1, Figure 7-1). Biologists have come to realize

TABLE 7 - 1

Intimacy and individuality: symbiosis and meiotic sexuality compared.

SYMBIOSIS	SEXUALITY
PARTNERS[a] (BIONTS): Two or more organisms, members of different species	PARTNERS (MATES, GAMONTS, GAMETES): Two organisms of complementary mating types, members of the same species
SYMBIOSIS (HOLOBIONT): Association[b] throughout a significant portion of the life history[c]	INDIVIDUAL: Association[b] throughout a significant portion of the life cycle (product of fertilization or meiosis in some protoctists, and in fungi, plants, and animals)[c]

SPATIAL RELATIONSHIPS[d]

OBLIGATE: One partner requires physical contact with the other throughout most or all of its life history. In "phoresy"[e] one partner physically "carries" the other; in "mutualism" both partners "benefit."	OBLIGATE: One partner or sex cell requires physical contact with the other to develop its body or produce its propagules.
FACULTATIVE: One partner can complete its life history in the absence of the other partner. In "commensalism" nutrient sources are shared; in "phoresy" one partner is borne or carried by another; in "mutualism" one partner "benefits" another.	FACULTATIVE: One partner can complete its life history or form propagules in the absence of the other complementary cell or partner ("asexuality").

TEMPORAL RELATIONSHIPS

ALLELOCHEMICAL: Chemical compounds produced by one partner evoke a behavioral or growth response in the other partner ("mutualism").	ALLELOCHEMICAL: Chemical compounds produced by one partner evoke a behavioral or growth response in the other partner ("sex pheromones").
BEHAVIORAL: The behavior of each partner is required for the establishment or maintenance of the association ("mutualism").	BEHAVIORAL: The behavior of each partner is required for the establishment or maintenance of the new individual "social relations," "mutualism," "cooperation," luminous mate attractants).
CYCLICAL: Physical association between partners is periodically established and disestablished.	CYCLICAL: Physical association between partners is periodically established and disestablished by fertilization and meiosis, respectively (obligate sexual reproduction).
PERMANENT: Physical association between partners is required throughout the life history of each (hereditary symbioses, "mutualism").	PERMANENT: Physical association between partners is required throughout the life history of each (permanent mating of schistosomes).

METABOLIC RELATIONSHIPS

METABOLITE: A product of metabolism (e.g., an amino acid, a carbohydrate, or a nucleotide derivative) of one partner becomes a semiochemical or a component of a semiotic product for other partner.[f]	METABOLITE: A product of metabolism (e.g., an amino acid, a carbohydrate, or a nucleotide derivative) of one sexual partner or cell becomes a semiochemical or a component of a semiotic product for the other partner (pheromones).

that these words refer to outcomes of the ecological relations of the partners and inhibit rather than foster scientific knowledge.

Before stating the criteria for considering the origin of organelles as permanent symbionts, this chapter examines the concept that cyclical (temporary) symbiosis is a form of parasexuality. The repetitive coming together and fusion of distinct individuals to form a complex entity is an essential feature of both cyclical symbiosis and meiotic sex (Figure 7-1). The fusion of bionts may be permanent and binding in that the loss of individuality and capacity for free living is irreversible (as in the mitochon-

TABLE 7 - 1 *(continued)*

SYMBIOSIS	SEXUALITY

METABOLIC RELATIONSHIPS *(continued)*

BIOTROPHY: One partner requires carbon, nitrogen, or some other nutrient that is a metabolic product of the other partner.

SYMBIOTROPHY, NECROTROPHY: One partner's nutritional needs are entirely supplied by the other partner, which (S) remains alive during the association ("mutualism," "parasitism") or (N) which is weakened or killed by the association ("parasitism," "pathogenesis").

BIOTROPHY: One mate or sex cell requires carbon, nitrogen, or some other nutrient that is metabolic product of the other in order to form a new individual.

SYMBIOTROPHY, Necrotrophy: One partner's or sex cell's nutritional needs are supplied by the other (e.g., nutritional social relations such as those for attached male angler fish or queen bee fed by drones).

GENETIC RELATIONSHIP

GENE-PRODUCT TRANSFER (PROTEIN, RNA): Protein or RNA synthesized off the genome of one partner is used in the metabolism of the other partner ("mutualism," "parasitism").

GENE TRANSFER: Gene(s) of one partner transferred to the genome of the other partner ("mutualism," "parasitism" involving horizontal gene transfer).

GENE-PRODUCT TRANSFER (PROTEIN, RNA): Protein or RNA synthesized off the genome of one partner is used in the metabolism of the other partner (obligate diploidy).

GENE TRANSFER: Gene(s) of the partner are transferred to the genome of the other partner (karyogamy, syngamy diploid formation).

[a]Partners: definitions with respect to only one partner. Biont: individual organism. Holobiont: symbiont compound of recognizable bionts.

[b]Association: physical contact between organisms (or their cells) that are members of different species (symbiosis) or same species (sexuality).

[c]Life history: events throughout the development of an individual organism correlating environment with changes in external morphology, formation of propagules, and other observable aspects. This refers to, but is distinguishable from, life cycle: events throughout the development of an individual organism correlating environment and morphology with genetic and cytological observations, e.g. ploidy of the nuclei, fertilization, meiosis, karyokinesis, cytokinesis (Margulis et al., 1990).

[d]See Margulis (1991) for discussions of these issues. Ecological relations (e.g., "parasitism," "pathogenicity," "mutualism") are given in quotation marks because only the outcome with respect to the relative growth rates of the partners can determine whether each term is appropriate in any given case.

[e]Traditional terms, given in quotation marks, may correspond to the relationships tabulated here.

[f]Semiochemical: chemical substance acting as signal (sense), i.e., capable of involving biological response. Allelochemicals, hormones, and pheromones are all examples of semiochemicals. Semiotic: meaningful, or the making of meaning, refers to chemical, verbal or other exchanges of signals or signs.

dria of animal cells), or it may be temporary and cyclical (analogous to meiosis–fertilization cycles) in that the bionts, living individually, are capable of returning again and again to the association (as in nitrogen-fixing bacteria and nodulating leguminous plants). The **holobiont** is the product, temporary or permanent, of the association between its constituent bionts (partners). The term **symbiont** in the literature is used both for each biont (especially the smaller one when the partners are unequal in size) and for the integrated holobiont. Since the meaning of "symbiont" in each instance can only be determined by context, we prefer the "biont–holobiont" terminology.

Both symbiosis and meiotic sexuality entail the formation of new individuals that carry genes from more than a single parent. In organisms that

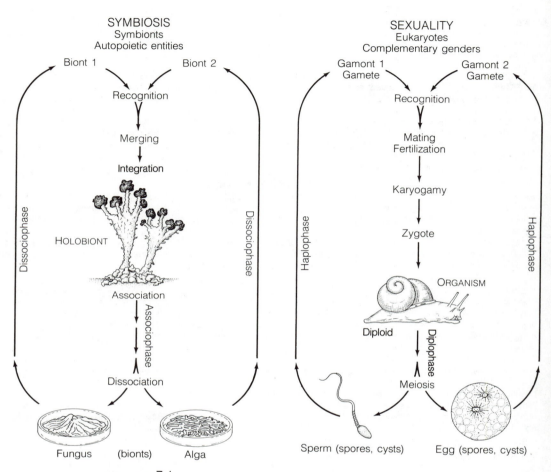

FIGURE 7-1

Comparison of symbiosis and sexuality. In cyclical symbiosis, new individuals are formed from "parent" individuals (bionts) of **different** species (in the example shown, alga and fungus in lichens), whereas in sexual association parents (and their eggs and sperm) are members of the **same** species (snails in the example shown). Both symbiosis and meiotic sex require recognition, merging and fusion of bionts, integrating mechanisms to maintain the association (holobiont or new individual). Later, disassociation (symbiosis) or meiotic reduction (sex) processes are required to again generate unassociated bionts.

develop from sexual fusion of cells from two parents, these parents share very recent common ancestors; partners in symbiosis have more distant ancestors. Meiosis occurs at some time in the life cycle after cells or nuclei fuse to form a diploid; it creates new individuals again capable of sexual fusion. The cyclical nature of symbiosis is less regulated: in some cases, the symbionts never return to the individual state; in others, there are definite and regular methods for ensuring alternation of the individual (biont) and symbiotic (holobiont) states.* In no symbioses are methods of restoring individual partners as elegant as the meiosis–fertilization cycles of tracheophytes and mammals. Geneticists define **parasexuality** as a process that leads to recombinant individuals (those with more than one parent) but bypasses the meiosis–fertilization cycle of biparental sex. The term was coined to describe, for example, bacterial conjugation or the production of recombinant monokaryons from dikaryotic fungi that bypass ascosporogenesis (Pontecorvo, 1958). Thus, cyclical symbioses conform precisely to the definition of parasexuality.

In any environment, symbiotic partnerships may be more fit than individual partners (bionts). Holobionts may persist and leave more offspring than do unassociated bionts. When the genes of symbiotic partners are in proximity, natural selection acts upon them as a unit. In the meiotic prophase of many if not most organisms, chromosome pairing and crossing-over occur. Genes from both parents are contained, at some stage of the life cycle, within the same nucleus; nucleotide sequences from the DNA of one parent are inserted into sequences of the other. In symbioses, the genetic relationship of the partners is generally less intimate and less well regulated through time; nevertheless, concepts of sexuality do apply to the analysis of symbiosis (Figure 7-1).

Ecologists and other field scientists, awe-struck by the diversity and complexity of species interactions, are frustrated by the paucity of information about natural communities. Given that growth rates differ, it would seem obvious that, in a constant environment and in the absence of interrelationships among organisms, one species should always predominate and outgrow all others. Especially in freshwater environments, however, observations show the opposite to be true; rarely does a single species exclude all others (Hutchinson, 1957). Why stable dynamic equilibria of many hundreds of species in the various niches of marine, freshwater, and terrestrial environments persist is not entirely understood. In spite of the nearly infinite biological potential for reproduction, balance is maintained. For each frog that lays 10,000 eggs or each mold that disseminates 1,000,000 spores in a season, that in the following season, only one frog lives and only one mold disseminates spores.

Natural selection acts relentlessly throughout all stages of the life cycles of all organisms, yet every organism is dependent on others for completion of its life cycle. Even in spaces as small as a cubic meter, no community of

---

*Endobiont or endosymbiont ("inside"), epibiont ("on"), endocytobiont ("inside the cell"), and similar terms refer to the topology of the bionts in forming the holobiont.

organisms is restricted to members of only a single species. Diversity, both morphological and metabolic, is the rule. Organisms depend directly on others for nutrients and gases. Only photo- and chemolithoautotrophic bacteria produce their organic requirements from inorganic constituents; even they require "food"—gases such as oxygen, carbon dioxide, and ammonia, which, although inorganic, are end products of the metabolism of other organisms. Heterotrophic organisms require organic compounds as food; except in rare cases of cannibalism, this food comprises organisms of other species, or their remains or exudates. Many heterotrophs are extraordinarily particular about their food sources; for example, some opisthobranchs choose only certain species of algae for their food and will starve rather than eat other, closely related algae (Edmundson, 1966). The lines between nutritional fussiness and dependency, parasitism, symbiosis, and other associations are fine; symbioses are distinguished by the continuity of physical contact between bionts. All biotic relationships are modulated by ambient conditions, especially with respect to water and nutrients. The obligate interdependence of biota illustrates the unity of life on Earth. If Darwin emphasized the common ancestry of living forms through time, Vladimir Vernadsky, Russian founder of biogeochemistry (1889–1945), recognized the interconnectedness of living forms through space (Lapo, 1987). The unity of life, the common chemistry of all autopoietic cell systems, in indisputable (see Chapter 1); symbioses are but one type of interrelationship.

Symbioses have been the subject of many studies (Lange, 1966; Lewis, 1973a, b; Cook, Pappas, and Rudolph, 1980; Richmond and Smith, 1979; Goff, 1983), yet for none are all the genetic, physiological, and biochemical details available. If my thesis is correct, cell biology and organellar genetics (rather than symbiosis research) will yield the most information about long-evolving, hereditary, intracellular symbioses; a beginning has been made with details of genetic control of mitochondria and plastids (Gillham, 1978). Metabolic pathways determined by organellar nucleic acids are relics of obligate endosymbioses. The origin of organelles as intracellular symbionts accounts for their "genetically autonomous" behavior.

Although often treated as exotic in the biological literature, symbiotic relationships abound; many of them affect entire ecosystems. For example,

In the tropics, many, probably most of the recurring animals have algal symbionts and these extend through most of the invertebrate phyla. . . . Quantitatively it is probably true that the symbiotic algae are more important than the phytoplankton and the free benthic algae in coral reefs and other shallow waters with calcareous sediments, and the productivity of such waters is essentially due to such symbionts, which are more usually known as zooxanthellae (Ferguson-Wood, 1967). [Most of these zooxanthellae are symbiotic dinomastigotes of the genus *Symbiodinium* (*Gymnodinium microadriaticum*) (Trench, 1979).]

Commonplace microbial symbionts are among the least known: surfaces of algae and leaves, tree bark, and animal skin have particular microbiota. For the conspicuous partner, the significance of these associations — number of symbionts tolerated, degree of specificity, nutritional and other exchanges — is not known. The microbiotic species have not been identified, since most ecologists ignore bacteria and protoctists. Dentists, confronted with the complexity of microbiota on the surfaces of teeth and gums, are acutely aware of limitations to knowledge, even of vertebrate symbionts. Only selected symbioses have been studied (Table 7-2).

Every kind of community contains symbiotic associations; partners come from all higher taxa. However, certain groups seem more prone than others to develop symbioses. Ascomycotes are the most likely fungi to enter lichen associations; of 13,250 species, more than 46 percent are mycobionts of lichens (Honegger, 1991). The dinomastigote genus *Symbiodinium* and the green algal genus *Chlorella* are more likely than other algae to become endocellular in coelenterates; *Nostoc* is more likely than other filamentous cyanobacteria to associate with ferns, cycads, and lichen fungi. The predisposition of the photosynthetic genera *Trebouxia* and *Pseudotrebouxia* (the algae of many lichens) to enter associations with a wide range of different ascobasidiomycotes must have had parallels: apparently, certain phototrophic prokaryotes became associated with many different hosts in the symbioses that evolved into algae, explaining why similar plastids are found in distantly related protoctists.

Why certain taxa are predisposed to enter into symbiosis can only be surmised in some cases, whereas in others the reasons are obvious. In many associations between luminous bacteria and marine animals (Figure 7-2), the selective advantage to the host is clear. The fish may use the light source as a flashlight, to illuminate potential prey; the light may be used to deter predators or in courtship behavior. The continuous bacterial light emission comes from the reaction of bacterial luciferase with luciferin (see Table 6-2). Some fish flash the light by using special tissues that form movable baffles. The uniqueness of the bacterial luciferases permits the symbionts to be identified in most cases, even when they cannot be cultivated outside the host (Nealson and Hastings, 1979). These marine symbioses exhibit a full range of intimacy, from casual associations of free-living luminous bacteria with the guts of marine organisms to closer associations in which the bacteria are cultured in special light-emitting organs inside the host but external to its cells; in at least one host, the tunicate *Pyrosoma*, the bacteria are intracellular symbionts. In some of these associations, luminous bacteria can be cultivated in the laboratory; in others they cannot, perhaps because of metabolic losses due to long association with the host. The partners can be identified, and the host animals are only distantly related, indicating that many of the associations evolved independently. In nature, the evolution of any single symbiosis has never

TABLE 7 - 2
Symbiosis literature.[a]

SYMBIONTS		REFERENCES
Ciliates	Microbial endosymbiont kappa particles and Gram-negative bacteria	Sonneborn, 1959; Preer et al., 1974; Preer (1975) correlates early genetic analyses of "cytoplasmic genes" with later morphological and biochemical work.
Ciliates	Anaerobic sulfide-oxidizing and other bacteria	Fenchel and Finlay, 1991
Ciliates	Intracellular chlorella (Figure 7-12)	Karakashian and Siegel, 1965; evolutionary significance of endosymbiosis in *Paramecium bursaria*, Karakashian, 1975; in *Coleps*, Esteve et al., 1988
Cnidarians, other animals	Intracellular dinomastigotes	Edmundson, 1966; Trench, 1979
Bacteria, fungi, and protists	Bacteria, fungi, and protists	Lange, 1966; Margulis, 1976; Cook et al., 1980; Smith and Douglas, 1989; Guerrero, 1991 (reviews of microbial predators and symbionts)
Fungi	Cyanobacteria and green algae (Figure 7-3)	Richmond and Smith, 1979; Ahmadjian, 1980; Honegger, 1991
Fungi, actinobacteria, and cyanobacteria	Plants	Lewis, 1973a, b; Prozynski and Malloch, 1975; Kendrick, 1991; Pirozynski, 1991 (reviews of mycorrhizal and other plant symbioses)
Bacterial-like microbes and "Blochmann bodies"	Cockroaches and other insects	Lanham, 1968; Schwemmler, 1984, 1991
Luminescent bacteria and animals	Teleost fish, cephalopods, mollusks, and nematodes	Harvey 1952; McFall-Ngai, 1991; Nealson, 1991
Algae and plants	Fungi, plants, and animals; translocation of carbohydrate from autotrophs to heterotrophs	Smith and Douglas, 1989; Cook et al., 1980
Bacteria and protoctists	Mulgulid tunicates, tubeworms, and clams; sulfur and nitrogen metabolism	Schwemmler, 1984; Saffo, 1991; Vetter, 1991
Protoctists	Bacteria; symbiotrophy	Kirby, 1941; Margulis, 1976; Preer, 1984
Cyanophora (protist)	Anomalous cyanobacteria (cyanelles); photosynthate transfer (Figure 7-5)	Trench et al., 1978; Trench, 1991

[a]See also the journals *Symbiosis* (Balaban Publisher, Rehovot Israel) and *Endocytobiology* (University of Tübingen, Germany); Nardon et al. (1989).

A

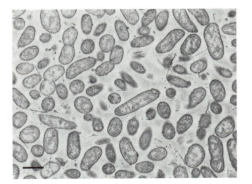

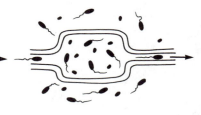

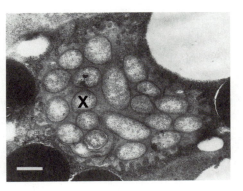

B

FIGURE 7-2

Symbioses of luminous bacteria. The diagrams depict schematically the various niches that the bacteria occupy. The unidentified symbionts in the photographs cannot be cultured. In all the photographs, bar = 1 μm. [From Nealson et al. (1981).] (**A**, **B** *above*; **C**, **D** *on following page*) (**A**) Casual association. Free-living and saprophytic *Photobacterium fischeri*. Biotrophic associations are neither species-specific nor permanent. (**B**) Extracellular gut symbionts. *Xenorhabdus luminscens* (*X*) in the pharyngeal cavity of a nematode. The luminous nematodes invade the tissues of caterpillars, rendering them luminous, which increases the chances that the caterpillars will be seen and eaten by birds, introducing the bacteria to their richest possible habitat. Bacteria excreted from bird guts may then be reingested by nematodes. (**C**) Extracellular symbionts in light organs. Top: *Photobacterium leiognathi* (*P*) surrounded by light-organ tissue of a pony fish; the light renders the fish cryptic to predators from below. Middle: unidentified symbionts (*L*) form the light organ of a flashlight fish, *Photoblepharon*; the fish shines the light to find prey and flashes it to communicate and to confuse predators. *n* = bacterial nucleoid. Lower left: unidentified symbionts of an angler fish, which uses their light to attract prey. Lower right: unidentified symbionts of the squid *Heteroteuthis hawaiiensis*; the squid squirts out the bacteria in a luminous "ink" to confuse predators in the dark. (**D**) Intracellular symbionts in light organs. Unidentified intracellular symbionts (*B*) of the luminescent tunicate *Pyrosoma*; the selective value to the tunicate is unknown, *M* = mitochondrion; *B* = bacteroids, which are modified symbiotic bacteria; *N* = nucleus of host cell.

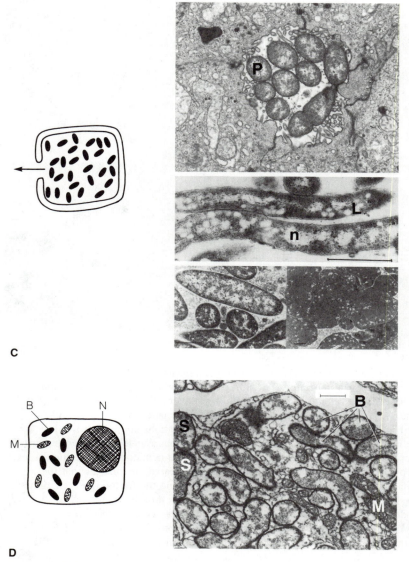

C

D

FIGURE 7-2 C, D

been documented, yet, in principle, because associations are numerous and varied, possible courses of evolution from autonomy to tight integration can be traced for many symbioses (Tebo, Linthicum, and Nealson, 1979; Smith and Douglas, 1989). McFall-Ngai (1991) presents strong evidence that the bacteria – fish association was important in the adaptive radiation of morids, macrourids, ceratoids, and especially leiognathid teleost fishes.

Jeon (1991) reports the establishment of a new symbiosis in the laboratory. The uniqueness of the amoeba-bacteria studies of Jeon are probably due to limited research support for symbiosis studies rather than to a low frequency of formation of new symbioses. When symbiosis literature is perceived as medically important [for example, in listeriosis or Legionnaire's disease (Portnoy et al. 1989)], it receives more funding and attention only if the term symbiosis is replaced with "pathogenesis" or "parasitism" (Margulis, 1990).

FROM SYMBIONTS TO ORGANELLES

Tous les êtres vivants, tous les animaux depuis l'Amibe jusqu'à l'Homme, toutes les plantes depuis les cryptogames jusqu'aux Dicotyledones, sont constitués par l'association, l' "emboîtement" de deux êtres differents. Chaque cellule vivante renferme dans son protoplasme des formations que les histologistes désignent sous le nom de "mitochondries." Ces organites ne seraient pour moi autre chose que des bactéries symbiontiques, ce que je nomme des "symbiotes." . . . Il est très naturel que les bactériologistes montrent quelque défiance à accepter d'emblée une notion aussi nouvelle et même aussi dangereuse pour leur science; il est très normal qu'ils exigent des preuves formelles et une interprétation convenable des faits. Il y a là en jeu une question de technique qui a une importance primordiale.*

P. PORTIER, 1918

*All living beings, all animals from amoeba to man, all plants from cryptogams to dicotyledons, are formed from an association, an engulfing of two different beings. Each living cell encloses in its cytoplasm bodies that cytologists designate by the name " mitochondria." These organelles are not anything other than symbiotic bacteria, so that I call them "symbionts." . . . It is very natural that bacteriologists show some hesitancy to accept a concept so new and even dangerous for their science; it is normal that they demand a formal proof and an interpretation fitting the facts. It brings up a question of technique that is of prime importance.

Cette expérience, qui constitue le point de départ
fondamental de la théorie des symbiotes, est, à notre
sens, tous à fait insuffisante pour entraîner une conviction.

Dire que ce microorganisme est indispensable à la vie
cellulaire, et que toute cellule le renferme, constitué une
généralisation abusive comme nous pensons pouvoir le
démontrer au cours de cet ouvrage.†

A. LUMIÈRE, 1919

The criteria required to establish the origin of organelles from microbial symbionts are listed in Table 7-3. These criteria have been applied amply to the examples noted at the bottom of the table. It is highly unlikely, based on these criteria, that certain organelles—the nucleus, flagellum, ribosomes, and others—were once free-living microbes.

In symbioses with a long history of physiological and morphological modification or involving three or more types of partners from different taxa, it is difficult to identify the bionts. However, in two classes of well-documented symbiosis—lichens and microbial symbionts of the ciliate *Paramecium*—partner identification is undisputed. Both symbioses were established by processes that can be related to the evolution of organelles.

Most lichens are morphologically very unlike either the phycobionts (algae) or the mycobionts (fungi) that compose them (Figure 7-3). They are remarkable examples of innovation, possessing many morphological, chemical, and physiological attributes absent from either fungus or alga grown independently (Honegger, 1991). Very few lichen phycobionts and mycobionts have been separated experimentally; the conditions of growth of each partner makes separation and pure culturing difficult (Ahmadjian, 1967, 1980).

In the late 1940s, a cytoplasmically inherited genetic phenomenon dubbed **kappa** (κ) was discovered in *Paramecium aurelia*, as Sapp (1987) describes. The presence of a nuclear gene K caused paramecia with κ to manifest a "killer trait" against non-κ-bearing strains. That is, κ-bearing paramecia incubated in the same medium as non-κ-bearing paramecia caused the sensitive paramecia to die. Because the nuclear genetic system of paramecia was understood and the transfer of cytoplasm during mating

---

†This experiment, which constitutes the fundamental point of departure of the symbiotic theory, is in our view, completely insufficient to support the contention.

To say that this microorganism is indispensable to cellular life, and that each cell contains it, constitutes an abusive generalization as we hope to be able to demonstrate in the course of this work.

TABLE 7-3

Organelles from symbionts.

CRITERIA FOR SYMBIOTIC ORIGIN	
Organelle at first must contain:	DNA messenger RNA Polymerases Ribosomes 5S, 16–18S, and 23–28S ribosomal RNA Ribosomal proteins Membranes with ion channels
Organelle must resemble:	An identifiable free-living microbe (more than it does the rest of the cell in which it resides)

STATUS OF HYPOTHESES RE ORIGINS	EXAMPLES
Established	Plastids from cyanobacteria (*Synechococcus*) Mitochondria from respiring eubacteria (*Paracoccus*) Kappa, mu, etc. particles from eubacteria (*Caedibacter*)
Hypothesized	Undulipodia from spirochetes (genera unidentified) Hydrogenosomes from *Clostridium* Peroxisomes from eubacteria (genera unidentified)
Unlikely	Rotary flagella Nucleus Endoplasmic reticulum Ribosomes

could be controlled, the inheritance pattern of κ was determined and found not to obey Mendelian rules of inheritance. The "cytoplasmic gene" for the killer trait was transferred from κ-bearing paramecia to non–κ-bearing mates only when cytoplasmic bridges were formed between the two mating paramecia and a substantial transfer of cytoplasm occurred. The killer ability was expressed only in paramecia that had both the cytoplasmic genetic determinant, κ, and the nuclear K gene. If κ was present in a cell with the genotype KK or Kk, the phenotype through many divisions was "killer." When the nuclear gene K was genetically transferred out of a κ-containing organism, producing the homozygous recessive kk diploid, κ was lost from the descent population. From the number of generations (N) required to lose κ after removal of the nuclear gene, the original number of κ genes (2^N) in the cytoplasm of the killer paramecia was estimated. The genetic behavior of κ was worked out by Sonneborn and his colleagues (Sonneborn, 1959). Cytoplasmic κ particles were eventually identified as Gram-negative rod bacteria capable of growth in complex media. The κ case is not unique; some twenty-different types of bacteria,

A

B

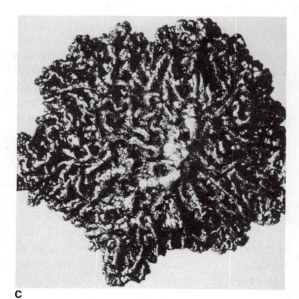

C

FIGURE 7-3

The lichen symbiosis.
(A) *Cladonia cristatella*, the British soldier lichen,
from deciduous forests of northeastern North
America. The fruiting bodies (the dark round
objects, actually red) are primarily fungal tissue.
[Courtesy of J. G. Schaadt.] (B) The isolated
mycobiont of *Cladonia cristatella*, growing on agar
in a Petri plate. Ascospore-bearing structures and
other complex morphologies are never expressed in
the fungus growing alone. (C) The isolated
phycobiont, the green alga *Trebouxia*, growing on
agar in a Petri plate. [B and C courtesy of V.
Ahmadjian.]

many of which harbor bacterial viruses, form tight or loose symbiotic relationships with *Paramecium* (Beale, Jurand, and Preer, 1969; Preer, Preer, and Jurand, 1974).

Although symbioses between large organisms can be recognized immediately, morphology alone is insufficient or misleading in distinguishing microbial symbionts. Other criteria are required. What properties of organelles derived from free-living symbionts can distinguish them from cell inclusions originating from direct filiation?

Microbial symbionts, loose and strict, are selected for as holobionts because, in a particular environment at a particular time, the integrated bionts manifest semes displaying advantages over those of the unassociated partners. In loose symbioses, when the growth conditions for one biont become more favorable than those for the whole complex, the partnership is likely to dissolve. Several factors — moisture, nutrients, temperature, and so forth — affect the establishment and dissolution of lichen symbioses, for example.

> Lichen-forming fungi and algae are not obligate symbionts. Given suitable conditions, each organism would remain in an independent free-living state. Appropriate fungi and algae must be forced by demands of their immediate environment into forming unions. A lichen which is kept under prolonged moisture conditions in a laboratory will soon dissociate into separate growth of the partners. Even under natural conditions such separations occur, either partially or wholly, dependent on the type and duration of changes in macro- or microclimates which surround these associations. Lichen development is a continuous balanced growth between fungus and alga. This fact helps explain why lichens are so abundant in barren dry areas and why they are so scarce in regions with large amounts of precipitation (Ahmadjian and Paracer, 1986).

The primary environmental variables that select for holobionts over unassociated bionts may be obvious: traits present in the complex are lacking in both free-living partners. Bacteria – plant symbioses for example, perform nitrogen fixation in nodulated legume roots; alone, neither partner is capable of fixing atmospheric nitrogen. Table 7-4 lists such metabolic novelties characteristic of microbial symbioses.

Any symbiont that originated as a free-living microbe must once have contained all minimal requirements for autopoiesis and reproduction (see Figure 5-1). These include DNA; messenger RNA colinear with the DNA; a source of ATP and other nucleotides; a cell membrane – synthesizing system; and a protein-synthesizing system composed of polymerases, transfer RNAs, their acetylating enzymes, ribosomes composed of many proteins, and at least three ribonucleic acids. Upon intracellular associa-

TABLE 7-4
Some novelties of symbioses.

LARGER BIONT	SMALLER BIONT	METABOLITES OR PATHWAYS CHARACTERISTIC OF NEITHER BIONT ALONE
Hydra viridis	*Chlorella* sp.	Mannitol
Cryptomonad mastigote *Peliaina*	Cyanobacterium	Starch synthesis (Lederberg 1952)
Leguminous plants	*Rhizobium* (bacteria)	Leghemoglobin (Verma et al., 1974)
Psychotria	*Klebsiella*	Nitrogen fixation (Silver and Postgate, 1973)
Ascomycotous or basidiomycotous fungi	*Trebouxia* (chlorophyte); *Anabaena* or other cyanobacterium	Depsides and depsidones (Miller, 1961; Hale, 1967; Ahmadjian, 1967)
Coelenterate gorgonians (for example, *Pseudoplexaura porosa*)	*Gymnodinium (Symbiodinium) microadriaticum*	"Gorgosterol" ($C_{30}H_5O$), a steroid derivative, and crassin acetate, a terpene derivative (McLaughlin and Zahl, 1966)
Paramecium aurelia	Lambda particles (bacteria)	Folic acid synthesis (Soldo, 1963)
Crithidia	Various bacteria	Lysine synthesized by the diaminopimelic acid pathway
Cockroaches and other insects	Inclusions (probably bacteria) in mycetocytes	Symbionts use nitrogenous waste products of the insects in synthetic metabolism (Lanham, 1968; Schwemmler, 1973, 1984)
Dry-wood and subterranean termites	Protists and bacteria	Cellulose digestion and nitrogen fixation (Breznak et al., 1973; Yamin, 1980)

tion, symbionts may lose none of these synthetic capabilities or all of them except those required to replicate their nucleic acid—DNA and complementary messenger RNA. Since a widely distributed enzyme, reverse transcriptase, permits the formation of DNA from a complementary RNA molecule, in principle after integration of the bionts a single informational macromolecule may be all that remains of an autopoietic intracellular system. If an organelle originated as a free-living microbe, its symbiont nucleic acid must reside somewhere in the host cell at every stage of the host's life cycle.

The growth rates of bionts must be approximately equal. Death or lysis of the host is a natural consequence of an endosymbiont growth rate that exceeds that of the host. If, on the other hand, the growth rate of the larger partner (the host) exceeds that of the symbiont, in each generation the number per host will decline, and offspring lacking endosymbionts will appear. If the symbiont-host complex is not of selective advantage in the particular environment, the association will dissolve. Even if the holobiont has great advantages over the free-living bionts, the partnership will not survive unless mechanisms evolve to ensure equal growth rates of the bionts.

In associations that are stable over long periods of time, redundancy is selected against. Traits dispensable to the partnership tend to be lost. Investigators have great difficulty separating closely associated bionts; if they are grown outside the partnership elaborate or exotic growth factors are often needed. Examples of such coevolution of partners include reduction of cell walls in algal endosymbioses of protoctists, flatworms, or corals. Symbiotic algae appear round and plastid-like (Figures 7-4 and 7-5; Trench, 1979); such morphological reduction often makes it difficult to identify the smaller partners. As Table 7-5 shows, symbiotic algae of marine animals and protoctists have a variety of origins. Their protection by the larger partner leads to the loss — sometimes reversible — of some traits that have functional significance only for the free-living algae (McLaughlin and Zahl, 1966). Symbionts released from radiolarians recover their typical dinomastigote form, with reappearance of the transverse and girdle undulipodia, which are not visible when the symbionts are inside. The changes in metabolic products, nutrition, pigmentation, or respiratory patterns that may accompany the shift from free life to symbiosis are not necessarily the expected ones. For example, although cell walls are reduced in most algal symbionts, in some cases they become more prominent. Nevertheless, expected changes do often occur: symbiotrophic worms tend to lose their sensory and motor appendages, becoming specialized for egg and sperm production; and endobiotic algae tend to lose genetic and synthetic capabilities, which are presumably relegated to the hosts — *Cyanophora paradoxa*, for example (see Figure 7-5) (Stanier and Cohen-Bazire, 1977; Trench, 1991). With the loss of capabilities needed for independent life, associations become progressively more obligate. Continual exposure to the environmental agents selecting for the partnership — alternating dryness and cold for lichens; anaerobiosis for protoctists harbored in animal tissue — probably increases the rate at which partners become interdependent, that is, new holobionts evolve.

Endosymbionts will be retained by their hosts only if each new host cell at reproductive division receives at least one copy of the endosymbiont genome. Any heritable change ensuring that the genome of the endosymbiont is distributed to host offspring will be of selective value to the evolving holobiont. A common, effective, but perhaps wasteful, way of retaining the valuable traits conferred by the symbionts on the holobiont is the retention of "copies," that is, many endosymbionts within each host cell. This increases the probability that each host offspring receives at least

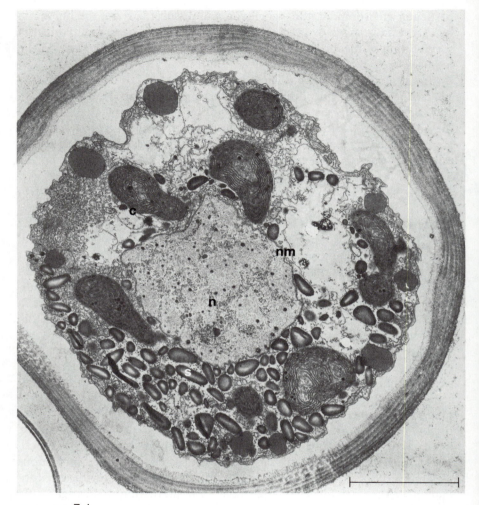

FIGURE 7-4

Glaucocystis nostochinearum. An "anomalous alga" with blue-green pigmentation but eukaryotic cell structures. The host is, apparently, a typical *Oocystis*-like chlorophyte harboring *Aphanothece*-like coccoid cyanobacteria that function as cyanelles. The presence of the thick wall around the complex suggests that the association is stable and that the cyanelles are permanent residents. *c* = cyanelle; *n* = nucleus; *nm* = nuclear membrane; *s* = starch. Transmission electron micrograph, bar = 5 μm. [Courtesy of W. T. Hall and G. Claus.]

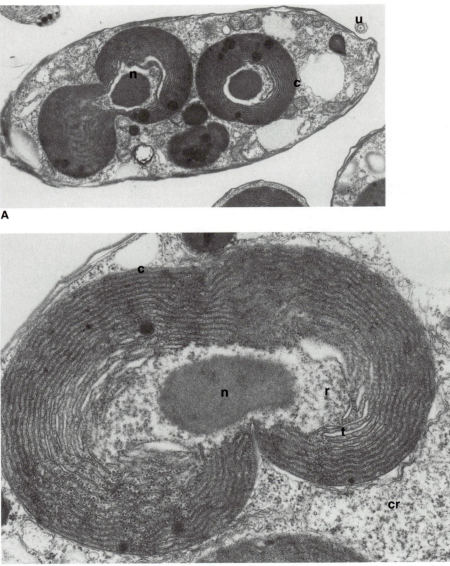

FIGURE 7-5

Cyanophora paradoxa.
(**A**) Whole complex: cryptomonad-like protist harboring cyanobacterium-like but nearly wall-less cyanelles. The association is obligate and there is exchange of metabolites and integration of gene products from both genomes (Trench et al., 1978). c = cyanelle; u = undulipodium; n = nucleoid of cyanelle. Transmission electron micrograph, ×11,500. (**B**) Cyanelle in division. c = cyanelle; cr = cytoribosomes of cryptomonad host; n = cyanelle nucleoid; r = cyanelle ribosomes; t = thylakoids. Transmission electron micrograph, ×35,000. [Courtesy of W. T. Hall and G. Claus.]

TABLE 7-5
Photosynthetic symbionts in protoctista, plants, and animals.[a]

HOST KINGDOM	HOST PHYLUM	PHOTOSYNTHETIC MEMBERS OF PHYLUM	SYMBIONTS (CLASSIFIED AFTER RELEASE FROM HOST)[b]	KINGDOM OF SYMBIONTS[b]
Protoctista	Actinopoda	Many radiolarians	(X) *Gymnodinium* (*Symbiodinium*) (dinomastigotes)	P
	Granuloreticulosa	*Globigerinoides ruber*	(X) Probably dinomastigotes	P
	Ciliophora	*Paramecium bursaria*	(C) *Chlorella*-like green algae	P
		Stentor polymorphus	(C) *Chlorella*-like green algae	P (?)
		Mesodinium rubrum	(Cr) Cryptomonad-like alga	P
	Bacillariophyta *Rhizosolenia*)	*Richelia intracellularis*	(BG) *Richelia* enclosed in five-layer envelope	M (?)
Animalia	Cnidaria (coelenterates)	*Paulinella*	(BG) "Cyanelles"	M
		Cladocora	(X) *Gymnodinium*	P
		Acropora	Unknown	
		Pacific Coast anemone	(X) *Peridinium*-like dinomastigotes	P
		Anthopleura xanthogrammica	(C) desmomonad-like	P
	Platyhelminthes (turbellarians)	*Convoluta*	(C) *Platymonas*[c] (mastigotes)	P
		Ophioglypha	Unidentified algae in epidermal plates	P
	Mollusca	*Tridacna*[d]	(X) *Gymnodinium*	P
		Tridachia[d]	(C) Chloroplasts, retained foreign organelles	M (?)
		Elysia[d]	(C) Chloroplasts, retained foreign organelles	M (?)
		Plachobranchus	(C) Chloroplasts, retained foreign organelles	M (?)
	Chordata	Didemnids (ascidians)	(C) *Prochloron*[e]	M
Plantae	Cycadophyta	All	*Nostoc* and *Anabaena*, cyanobacteria in roots[f]	M

[a]After Lange (1966); Taylor (1974); Lewin (1976); Margulis (1976); Trench (1979).
[b](X) = zooxanthella; (C) = zoochlorella; (Cr) = cryptomonad; (BG) = blue-green organism; M = Monera; P = Protoctista.
[c]See Figure 7-8.
[d]See Figure 7-7.
[e]See Figure 5-6.
[f]See Figure 12-3.

one endosymbiont genome. The behavior of centrioles, mitochondria, chloroplasts, and intracellular bacterial symbionts at cell division can be viewed as a way of ensuring organellar or symbiont genetic continuity (see Tables 10-6 and 11-1).

Mitotic organisms, which contain elaborate mechanisms for the distribution of their nuclear genes, tend also to retain endosymbionts effectively. For example, some endosymbionts divide synchronously with their host cells; often, eggs are provided with packages of symbionts, ensuring the genetic continuity of the partnership (Schwemmler, 1991). Subtle chemical mechanisms are probably involved. For example, gametogenesis of some hypermastigote protoctists (Phylum Zoomastigina)—the wood-eating symbionts in cockroaches—is controlled by a hormone, ecdysone, produced by *Cryptocercus* (the insect) just before molting. The molting wood-eating cockroach sheds its hindgut with its microbial associates; ecdysone induces the protists to form dormant, resistant cysts, including gametes in gametocysts that will be safely ingested later by other nest mates (Cleveland, 1951).

Since organelles originating by symbiosis did not evolve by accumulation of mutations, no living or fossil organism contains intermediate stages in the development of these organelles. The metabolic capabilities of the microbial symbiont are "packaged" as a unit and acquired by the host in one step. As documented in the literature of parasitology, loss of traits and dedifferentiation of the endosymbiont (rather than gain and complexification) represent evolutionary "advancement."

When first established, a holobiont or symbiotic complex has two or more unrelated ancestors rather than a single direct ancestor. No photosynthetic ciliate, for example, is directly ancestral to *Paramecium bursaria*, nor do lichens necessarily have lichen ancestors. However, strong selective pressures that maintain new holobionts may lead to many simultaneous modifications of both symbiotic partners. Such coevolution may obscure the identities of the original partners. Because it is easy to separate *P. bursaria* from its intracellular chlorellas (Karakashian and Siegel, 1965; Karakashian, 1975), it may be inferred that the *P. bursaria* symbiosis is more recent than that of most lichens; in both complexes, however, permanent hereditary changes have occurred. Complex life histories of endosymbiotic organisms parallel so ingeniously and closely the development of their hosts (for example, the apicomplexan *Plasmodium*, the causative agent of malaria, or the flatworm *Cryptocotyle*, with three intermediate hosts—seagulls, cunha fish, and snails) that early investigators found it hard to imagine anything short of divine—or demonic—intervention to account for the origin of these organisms.

Symbioses can form and later dissolve; stability is, in part, a function of time and the intensity of environmental selective pressure. To counter a threat to its own existence, the larger partner may take advantage of its symbionts; for example, it may digest them. Thus, under certain conditions, the presence of the symbiont is gratuitous. Analogously, in heterotrophic media containing minimal amounts of phosphate, *Euglena gracilis* may lose its plastids permanently by digesting them (Epstein and Alloway, 1967). When intracellular symbionts are lost, any metabolic functions that depend on the presence of their genomes are also lost. If each partner contributes to a metabolic pathway, the removal or impairment of one of the partners may cause certain telltale metabolic intermediates to accumulate.

Initially, each biont has its own genes, gene products (RNA and proteins), and metabolic products, including wastes. The possible ways in which different bionts may be integrated into a new whole, the holobiont, are many. Lewis (1973a, b) has analyzed the utilization of photosynthate of heterotrophs symbiotic with autotrophs—mainly fungi and plants. One partner may utilize and remove waste from the other. In the *Methanobacillus omelianskii* partnership, for example, hydrogen, an end product of fermentation in the "S" bacterium, is removed for utilization as substrate by a methanogen (Bryant et al., 1967; Reddy, Bryant, and Wolin, 1972). One partner may produce a coenzyme, the other partner, the apoenzyme. In nodules of legumes, for example, *Rhizobium* bacteroids produce the heme of leghemoglobin, while the plant produces the globin protein. Although free-living *Rhizobium* does not fix nitrogen, it may be induced to do so in pure cultures on media containing arabinose. Synthesis and accumulation of arabinose by plants apparently derepresses the bacterial nitrogenase system. Thus, symbioses may operate simultaneously at more than one level: gene product and metabolite in this case. The genes of one partner may be transferred to the genome of the other. The details of such transfers are still obscure. In the best known case, the genes of *Agrobacterium* are carried by bacterial plasmids into the cells of susceptible plants (Schell et al., 1979).

Metabolic innovations appear in symbiotic partnerships (Table 7-4). The substrate for a biosynthetic reaction in one partner may be the product of biosynthesis in the other; the second member, metabolizing the substance further, may then produce some compound uniquely characteristic of the partnership. If this product is of high selective value, it will stabilize the relationship. Alternatively, the final product may represent the total catabolism of a food source.

Levels of partner integration are (1) behavioral, (2) metabolic, (3) gene-product (protein or RNA), and (4) genetic; Figure 7-6 diagrams the last two. The more closely associated and dependent the genetics and metabolism of symbionts become, the more obligate and refractory to dissociation the holobiont. Well-integrated symbioses are more difficult to study than casual ones. The deepest insight into the mechanisms of gene transfer in symbioses comes from work on organellar–nuclear relations (Gillham, 1978).

A symbiont can never be produced by the action of its host's genes. Once lost, a symbiont cannot reappear unless it is reacquired, for example by ingestion. Unless reingestion follows loss quickly, geologically speaking, precisely the same symbiont will not be reacquired. The experimental

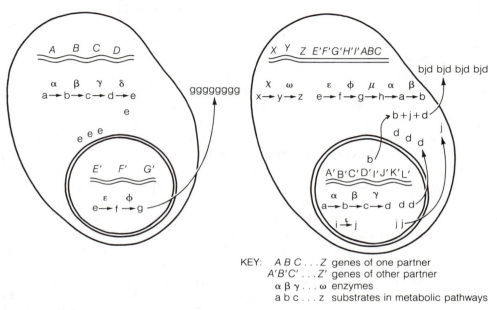

FIGURE 7-6

Partner integration in symbiosis. The proximity of the endosymbiont demonstrates integration at the behavioral level. In the transfer of gene products, either a primary gene product (RNA or protein) or a secondary product (a metabolite) of the genome of one partner is taken up by the other partner for immediate use or for further metabolism into a product (*g*).

With the transfer of genes, a symbiosis becomes more closely integrated. Part of the genome of one symbiont is transferred to the genome of the other. The new genome may underlie metabolic pathways leading to an advantageous product (*bjd*) that neither partner was capable of producing alone.

Key: *A*, *B*, *C* . . . *Z* = genes of one partner; *A'*, *B'*, *C'*, . . . *Z'* = genes of other partner; *α*, *β*, *γ*, . . . *ω* = enzymes; *a*, *b*, *c*, . . . *z* = substrates in metabolic pathways.

induction of loss and the reacquisition of "foreign" chlorella endosymbionts by *P. bursaria* and the green freshwater hydra have been described (Karakashian, 1963; Fracek and Margulis, 1979; see Figure 7-13). Hydras are remarkably good at recognizing and reacquiring lost chlorella symbionts.

Analogous loss and regaining of different symbionts was hypothesized in botanical literature of the early twentieth century to explain the evolutionary origin of certain anomalous algae. *Glaucocystis nostochinearum*, an alga, should be considered a product of "reingestion." As its generic name implies, *G. nostochinearum* has "host" features that relate it unambiguously to the Oocystaceae family of chlorophytes; however, it contains cyanobac-terium-like chromatophores rather than green chloroplasts. Likewise, ac-cording to Fritsch (1935), *Gloeochaete*, an organism "long referred to as an anomalous genus of the Myxophyceae [cyanobacteria]," . . . is "now known to represent a colorless Tetraporaceous [green algal] form in which the blue-green chromatophores are symbiotic Blue-green Algae." Fritsch suggested that the loss of green chloroplasts in these organisms was fol-lowed by reingestion of cyanobacteria that became endosymbionts. Ultra-structural studies of *G. nostochinearum* have confirmed and clarified Fritsch's original explanation (see Figure 7-4; Hall and Claus, 1967).

The *Cyanophora paradoxa* story is similar. *Cyanophora*—common in algal blooms in warm, fresh water—is a euglenoid mastigote harboring blue-green **cyanelles**, structures that resemble coccoid cyanobacteria that have retained vestiges of their penicillin-sensitive walls (see Figure 7-5). The work of Trench (1991) suggests that *Cyanophora* is a unique, dead-end symbiotic complex in which cyanelle pigment synthesis is under at least the partial control of the host's ribosomes. Are the cyanobacteria-like struc-tures blue-green endosymbionts, cyanelles, chromophores, or plastids? The name is really a matter of taste; the morphological affinities of these photosynthetic units with free-living cyanobacteria are obvious. Since the cyanelles import 80 percent of their protein from host cytoplasm and their circular genomes are only about 130 kilobases long, it is not surprising that they have defied cultivation in vitro.

The cell walls of *Cyanophora* and *Glaucocystis*, the hosts, restrict phago-cytosis and pinocytosis. If the photosynthetic symbionts of these organisms were acquired by feeding, then the hosts' cell walls must have arisen after the associations formed, and they ought to differ from those typical of chlorophytes.

The problem of identifying endosymbionts has been exacerbated since it has been found that bona fide organelles, such as chloroplasts ingested along with food, can remain functionally autotrophic for weeks inside animal tissues. In several marine slugs and marine ciliates, what were thought to be symbiotic algae are not the entire algae but only the chloro-plasts (see Table 7-4); this phenomenon (Figure 7-7) has been called *"foreign organelle retention"* (Taylor, 1974, 1979) or even *"chloroplast en-slavement"* (Stoecker and Silver, 1990).

Cyanophora is a model for more than just the equation **heterotrophic protoctist + photosynthetic moneran = alga**; it is a model for genomic reduction as well. Although each cyanelle has about 40 copies of its own DNA, cyanelles contain less than 10 percent of the DNA found in typical cyanobacteria (Herdman and Stanier, 1977). This tendency of obligate endosymbionts toward a change in genome number (from one or two, to

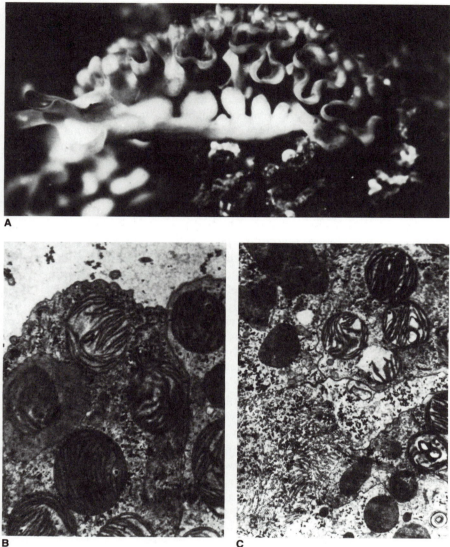

FIGURE 7-7

Foreign organelle retention: photosynthetic mollusks.
(**A**) A saccoglossan mollusk, *Elysia viridis*, that is functionally photosynthetic owing
to the chloroplasts harbored in the diverticulum of the gut. Light penetrates the
epithelium and the gut wall. This gastropod can survive more than three months of
starvation if adequate light is present. (**B**) Digestive cell of the photosynthetic
mollusk *Tridachia crispata*, showing a healthy chloroplast in a cell with others being
digested in a phagocytic vesicle. Transmission electron micrograph, ×8200. (**C**)
Digestive cell of *Elysia viridis*, showing healthy chloroplasts derived from
siphonaceous food algae. Transmission electron micrographs, ×8000. [Courtesy of
R. K. Trench.]

zero or hundreds of genomes/symbiont), as well as toward a reduction in total genome, should be recalled later, in the discussions of undulipodia, mitochondria, and plastids.

Symbiont loss must be distinguished from dedifferentiation. When a symbiont confers on a host metabolic capabilities that are not required at all stages of the host's life history, selection leads to hosts that maintain dedifferentiated, nonfunctional symbionts during the parts of the life history in which they are dispensable, blurring the distinction between genetic loss and phenotypic dedifferentiation. For example, dedifferentiated mitochondria in anaerobically growing yeast are not always visible, even with the electron microscope. However, from a genetic point of view, dedifferentiation and loss are radically different. Loss, like death, is permanent: when functional endosymbiont nucleic acid is lost from the holobiont, the host loses the potential to form any phenotypic manifestations of the endosymbiont or its genome. In dedifferentiation, on the other hand, the host retains the potential to reform the symbiont or organelle under appropriate developmental and environmental conditions—that is, the host retains the endosymbiont genome, if only as a single molecule of nucleic acid. Organelles derived by differentiation from nuclear genes or metabolic pathways that produce some recognizable structure—for example, secretory granules, cell-wall materials, and lysosomes—if lost or dedifferentiated, may be replaced, in the proper environment, by the interaction of the nuclear genes with the cytoplasm. But organelles that originated as microbial symbionts, if lost, can never be replaced by nuclear genes, just as no spontaneous generation of cells occurs today.

Any microbial symbiont transmitted from a host to its progeny must have replicated its own genes between divisions of the host. Potentially, then, the metabolic or morphological presence of the symbionts may be correlated with the genetic traits that they confer upon the host. This approach established that κ particles are bacteria. Most, probably all, nonnuclear (non-Mendelian) eukaryotic genetic factors, many known to be inherited uniparentally, have symbiont or viral origins. Examples include variegated plant plastids (Chapter 11; Correns, 1909), the *Drosophila* sex-ratio "gene," now known to be symbiotic bacteria (Oishi, 1971), μ particles in paramecia (Preer, Preer, and Jurand, 1974), the genetic determinant of lysine biosynthesis in *Crithidia oncopelti*, γ particles in *Blastocladiella* (Cantino and Myers, 1973), and perhaps even the B chromosomes of maize (see Margulis, 1976, for review). The correlation of feeding patterns with ultrastructural morphology has been used to infer an endosymbiotic origin for trichocysts in the protist *Cyathomonas truncata* (Schuster, 1968); whether hydrogenosomes (organelles which may lack DNA) are degenerate clostridial symbionts (Müller, 1988), aberrant mitochondria (Fenchel and Finlay, 1991), or neither is unresolved.

The symbiotic theory predicts that the phenotypic trait coded for on an organellar genome must be associated consistently with the parent donating it. If parental respiratory phenotypes can be distinguished in sea-urchin mitochondria, then one particular blastomere of the 32-cell stage of the developing larva should display the paternal phenotype; the 31 other blastomeres should display the alternate maternal phenotype. This prediction is based on the observation that, at the 32-cell stage of embryonic development in sea urchins, all the paternal mitochondria are contained in only one blastomere (Wilson, 1959).

Because of the conservatism of evolution, especially at higher taxonomic levels, free-living counterparts of the organelles are predicted to exist. The photosynthetic symbionts of the marine platyhelminth *Convoluta roscoffensis* are a striking example. These worms resemble green seaweeds (Figure 7-8). Each worm contains platymonad green algae (*Platymonas convoluta*), rounded and barely recognizable inside the tissues of the animal hosts. Released into sea water, the platymonads grow undulipodia, recovering their typical prasinophyte morphology. Swimming toward the worm eggs, they are eaten by young worms and not digested. Both intra- and intercellular, the algae invade the developing tissues of the worms, growing inside. The holobionts appear to be green seaweeds, but actually they are partnerships between marine worms and both intra- and intercellular platymonads. *Convoluta paradoxa*, a second photosynthetic species, harbors diatoms, which give the worms a yellowish color. In the *C. paradoxa* association, unlike others the symbionts can be identified. Even if unidentified, symbionts are expected to have genetic and physiological characteristics similar to those of known taxa.

Harmoniously growing partners in symbiotic relationships often grow more slowly than their free-living counterparts. Lichens, extremely slow-growing holobionts, are composed of fungi and algae, bionts that grow at typically rapid rates when unassociated (Ahmadjian, 1967). The characteristically slow doubling rate of eukaryotes (which is measured in days and hours) compared to that of prokaryotes (generally measured in minutes) is due probably more to their multiple symbiotic origin than to their size; holobionts grow more slowly than the unassociated bionts that comprise them.

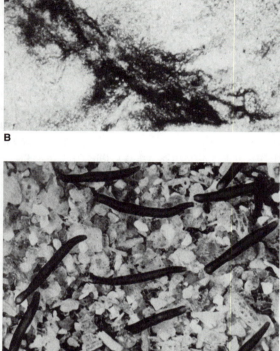

FIGURE 7-8

Convoluta roscoffensis, a photosynthetic worm.
(A) Patches of "seaweed" as seen in drainage ditches on the beaches of Brittany and the Channel Islands. **(B)** Close-up of drainage ditch. The "seaweed" tends to enter the sand at low tide. **(C)** Although fully photosynthetic, the "seaweed" is seen on close inspection to be a symbiotic complex of platyhelminth worm (*Convoluta*) and a prasinophyte green alga (*Platymonas*). [Courtesy of D. C. Smith.]

MIXOTRICHA MOTILITY SYMBIOSIS

I was raised in the belief that these [mitochondria] were
obscure little engines inside my cells, owned and
operated by me or my cellular delegates, private,
submicroscopic bits of my intelligent flesh. Now, it
appears, some of them, and the most important ones at
that, are total strangers.

L. THOMAS, 1974

Mixotricha, a motility symbiosis of protists and bacteria, is a model for the origin of undulipodia.

Mixotricha paradoxa is a parabasalian trichomonad found in the intestine of *Mastotermes darwinensis*, a distinctive Australian termite of the family Mastotermitidae. Because this protist is covered with thousands of moving, hairlike projections, it was thought at first to be a specialized hypermastigote (Phylum Zoomastigina). Closer inspection, however, revealed that the hairlike projections were swarms of spirochetes. Like other trichomonads, which are also zoomastiginas, *Mixotricha* has only a few undulipodia (four). Yet its straight-line movement through the termite intestine is not typical of termite protists. Cleveland and Grimstone (1964) described it in this way:

> Except when obstructed, *Mixotricha* glides along uninterruptedly, at constant speed and usually in a straight line. These characteristics, which have been repeatedly observed in living material and are shown very clearly in our ciné-films, are markedly different from those of other large flagellates [mastigotes], such as any species of *Trichonympha*, which swim at varying speeds, turning from side to side, changing directions, and sometimes coming to rest. The extreme smoothness of the movement is also in contrast to the jerky advance of other flagellates [mastigotes] which, like *Mixotricha*, have only a few anterior flagella [undulipodia]. The movement of *Mixotricha* is apparently unrelated to the activity of its flagella [undulipodia] and may continue even though the latter are motionless. The flagella [undulipodia] cannot, therefore, be responsible for locomotion, a conclusion which is scarcely surprising in view of the great disparity in size of the flagella [undulipodia] and the whole cell. It seems likely that their only function is to alter the direction of movement, and we have occasionally seen them do this.
>
> *Mixotricha* is propelled not by its flagella [undulipodia] but by the undulations of its attached spirochetes. The chief evidence for this assertion is that in all actively swimming individuals the spirochetes undulate vigorously and are well-co-ordinated, while in moribund individuals, in which movement is slow or absent, the spirochetes cease to undulate or do so in a slow and poorly co-ordinated manner. There is, in general, a clear correlation between the degree of activity and coordination of the spirochetes and the speed of movement of their host. In spite of the closest study no other possible means of locomotion has been detected.

Like most protist symbionts in dry wood–eating termites, *Mixotricha* cannot be cultured. Its taxonomic status has been clarified by ultrastructural observations, proving that *Mixotricha* is a bizarre symbiotic complex (Cleveland and Grimstone, 1964). *Mixotricha* has one forward-directed undulipodium, three trailing ones, and other structures typical of trichomonads (Figure 7-9A). Its peculiar movement is due to the treponeme-like

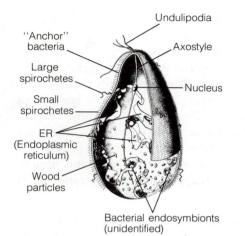

A

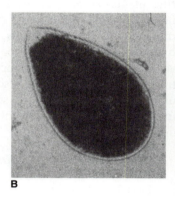

B

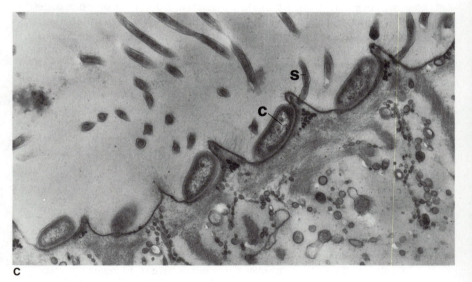

C

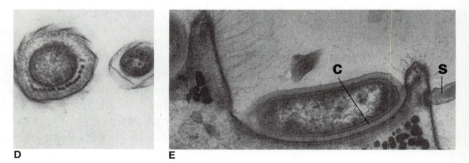

D

E

FIGURE 7-9

Mixotricha paradoxa.
(A) Drawing of *Mixotricha paradoxa*, motility symbiont complex found in the hindgut of *Mastotermes darwinensis*, the Australian termite. Three cortical symbionts (two types of spirochetes and one eubacterium) and one type of endosymbiotic bacterium are regularly found associated with *Mixotricha*, making it at least a pentagenomic organism. [Drawing by Christie Lyons.] **(B)** Live *Mixotricha*. Light micrograph from film, ×900. [From Cleveland, 1956]. **(C)** Cortex of *Mixotricha*, showing the alignment of cortical bacteria (*c*) and the small spirochetes (*s*) responsible for the forward movement of the protist. Transmission electron micrograph, ×45,000. **(D)** Cross sections through the large spirochete casually associated with the cortex of *Mixotricha* and the smaller treponeme-like spirochete responsible for the movement. Transmission electron micrograph, ×270,000. **(E)** Close-up of a protruding portion of the cortex of *Mixotricha*, a spirochete (*s*), and cortical bacterial attachment site (*c*). Transmission electron micrograph, ×78,000. [**C**, **D**, and **E** courtesy of A. V. Grimstone.]

attached spirochetes arranged regularly on its cortex. The four standard undulipodia are tiny compared to the hypertrophied cell. Acting as rudders, they cause occasional changes in the direction of movement (Figure 7-9A, B). About 500,000 small spirochetes, each connected to a specific site, are attached to each *Mixotricha* at a protruding portion of the host. An extracellular, symbiotic, rod-shaped bacterium is also attached to the protrusion (Figure 7-9C, E). Other, large, spirochetes, probably members of a new genus, are attached casually to the surface of *Mixotricha*; although they are motile, apparently they do not serve to move the complex.

Living, like other trichomonads, under the low oxygen tensions of the termite gut, *Mixotricha* lacks mitochondria. Each *Mixotricha* harbors, in addition to the surface symbionts, intracellular symbiotic bacteria, each surrounded by host endoplasmic reticulum and ribosomes. These intracellular symbionts, Cleveland and Grimstone (1964) suggest, may be functional equivalents of mitochondria. Perhaps they remove lactate or pyruvate and act to produce ATP.

Thus, each *Mixotricha* "cell" is a complex of at least five heterologous types of organisms: the protist trichomonad host; three different prokaryotic surface symbionts—large spirochetes, small spirochetes that produce the anomalous movement (Figure 7-9A, D), and their associated "anchoring" bacteria (Figure 7-9C, E); and the fourth symbiont, an endosymbiont. Furthermore, *Mixotricha* itself is a symbiont restricted to the intestine of its insect host.

Although *Mixotricha* is a well-studied motility symbiosis between spirochetes and other cells, it is not the sole example. Other, unidentified, protists associated with spirochetes from the intestine of the Florida termite *Kalotermes schwartzi* are shown in Figure 7-10. The peritrichously flagellated bacteria that form motility symbioses with devescovinids have been

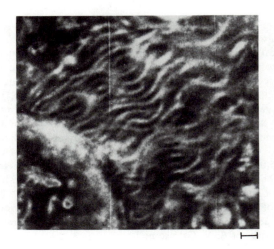

FIGURE 7-10

Spirochete–protist motility symbiosis. Both partners from the dry-wood-eating termite *Kalotermes schwartzi* are unidentified. Light micrograph from film of live organisms; bar = 1 μm.

FIGURE 7-11

Motility symbiosis between a protist and bacteria. This unidentified devescovinid protist "*Rubberneckia*" with its undulipodiated rod and fusiform bacteria is from the subterranean termite *Cryptotermes cavifrons* (Tamm, 1982).
(A) The devescovinid (*d*), which is about 100 μm in length. The parallel rows of rod bacteria run obliquely on the body surface (*b*) and more or less transversely on the head surface (*h*). Between the head and body is a belt of bacteria-free membrane (shear zones). Fusiform bacteria are not visible. *u* = undulipodium. Phase-contrast light micrograph. **(B)** Transverse section through the cortex of a devescovinid body. Rod bacteria (*r*) lie in specialized pockets that are coated with dense fuzz, while fusiform bacteria (*f*) are attached along their length to ridges of the host's surface. The exposed surface of each rod bacterium bears flagella; the insertion of one is seen on the rod bacterium on the left, with fuzzy glycocalyx. Flagella are missing, and the glycocalyx is reduced or absent on the part surrounded by the devescovinid pocket. Cell walls on the fusiform bacteria are grooved to match the host's ridge — like a tongue-and-groove joint. The bacteria divide in situ on the surface of the protist. Transmission electron micrograph, ×42,400. **(C)** Negative-stain preparation showing bacterial flagella (*f*) projecting from the edge of the whole-mount devescovinid. Transmission electron micrograph, ×10,000. **(D)** Detergent-isolated rod bacterium from the surface of the protist. Each rod bacterium bears about a dozen flagella (*f*), but only on the side that was not covered by the pockets in the surface of the protist. The flagella are about 4 μm long. The rod bacteria are about 3 μm long and about 0.75 μm in diameter. Undulipodial microtubule doublets (*mtd*) from the devescovinid are shown for comparison. [Courtesy of S. L. Tamm.]

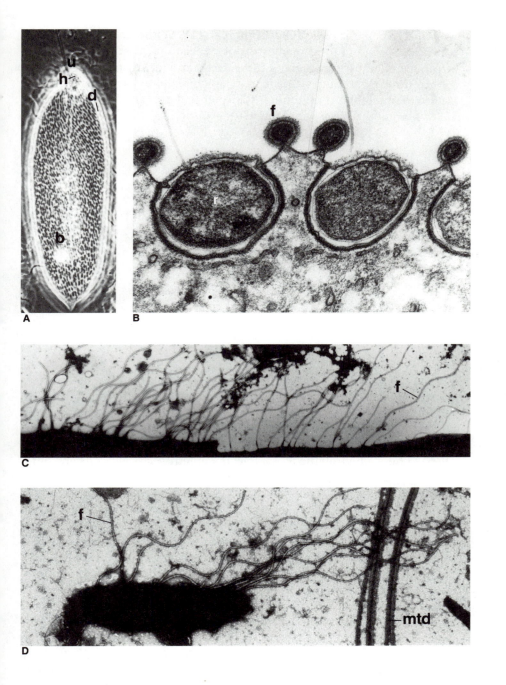

well analyzed by Tamm (1982) (Figure 7-11). Motile bacteria, including spirochetes, tend to form many sorts of associations, from loose to highly integrated. Complete sequences from irregular attachments with a protist surface through motility symbioses to "emboîtement" have been observed in termite microbial communities (Chapter 9; To, Margulis, and Cheung, 1978; Margulis, Chase, and To, 1979).

Photosynthetic ciliates

Chlorophyll bodies grow, are nourished, reproduce, synthesize proteins and carbohydrates, hand down their characteristics — all independent of the nucleus. In a word, they behave like independent organisms and therefore should be examined as such. They are symbionts, not organs.

K. S. MERESCHKOVSKY, 1910

(in Khakhina, 1979)

Paramecium bursaria and *Hydra viridis* show how heterotrophs can stave off starvation by association with algae, analogous to the relationship of plastids with the rest of the cell in which they reside. *Paramecium bursaria*, an algal–ciliate symbiosis, has been studied since the 1920s, when Pringsheim isolated and developed an inorganic culture medium for it (Lange, 1966). These photosynthetic paramecia contain relatively constant numbers of algal symbionts. In the light, the symbionts divide at the approximate rate of the host, maintaining fairly constant symbiont/host ratios — from about 60 to more than 2000, depending on the particular strain (Figure 7-12; Karakashian, Karakashian, and Rudzinska, 1968). When released from the paramecium these photosynthetic symbionts resemble free-living chlorellas. They may be cultured axenically on relatively simple media. *P. bursaria* fed on bacteria and kept in the dark tends to outreproduce its symbiont chlorellas; eventually, dark-grown paramecia produce offspring cells that lack symbionts. If fed and cleaned regularly, these "aposymbiotic" paramecia can be cultured indefinitely in the absence of chlorellas. In culture, the paramecia without symbiont retain morphological characteristics of *P. bursaria*, which in nature has never been found without the algae. When food is scarce and light plentiful, the growth rate of paramecia containing intracellular algae is significantly greater than that of the same strain experimentally deprived of the symbionts. Thus, in

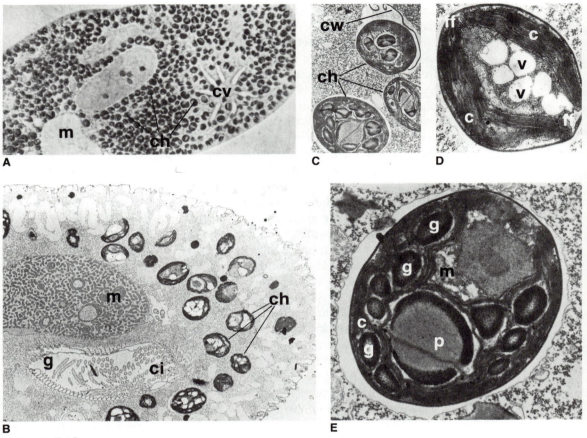

FIGURE 7-12

Paramecium bursaria, the photosynthetic ciliate.
(A) Live *Paramecium bursaria*. Hundreds of symbiotic chlorellas reside in the
cytoplasm. *ch* = chlorellas; *cv* = contractile vacuole; *m* = macronucleus. Light
micrograph, ×750. **(B)** Thin section showing the symbiotic chlorellas (*ch*) in the
paramecium gullet. *ci* = cilia; *m* = macronucleus; *g* = gullet cilia. Transmission
electron micrograph, ×2500. **(C)** The cell wall (*cw*) of a symbiotic alga is peeling
away into the algal vacuole, the membrane-bounded space surrounding the alga.
ch = chlorellas. Transmission electron micrograph, ×5600. **(D)** Symbiotic chlorella
in an early stage of division in its vacuole. The chloroplast (*c*) has divided already,
the pyrenoid (the center of starch deposition) has regressed, there is a fission furrow
(*ff*) at each end of the cell, and the vacuoles (*v*) lie on the incipient fission plane.
Transmission electron micrograph, ×15,300. **(E)** Mature chlorella in its vacuole.
c = cup-shaped chloroplast; *g* = ellipsoidal starch grains; *m* = mitochondrion;
p = pyrenoid. Transmission electron micrograph, ×14,900. [Courtesy of S. J.
Karakashian.]

nature, food scarcity probably selects for the symbiosis over the separate partners (Karakashian, 1963).

The relationship between each strain of *P. bursaria* and its chlorellas is specific. Paramecia are able to distinguish their endosymbiotic algae from free-living chlorellas and from chlorellas released by different paramecia or hydras. When aposymbiotic *P. bursaria* that has outgrown its chlorellas is placed in cultures containing chlorellas from various sources (including themselves), invariably it ingests its own chlorellas. It fails to establish symbiosis with certain other chlorellas, ejecting them or refusing to ingest them. Symbiosis between previously adapted partners, however, is reestablished easily—even if the previous adaptation was to a different type of host. The number of undigested chlorellas in an individual paramecium rises after reingestion; the chlorella are sequestered in vacuoles and resist digestion. Then the growth rate of the chlorellas tapers off to equal that of the host. Will aposymbiotic *P. bursaria* reestablish symbioses with chlorellas derived from sponges, hydras, or different *P. bursaria* strains? Generally, reestablishment occurs most easily with the original symbionts. However, for unknown reasons, in some experiments fast-growing new symbioses were established between paramecia and chlorellas released experimentally from green hydras (Karakashian, 1963; Karakashian, Karakashian, and Rudzinska, 1968). Experiments in separating and reestablishing the symbiosis of some strains of the freshwater coelenterate *Hydra viridis* with chlorellas similar to those of *P. bursaria* have also been performed (Cooper and Margulis, 1977; Cook, 1980). In some cases, heterologous algae both from paramecia and from distant strains of green hydras reestablished stable symbioses with hydras (Fracek and Margulis, 1979).

Genetic differences between symbiotic and free-living forms of both paramecia and chlorellas have been documented (Karakashian and Siegel, 1965). Sexuality in *P. bursaria* requires the presence of chlorellas. The complex ciliate micronuclear sexual cycle, in which meiosis to form gametic nuclei is followed by autogamy or conjugation, has never been observed or induced in *P. bursaria* that have been "cured" of their endosymbiotic algae, although sexuality in green forms is studied easily. Permanent genetic changes have occurred in both partners as a result of the symbiotic association; ultrastructural observations have corroborated this conclusion. The endosymbionts are now an integral part of the genetic endowment of the paramecia. Even if they could be made, careful distinctions between chlorella and host contributions to the genetic system are not necessary, as Karakashian and Siegel (1965) claim in a carefully reasoned paper. The intimate association between the host and symbiont is shown in Figure 7-12.

Because of the symbiosis, each cell in the green forms of *P. bursaria* should contain at least two different DNAs, one ciliate-specific and one chlorella-specific. Lysates of green paramecium cultures were sedimented to equilibrium in a cesium chloride density gradient (Margulis, 1963, unpublished results) [the buoyant density of DNA bands is expressed as the percentage of guanine–cytosine (GC) nucleotide pairs]. Two conspicuous bands were found: one sedimented at a position representing about 60 percent GC and the other at approximately 20 percent GC. The latter value is close to that of nuclear DNAs of other paramecia (Schildkraut et al., 1962), and the former is close to published data for chlorophytes (Sueoka, 1961); thus, the two DNAs were attributed to the chlorella nuclei and the ciliate nuclei. Perhaps 30 percent of *P. bursaria*'s DNA was from chlorellas. The large quantity of chlorella DNA may reflect the newness of the association. The fact that the cyanelles of *Glaucocystis nostochinearum* have only about 10 percent as much DNA as the free-living coccoid cyanobacteria from which they were derived may be interpreted as evidence for the same trend: genomic reduction in symbioses with time (Herdman and Stanier, 1977). Direct studies of DNA help in dating the symbioses, thereby estimating the degree of their interdependence (Edelman et al., 1967; Searcy, 1970; Herdman and Stanier, 1977). The difference between cultivable green algal symbionts and the uncultivable chloroplasts of algae and plants must be due, in part, to age; the chloroplasts have been obligate symbionts for far longer.

The hereditary endosymbiosis of *P. bursaria* has produced motile photosynthetic protists. Like *Convoluta roscoffensis* and its *Platymonas convolutae*, the *Hydra viridis*–chlorella association is a photosynthetic animal (Figure 7-13). Labeled carbon dioxide taken up by the chloroplasts of the algae can be detected quickly in the proteins of the heterotrophic hosts. The nutritional exchange from chlorellas to host is analogous to metabolite exchange between plastids and the nucleocytoplasm of algae and plants (Karakashian, 1963; Trench, 1979; Thorington, 1980; Smith and Douglas, 1989).

Searching among cyanobacteria for an ancestor to algae makes sense only within the direct-filiation paradigm. The symbiotic paradigm raises other questions: What autotrophs are ancestors of the chlorophyte plastids? By what heterotrophs were photosynthetic symbionts first acquired? How were hosts and symbionts modified after the establishment of the partnerships? Do hosts and symbionts preserve their genetic and metabolic autonomy? How many different times and with what partners did heterotrophs and phototrophic prokaryotes form symbioses?

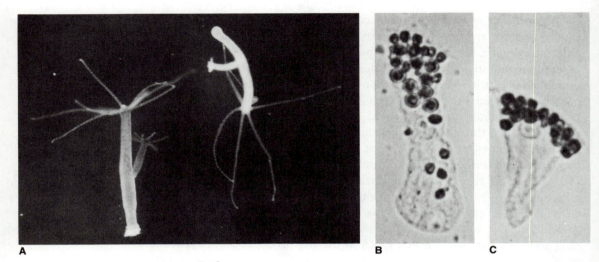

FIGURE 7-13

Hydra viridis, the photosynthetic hydra.
(A) Green freshwater hydra as found in nature (left) compared with white hydra "cured" of its intracellular chlorellas (right). If green hydras are exposed to intense light in the presence of photosynthesis inhibitors, they expel their symbionts. Light micrograph, ×30. **(B)** Gastrodermal cell from a bleached *H. viridis*, Carolina strain, about five hours after the hydra was fed chlorellas. If bleached hydras are fed a slurry of appropriate chlorellas, they ingest the algae, taking them into the gastrodermal cells. The algae migrate to the distal end of each cell. This migration is sensitive to drugs that inhibit microtubule polymerization (Cooper and Margulis, 1977; Fracek and Margulis, 1979). Light micrograph, ×3200. **(C)** Gastrodermal cell from *H. viridis*, showing the normal distribution of intracellular symbiotic algae at the peripheral end of the cell, where the light intensity is maximum. Light micrograph, ×3200.

PREDATORY PROKARYOTES

Predator-prey interaction is a mode of nutrition that implies a victor and a . . . victim. Because of the small size of the bacteria that display this mode . . . most biologists biologists are unaware of the ancient history of predatory-prey relationships prior to the existence of animals and even of protoctists.

R. GUERRERO, 1991

Predatory prokaryotes and other respiring eubacteria are models for mito-chondrial origins. Until the 1960s prokaryotic cells were considered impenetrable by other bacteria. Now, predatory prokaryotes are known. Penetration of their bacterial victims provides a model for the entrance into prokaryotes of other prokaryotic symbionts that became organelles. Descriptions of at least three types of predatory prokaryotes are relevant to the origins of organelles: bdellovibrios, capable of entering the periplasm of Gram-negative victims; daptobacteria, which can penetrate the plasma membrane and divide inside the cytoplasm; and vampirococci, which extract nutrient from their *Chromatium* hosts by attachment to the outer surface of the cell wall. Not only do these bacteria – bacteria interactions span the range of topological possibilities (periplasmic, cytoplasmic, and external), but they show different responses to ambient oxygen. *Vampirococcus* is an obligate anaerobe (and obligate predator) and *Daptobacter* is a facultative anaerobe (and facultative predator), whereas the various bdellovibrios (only remotely related to each other) are obligate aerobes (with high respiration rates) and obligate predators. These strategies for bacteria – bacteria interactions have implications for the origins of cell organelles, as described by one of their discoverers (Guerrero, 1991): Topologically, undulipodia resemble daptobacters most, since they are cytoplasmic, whereas mitochondria, which face the external side of the membrane, are topologically most like bdellovibrios. Since vampirococci are extracellular, their topology does not serve as a direct model for mitochondrial origins. However, they demonstrate the fact that obligate bacterial-bacterial symbiotrophic physical associations exist in anoxic environments, even today. The observation that bacteria are preyed upon by distinguishably different bacteria in extant microbial communities simply emphasizes the mechanisms for bacterial interaction available to an evolving biosphere as symbionts became organelles.

Bdellovibrio is an example of prokaryote – prokaryote cell "emboîtement" without phagocytosis. Like viruses, these tiny, flagellated bacteria form plaques when placed on lawns of Gram-negative bacteria. Their plaques, caused by lysis of host bacteria, resemble those of bacteriophages and, in fact, were discovered in the course of virological manipulations. Unlike those of phage on virus, however, bdellovibrio plaques keep growing until they have cleared the entire Petri plate. Each bdellovibrio cell, about 0.5 μm in diameter and vibrio-shaped, bears a single polar flagellum. Bdellovibrios have been found in soil, marine, and freshwater environments. Three types have widely varying GC percentages (28, 38, and 51 percent), indicating convergent evolution of the predatory habit (Stolp, 1979). Wild-type swimming bdellovibrios collide with, attach to, and penetrate their larger bacterial prey. Victims may be coliform bacteria, spirilli, or other eubacteria (Starr, 1975).

Once they have penetrated, bdellovibrios repair the outer membrane of their victim with lipid sealers and proceed to break down the cell, inducing

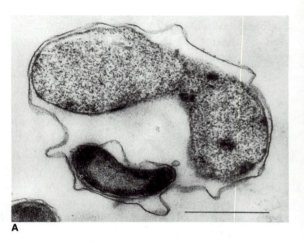

A

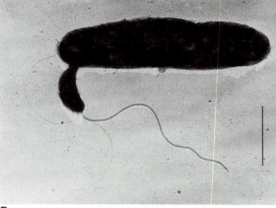

B

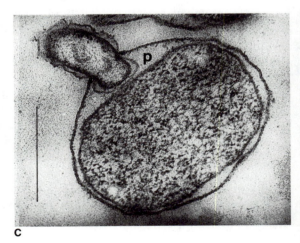

C

FIGURE 7-14

Bdellovibrio, a bacterial predator. In all
photographs, bar = 0.5 μm.
(**A**) A single parasite ensconced in the periplasm
of its host. Transmission electron micrograph. (**B**)
Negative-stain preparation of a *Bdellovibrio*
attached to a flagellated host rod. Transmission
electron micrograph. (**C**) A *Bdellovibrio* that has
penetrated partly into the host periplasm (*p*).
Transmission electron micrograph. [Courtesy of
H. Stolp.]

the host to form spheroplasts. They digest the peptidoglycan layer of the host cell wall and then reproduce, surrounded by the host's outer membrane, topologically outside the cell but inside the periplasm — the space between wall and membrane (Figure 7-14). Several bdellovibrio nucleoids and associated cytoplasm form, making long filaments which are then cut into standard-size bdellovibrios by the formation of cross-walls. The host lyses. Although mutant bdellovibrios have been isolated and grown on host-free media, their growth rate is far lower than the optimum achieved inside a host. Evidently, their nutritional requirements are complex. Since neither pinocytosis nor phagocytosis is known in any prokaryote, the entry of these respiring heterotrophic bacteria into the periplasmic space of other bacteria is an important model for the acquisition of precursors to undulipodia, mitochondria, hydrogenosomes, and even other organelles, prior to the evolution of the types of cell motility absolutely required for ingestion (for example, phagocytosis, pinocytosis, cyclosis, and endocytosis).

The most likely candidate for the ancestors of mitochondria is *Paracoccus dinitrificans*, a free-living, respiring, facultatively aerobic bacterium (John and Whatley, 1977b). The argument rests on the detailed similarity of the respiratory system of *Paracoccus* to that of animal and yeast mitochondria (Table 7-6; John and Whatley, 1977a). Not only are the quinones and cytochromes of the electron transport chain similar (Figure 7-15), but the three-dimensional spatial configuration of the cytochromes and the nature of the membrane F_0F_1-ATPase make *P. denitrificans* more similar to mitochondria than are other bacteria. Three conspicuous differences

TABLE 7-6

Mitochondrial features of *Paracoccus denitrificans*.[a]

RESPIRATORY CHAIN
  Succinate and NADH dehydrogenases
  Transhydrogenases and FeS proteins
  Two *b*-type and two *c*-type cytochromes, easily distinguished by spectroscopy
  Ubiquinone-10 is the sole quinone
  Cytochromes *a* and a_3 act as oxidase
  Sensitive to low concentrations of antimycin
  NADH oxidation is inhibited by rotenone and piericidin A
  Succinate oxidation is inhibited by carboxin and thenoyltrifluoroketone

OXIDATIVE PHOSPHORYLATION
  H^+ : O ratio of 8 with NADPH
  Respiratory control is released by ADP or by uncouplers of oxidative phosphorylation
  ATPase has tightly bound nucleotide, exchangeable on energization
  ATPase is inhibited by venturicidin and 7-chloro-4-nitrobenzo-2-oxy-1,3-diazole
  ATP synthesis, but not ATP hydrolysis, is inhibited by aurovertin

MEMBRANE PHOSPHOLIPIDS
  Phosphatidyl choline is the main constituent
  All fatty acids are straight-chain saturated and monounsaturated

[a]From John and Whatley (1977a, b). Many of these features are found in other bacteria but the complete list is unique to *P. denitrificans* and mitochondria, as far as is known.

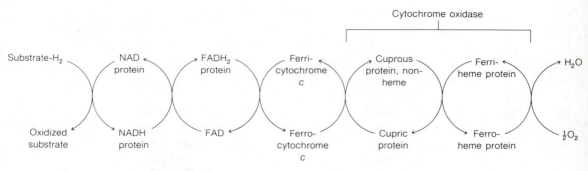

FIGURE 7-15

Electron transport chain common to mitochondria and *Paracoccus denitrificans*.

between mitochondria and *P. denitrificans*: the bacteria have cell walls, reduce nitrate and nitrite to nitrogen and nitrous oxide under anoxic conditions, and lack an ATP transport system. In mitochondria, the ATP system transports intramitochondrially produced ATP, releasing it to the rest of the cell. All three differences are explained easily if mitochondria evolved from *Paracoccus*-like ancestors. Cell walls and the capacity for nitrate reduction were lost because they were gratuitous in the oxidizing, salt-regulated, nitrogen-rich intracellular milieu. The release of ATP by *Paracoccus* into its surroundings would be analogous to throwing cash into the streets. The complex mitochondrial ATP transport system must have evolved with symbiosis. Based on the 16S ribosomal RNA gene, the ancestors of mitochondria belong to the "proteobacteria" [or the new group of purple bacteria as described in the introduction of Balows et al. (1991)] but have respiratory features like those of *P. denitrificans*. With origins in various lineages of respiring eubacteria, the probability of the polyphyly of protoctist mitochondria is high (Margulis, 1988). Molecular details of the respiratory system of bdellovibrios are not known. Perhaps if they were, some bdellovibrios would also be considered probable ancestors of some lineages of protoctist mitochondria.

GENETIC ANALYSIS OF SYMBIOSIS

The clearest and most direct interpretation of the [5S RNA, ferredoxin, cytochrome *c*] sequence data is provided by the symbiotic theory for the origin of eukaryotes.

R. M. SCHWARTZ and M. O. DAYHOFF, 1978

In view of the polygenomic nature of eukaryotic cells relative to prokaryotes, Taylor (1974) asserted that the word **cell** is inadequate. This simple monosyllable obscures the fact that prokaryotes are single protein-synthesizing units, whereas eukaryotes are multiple. Taylor suggested that "cell" be specified as **monad, dyad, triad, quadrad**, and so forth, depending on the number of different (heterologous) kinds of gene-directed protein-synthesizing units behaving as individuals in association (Figure 7-16). Most bacteria are units and rightfully would be called monads. The word **multicellular**, referring to organisms composed of many copies of unit cells, is also inadequate. It might also be replaced by a series of terms: **polymonad, polydyad, polytriad**, and so on. For example, if nucleocytoplasm, mitochondria, and plastids are each considered monads, plant cells should be referred to as triads and whole plants as polytriads. Multicellular bacteria —for example, myxobacteria and filamentous cyanobacteria—would be polymonads.

The traditional, and inadequate, classification of symbioses has focused on the relative growth rates of the partners (parasitisms, pathogenicities, mutualisms, and commensalisms) (Margulis, 1990). Some authors have classified symbioses by the nutritional modes of the partners and their mutual effects (necrotrophisms, biotrophisms, and so forth; Lewis, 1973a, b). However, relative growth rates and even nutritional modes are sensitive to environmental modulation. For example, in studies of amoebae, yesterday's bacterial pathogen became today's benign symbiont (Jeon, 1991; see Figure 10-3). Illuminated green hydras get their nutrients from *Chlorella*, photosynthetic symbionts; both in the dark and in the light, however, nutrients move from the hydra to the algae (Thorington and Margulis, 1981). Furthermore, in intense light, green hydras reject their symbionts (both the green algae and the symbiotrophic aeromonads; Berger, Thorington, and Margulis, 1979); in only a few days genetic and metabolic aspects of their associations change (Margulis, Chase, and To, 1979). Traditional terms are inadequate to describe such changing subtleties of real interrelationships.

Both polygenomic cells and symbioses are grouped easily by the number of their heterologous gene-directed, protein-synthesizing systems (Figure 7-17). Moreover, in Taylor's scheme, the description of symbioses raises questions amenable to solution by molecular biological techniques.

The polygenomic nature of eukaryotic cells is conspicuous more because of our familiarity with them than because of their intrinsic uniqueness; indeed, polygenomic, bounded individuals are rampant in nature. A continuum extends from loose ecological and functional associations entailing no direct contact, through temporal and spatial interactions so tight

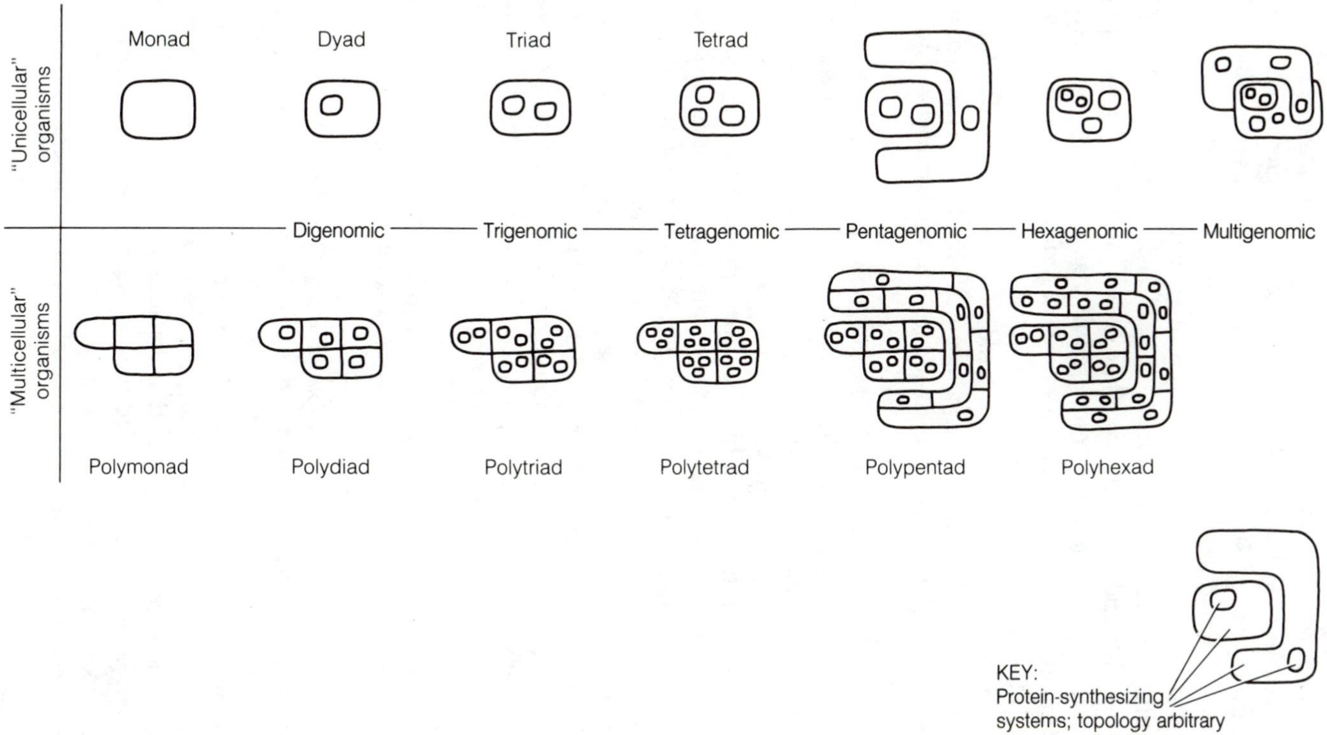

KEY:
Protein-synthesizing
systems; topology arbitrary

FIGURE 7-16

Terms for analyzing symbioses. [After Taylor (1974).]

	MONO-	DI-	TRI-	TETRA-	PENTA-	HEXA-	SEPTA-	POLYGENOMIC
MONADS	b *Bacillus*	b + b *Pelochromatium*	n + pls + mi *Chlamydomonas*	n + b + b + sb *Mixotricha*	n + mi + a *Paramecium bursaria*	a + a *Brennekella?*		n + mi + b + sb + b + pro *Reticulitermes*
POLYMONADS	b *Actinoplanes*	inv *Drosophila*	inv + b *Eubostrichus*	pla + b *Psychotria bacteriophila*	inv + a *Cassiopeia*	pla + pla *Cuscuta*	a + f + f *Sargassum natans*	
SUPERNUMERARY GENOMES	b + v *Escherichia coli*	cy + b *Lyngbya*	ve + v *Homo sapiens*	pro + b + v *Paramecium aurelia*	pla + ? *Zea mays,* *B chromosomes?*	n + mi + pls + pro + v *Peridinium balticum*	inv + sb + v + v *Drosophila*	

KEY:

Number of genomes, monads	Abbreviation	Type of organism
1,1	b	bacteria
1,1	sb	spirochete bacteria
1,1	cy	cyanobacteria
1,1	n	nucleocytoplasm
1,1	pls	photosynthetic plastid
1,1	mi	mitochondria
1,0	v	virus
2,2	ve	vertebrate animal (n-mi)
2,2	inv	invertebrate animal (n-mi)
2,2	f	fungi (n-mi)
3,3	a	algae (n-pls-mi)
3,3	pla	green plants (n-mi)
2,2	pro	protist

FIGURE 7-17

A matrix of symbioses.

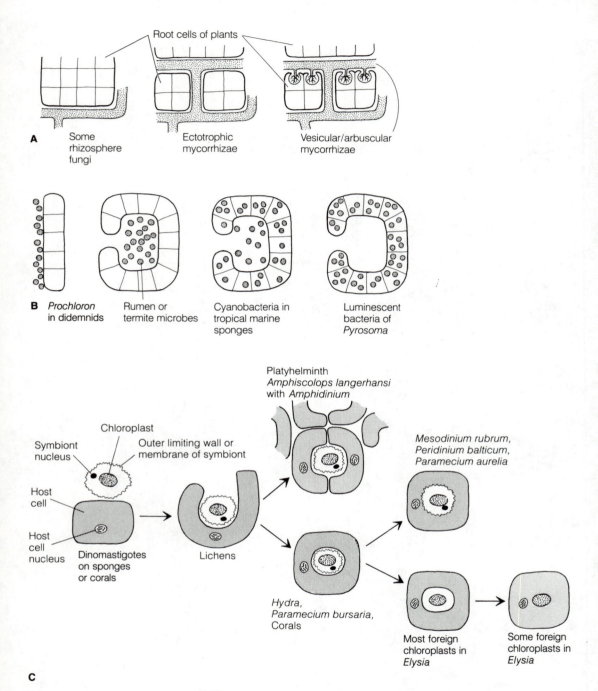

FIGURE 7-18

Ranges of intimacy in symbioses.
(A) Fungal (gray) and plant symbionts. (B) Protist or moneran (gray) symbionts with animal or plant hosts. (C) A hypothetical series of increasing intimacy, illustrated by actual symbioses. [Modified from Smith, 1979.]

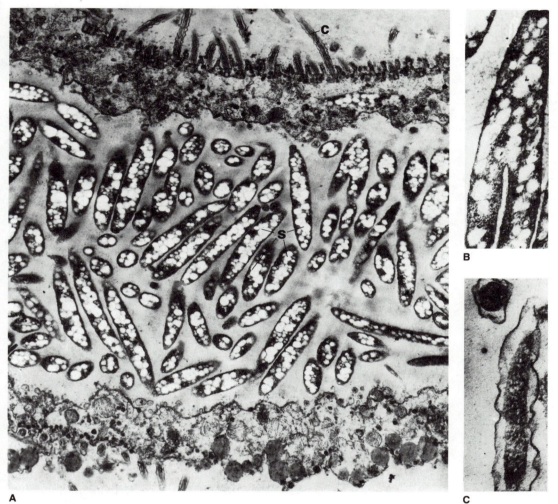

A   **B**   **C**

FIGURE 7-19

Kentrophorus, a ciliate culture chamber for anaerobic bacteria.
(A) The dorsal portion of the body of the ciliate *Kentrophorus fistulosum* Fauré-
Fremier facing the lumen of a tube in which symbiotic purple photosynthetic
bacteria (*s*) are cultured. The ventral cilia (*c*) are seen toward the top of the photo.
The attached symbiotic bacteria lack cell walls; they are engulfed gradually by
pseudopods on the dorsal portion of the ciliate. Transmission electron micrograph,
×7000. **(B)** Longitudinal division of a purple sulfur photosynthetic bacterium
cultured by the ciliate. Transmission electron micrograph, ×15,000. **(C)** Longitudinal
and transverse (upper left) section of a spirochete-like bacterium found attached to
the surface of *Kentrophorus*. These bacteria have also been seen attached to the
sulfur bacteria within the tube formed by the ciliate (Raikov, 1982). Transmission
electron micrograph, ×52,000. [Courtesy of I. B. Raikov.]

that their origins are obscured. The range of spatial associations found in heterotrophic – photosynthetic symbioses is diagrammed in Figure 7-18. An obscure, unexpected ciliate, at least a dyad, rolled into a tube inside which it cultivates phototrophic food bacteria, is shown in Figure 7-19.

The merging of genetic systems when symbiotic complexes form is, in many ways, analogous to sexual processes, which unite in a single individual genes that were once in separate genomes. The parasexual aspect is obvious in cyclical associations, in which the partners enter and reemerge regularly: parasitic flowering plants; legumes and their nitrogen-fixing rhizobia; the fungi and algae of lichens; the photosynthetic flatworm *Convoluta* and its alga *Platymonas*. Once bounded by a membrane, integument, or other continuous covering, autopoietic systems at more complex levels of organization appear. New individuals, evolutionary innovations that are the products of the merging of once independent genetic systems, then form. Natural selection, acting on the complex, retains and modifies the partnership, which leaves more offspring than do the unassociated bionts. If the association becomes permanent, it is an example of the origin of evolutionary novelty through symbiosis — exactly Mereschkovsky's (1909) "symbiogenesis" or Wallin's (1927) "symbionticism." But if dissociation occurs periodically, the reestablishment of the symbiosis is directly comparable to fertilization, and its dissolution, to meiosis. The only difference here between the meiosis – fertilization cycle and cyclical symbiosis is the length of time since the "parent" diverged from a common ancestor. In sexuality it is very short, from a few generations to some thousands. In symbiosis it is generally much longer; the "parents" may have diverged greatly. Eukaryotes and other stable polyads can be compared to schistosomes. These worms spend most of their life cycles mated — female stuck to male in the copulatory position. With rare exceptions, such as algae without plastids and anaerobic ciliates that have lost functional mitochondria, eukaryotes do not periodically lose their organelles. Even more than schistosomes, the eukaryotic cell lives and dies in permanent copulatory fusion!

Symbiosis, like sexuality, is a powerful force that brought together preadapted genetic combinations in the absence, at least initially, of mutation. Symbiotic complexes are major sources for genuine evolutionary innovation. Symbiont integration underlies adaptive radiation. In spite of their zealous overstatement, Famintzyn (1891), Mereschkovsky (1909), Portier (1918), and Wallin (1927) all realized correctly that symbiosis has played a crucial role in the origins of new species and through them, eventually, new higher taxa (Khakhina, 1979).

NAMES: BINOMIAL NOMENCLATURE AND PHYLOGENIES FOR SYMBIONTS

Generally things are ancienter than the names whereby
they are called.

HOOKER, 1597

Since the time of Linnaeus in the eighteenth century, biologists have used a two-name (*Genus* and *species*) Latin system for identifying organisms: *Homo sapiens* (human), *Canis lupus* (wolf), *Canis familiaris* (dog), *Canis latrans* (coyote), *Taraxacum officinale* (dandelion), *Ixodes dammini* (deer tick), and so forth. Geologists employ the same binomial system for fossils, although, in the cases of worm burrows, stromatolites, reef communities, and other trace fossils, the reader is made aware, by the term "form genus" or "form species," that the given "*Genus species*" binomial refers to the consortia, the community, or the traces of former behavior of the organism rather than to the body itself of an individual organism.

The irony of biological naming is the subliminal fact that, since all eukaryotes are coevolved communities (consortial structures of organisms derived polyphyletically from more than a single ancestor), all binomial nomenclature of protoctists, fungi, animals, and plants is "form" nomenclature. The attempt to distinguish "form fossils" from "body fossils" or "real genera" from "form genera" reverts to the specious.

Bifurcating dendrograms, genealogical trees, except in the case of prokaryote genera or when specified as "partial phylogenies" (that is, phylogenies for single classes of molecules), are, in principle, inadequate two-dimensional projections of the pluralistic ancestry of eukaryotes. Bushes with anastomosing branches would be more accurate representations. As soon as the symbiotic community in question becomes bounded by a common membrane of its own making — that is, as soon as it develops autopoiesis at a higher level of organization — the entity deserves, and by tradition is endowed with, its own name. The red-tide ciliate *Mesodinium rubrum* is an example (Lindholm, 1985). This ciliate has at least six heterogenomes: (1) macro-micronucleus of the *Mesodinium* ciliate; (2) nucleus of the cryptomonad; (3) nucleomorph of the cryptomonad-plastid complex; and nucleoids of (4) the plastid, (5) the tubular mitochondria of the *Mesodinium*, and (6) the flattened mitochondria of the cryptomonad.

Species names are changed as species differences become marked enough to be distinguished by the investigator. To me, the likelihood that the name changes are directly proportional to the acquisition of newly integrated heterogenomes [essentially a point realized already by Wallin (1927)] is immense. "Speciation" seems Darwinian and gradual to botanists and zoologists because microbial genomes are acquired, transferred, and integrated subvisibly. Furthermore, the integration of bacterial genomes by eukaryotes (unlike plasmid- or viral-level gene trading in bacteria) is irreversible. This irreversibility of integration confers legitimacy on the naming process and gives evolution through time its unidirectional property. By tradition, species binomials — biological names — refer nearly always to the most conspicuous member of the consortium: the member most likely to be encountered first by the scientist. Almost all lichens bear the name of the fungus, and ferns that of the sporophyte. Yet *Ascophyllum*, the "inside-out" lichen, is named for the alga. Of course, the cow (*Bos taurus*), rather than its cellulose-digesting, entodiniomorph ciliates, contributes the name, just as the mesodinia in *M. rubrum* do. We people tend to see and name only the most superficial or largest members of the consortium; we then act upon the self-deceptive construction that the consortium is an independent "individual."

CHAPTER 8

NUCLEI, MITOSIS, AND UNDULIPODIA

NUCLEAR ORIGINS

There is not so much Life as talk of Life, as a general thing.
Had we the first intimation of the definition of Life
The calmest of us would be lunatics.

E. DICKINSON, c. 1880
(in Sewall, 1974)

Eukaryotes can be defined only by the ubiquity of nuclei. No other class of organelles (mitochondria, plastids, peroxisomes, or undulipodia, for example) are universal in eukaryotes. How did this unequivocal symbol of eukaryotic status evolve? Nuclei, like all other organelles, originated either by differentiation (direct filiation) or exogenously, by symbiosis.

Double-membrane-bounded structures, extant nuclei bear octagonal pore complexes composed of several kinds of proteins. Spaced at some 11 pores per square micron, these pore complexes (with molecular weights of 50 thousand to 100 thousand kilodaltons) regulate the passage of macromolecules into and out of the nucleus through the aqueous pores at their centers. Evolution of the eightfold pore complexes from a membrane-attached genome, the number of which complexes is augmented by polyploidy, is outlined by Maul (1977). Nuclei are buttressed on the inside by a meshwork of intermediate filaments composed of no fewer than three kinds of nuclear lamin proteins (A, B, and C, with molecular weights from 60 to

70 kilodaltons), at least in mammals, where they have been studied most closely. The fenestrate nature and lability of the nuclear membrane is well known from its behavior throughout the mitotic cycle. Either the nuclear membrane disintegrates entirely at the beginning of mitosis to re-form later at telophase by interaction of the membranous materials at the surfaces of chromosomes, or, at the other extreme in many protoctists (those with closed mitosis), it remains intact throughout karyokinesis.

The argument that the nucleus was acquired as a symbiont makes no sense to me because, by itself, the nucleus falls far short of being an autopoietic system: it lacks any wherewithal for protein synthesis. To have originated symbiotically, a nucleus would have had to lose its accompanying protein-synthesizing system and gain that of its host — whereas the host would have had to lose its DNA replication apparatus, genophore, or nucleus. In this scenario one begins almost with the thing one is trying to evolve. If we apply the criteria for a symbiont origin of organelles (see Table 7-2) to the nucleus, we must reject unequivocally the hypothesis of an exogenous origin of that double-membraned enigma.

Nuclei contain two types of nucleic acid — chromosomal DNA and several RNAs. Cytoplasmic syntheses under nuclear control make possible the duplication of DNA in the nucleus and the growth of the cell prior to the next division. Messenger RNA, produced in the nucleus and matured by many editing processes, is transported through nuclear pores to the cytoplasm, where it determines the sequence of amino acids in proteins. Numerous chromosomal loci, some separated widely, control the production of both messenger and ribosomal RNAs. The RNA products of ribosomal genes generally accumulate at nucleoli and, at least in some amphibian eggs, are stored in large quantity (Ursprung et al., 1968). Thus the nucleus and cytoplasm of the eukaryotic cell are part of a single, integrated, gene-controlled, protein-synthesizing system: the nucleocytoplasm. Whereas I see this as an entity, because of molecular sequence data Sogin proffers an alternative: A protoeukaryote engulfed an archaebacterium, and the protoeukaryote plus the archaebacterial DNA formed a nucleocytoplasmic chimera. The archaebacterium provided most protein-coding DNA, whereas the translation apparatus and cytoskeleton derive from the protoeukaryote (Sogin, 1993).

Since Mereschkovsky (1909), the vague suggestion of a symbiotic origin of nuclei has not been developed seriously. By contrast, the various theories of endogenous origin bear scrutiny. Just the established fact that the outer nuclear membrane is continuous with the endoplasmic reticulum, which itself is a continuation of the rest of the endomembrane system, including the plasma membrane, argues for differentiation; that is, for direct filiation of membrane-wrapped DNA as nuclei from the original prokaryotic host cells.

Membranous structures that partition the genophore and other cytoplasm from ribosome-rich cytoplasm are evident in *Prochloron* (Whatley, 1977), as well as in several eukaryotes—for example, the nuclear cap containing the ribosomes of *Blastocladiella* and the chloroplast endoplasmic reticulum of many brown algae. In *Pelomyxa*, special membranous tubules, apparently continuous with the endoplasmic reticulum, extend from the membranes of perinuclear endosymbiotic bacteria to the nuclear pores of the host. These tubular elaborations probably ensure proper distribution of the bacterial symbionts when the nuclei divide (Whatley, and Chapman-Andresen, 1990). Together, such observations suggest convergent evolution of internal membrane elaborations, including those of membrane-bounded nuclei. The nucleocytoplasm may have evolved together as a reticulating unit, the nuclear membrane itself appearing along with the rest of the endomembrane system.

The tendency of prokaryotes to proliferate internal membrane systems is indisputable: autotrophs produce surface membranes reversibly in response to the availability of light and metabolites; endospore-forming bacilli and clostridia engulf and wrap the genomes protected in membranes during sporulation; and, most dramatically, an obscure eubacterium, the freshwater, non–spore-forming, heterotrophic *Gemmata obscuriglobus*, wraps its entire nucleoid in a continuous membrane structure, making the wrapped nucleoid closely resemble the bona fide nucleus of eukaryotes (Fuerst and Webb, 1991). Of course, the nucleoid membrane of *Gemmata* lacks the inner-membrane nuclear lamina structure and accompanying octagonal pore complexes. The *Gemmata* genophore is certainly not organized into histone-bearing chromosomes; no evidence suggests that *Gemmata* is anything but a standard prokaryote, emphasizing to me the probability of convergent evolution of nuclei in general.

In aged cultures of *Oscillatoria* (sheathed, filamentous cyanobacteria), Jensen (1989) noted a strong tendency of the thylakoid membranes to embrace the nucleoids and interpreted this wrapping behavior as evidence for the direct-filiation mode of nuclear origin. Jensen (parting company with me, M. W. Gray, and others) even sees internal membrane proliferation with its DNA association as a preadaptive behavior for the endogenous origin not only of nuclei, but also of other organelles (plastids and mitochondria).

A second model for the endogenous origin of the nucleus is presented by Gould and Dring (1979). These authors describe the engulfment of its own bacterial cytoplasm during endosporulation in *Clostridium*, *Bacillus*, and *Thermoactinomyces* and suggest that endospore formation is a legacy of self-engulfment in bacteria that led, in some lineages, to eukaryotes.

Although a most problematic aspect of the origin of eukaryotes is this detail of the history of the nucleus, it appears quite likely to me that the tendencies toward membrane-wrapping and engulfment do exist in pro-karyotes. Membrane hypertrophy, probably aggravated by infection and predatory attack, was supplemented by heredity symbiosis, in my opinion. Evolution of the mitotic nuclei of eukaryotes required much more than membrane wrapping of bacterial DNA genophores. Chromatin is coiled into chromosomes, which are propelled internally by the mitotic spindle – microtubule system, the subject of this chapter and the next.

My current "best scenario" is that nuclei are polyphyletic in protoctists; in some amoebae and other anaerobic sexless protists, the nucleus evolved just as it did in the prokaryote *Gemmata*. In most eukaryotes, however, attack by surface spirochetes that became the mitotic apparatus led to dynamic membrane hypertrophy and the karyomastigont systems.* With time, the various components of the endomembrane system (outer mem-brane of the nucleus, endoplasmic reticulum, outer membrane of other organelles, and the Golgi apparatus), differentiated from the DNA-asso-ciated plasma membrane in the adaptive radiations of early eukaryotes (the first protoctists). Mitosis, then, is a chimeric process, the establishment of which involved symbiosis (in the origin of certain elements of the motility system such as centrioles and microtubules) as well as direct filiation (as in the differentiation of nuclear membranes).

Variations on the mitotic theme in extant protoctists emphasize the protracted evolution of the mitotic process. Mitochondria and plastids were acquired during and after the evolution of mitosis and meiotic sex in heterotrophic protists. The integration of multi-component microbial sys-tems (which process formed the eukaryotic cell and, indeed, is **eukaryosis**) occurred in a specific order, with corresponding identifiable products: first, the two-component system (undulipodia – mitotic microtubules/mitochon-dria) integrated to become anaerobic zoomastigina mastigotes and amoe-bae; then followed integration of the three-component system with forma-tion of the nucleus (undulipodia – mitotic microtubules/nucleocytoplasm/ mitochondria) to produce ciliates, foraminifera, amoebomastigotes, and so forth; finally, integration of the four-component system (undulipodia – mitotic microtubules/nucleocytoplasm/mitochondria/plastids) led to the formation of most algae.

The genetic, metabolic, and behavioral integration of two or more membrane-bounded autopoietic entities (bionts, free-living cells) into a membrane-bounded autopoietic "individual" (holobiont, heterogenomic individual, of greater biological complexity) is the crucial distinctive fea-ture of the origin of eukaryotes. The integrative process differentiates a community from an individual as it does a chromonemic prokaryote from a chromosomal, polyheterogenomic eukaryote. Eukaryotes are not distin-

---

*Organellar complex including kinetids and their undulipodia, nuclei and often other fibrous and tubular structures (in parabasalians, *Giardia* and protists).

TABLE 8 - 1

Correlation of species with symbiont acquisition in a marine flatworm.

GENUS	SYMBIONT	COLOR	SPECIES
Convoluta	*Prasinomonas*	green	*C. roscoffensis*
Convoluta	diatom	yellow	*C. paradoxa*
Convoluta	none	translucent	*C. convoluta*

guishable simply because of their membrane-bounded nucleus or their unique 16S ribosomal RNA gene sequences. Rather, because they resulted from symbiogenesis, they are all products of an irreversible integration of prokaryotes into a new individualized symbiotic unity. Speciation, a property of eukaryotes only, results from the integrative process itself (Table 8-1). In their single-genome systems, prokaryotes, autopoietic entities that flaunt their reversible sexuality, differ in principle from eukaryotes. Bacteria cannot form stable species or isolated lineages; therefore, in their evolutionary history, they have suffered no extinction and, until some of them became eukaryotes, they demonstrated very little directional evolution through time (Sonea, 1991).

PROTOCTIST CELL DIVISION

Dans le cytoplasme de plusiers espèces de cellules on a décrit depuis long temps un ou plusiers petits corps arrondis en forme de bâtonnet, réfringents, plus dense que le cytoplasme et se colorant plus fortement sons l'influence des réactif qui colorent ce denier: ce sont les *centrosomes* ou *centrioles*. . . . Au moment de la division de la cellule, le centrosome et sa sphère attractive jouent un rôle important comme on le verra plus loin. . . . Dans les cellules amiboides, telles que les leucocytes, le centrosome occupe le centre de la cellule et de ce centre partent des lignes rayonnants nombreuses s'étendant jusqu'à la périphérée; dans les cellules á clls vibratiles, á la base de chaque cil se trouve un centrosome (corpuscle basal, blépharoplaste); dans le spermatozoide le centrosome est suité entre la tête et la queue.*

L. F. HENNEGUY, 1923

No one who sees mitosis fails to marvel at its living elegance. Lacking physicochemical explanation, mitotic phenomena is fascinating because of its intrinsic logic.

Recombination of DNA molecules evolved in prokaryotes. The enzymatic repair of DNA ruptured by ultraviolet light may have been a preadaptation for the various forms of bacterial genetic recombination, such as phage transduction and conjugation. Mitosis, "the dance of the chromosomes," evolved in protists, as did meiosis, a variation that evolved in organisms that were already mitotic.

The nuclei of most animal and plant cells contain two full, homologous, but not identical sets of chromosomes with 10^4 to 10^6 genes per set. Cells whose nuclei contain two sets of chromosomes are **diploid**; cells whose nuclei have a single set are **haploid**. In mitotic division, each offspring cell receives a copy of each of the parent cell's chromosomes. Thus, cells produced by mitosis are either diploid or haploid depending on the ploidy level of the parent cell. Meiosis distributes the chromosomes so that each offspring cell has exactly one member of each of the homologous chromosome pairs originally present in the parent. Thus, cells produced by the meiosis of a diploid cell are always haploid. In fertilization, the fusion of two haploid cells (each the product of meiosis) reestablishes diploidy. Thus, meiosis and fertilization lead to new assortments of chromosomes. Meiosis may include crossing-over, the exchange of DNA between homologous chromosomes, and reassortment can take place even without crossing-over. However, any variation of mitosis that leads to imprecise segregation of chromosomes or loss of genetic material is selected against immediately.

Mitosis requires the close coordination of two cell systems: one doubles the quantity of the chromosomal DNA by synthesis, and the other forms the mitotic spindle that moves chromosomes to the opposite sides of the dividing cell. The spindle is composed of spindle fibers, centrioles, centriolar plaques or other "mitotic centers," and spindle attachment sites, **kinetochores**, generally on chromosomes. The details of mitosis and meiosis

---

*In the cytoplasm of several species of cells, one or several small round rod-shaped refringent bodies have long been described, denser than the surrounding cytoplasm and more strongly staining: these are the *centrosomes* or *centrioles*. . . . At the time of cell division the centrosome and its surrounding sphere plays an important role as we shall see later. . . . In amoeboid cells, such as leucocytes, the centrosome occupies the center of the cell and from its numerous radiating lines extend to the periphery; in cells with motile cilia, a centrosome (basal body, blepharoplast) is found at the base of each cilium; in sperm the centrosome is found between the head and the tail.

vary considerably, especially in protoctists. Their functions, however, do not vary: mitosis ensures equal distribution of genetic material to offspring, and meiosis produces haploid cells from diploid ones. Fertilization, the fusion process that reestablishes diploid chromosome numbers, must intervene between meioses in any stable life cycle. The diploid–haploid cycle, detected by the observation of meiosis and fertilization, is the surest indication that standard mitosis has evolved in a group. Mitotic organisms may be recognized by their typical genetic patterns, called Mendelian, whether or not direct observations of chromosome behavior exist.

Microscopic studies of live cells in division suggest that meiosis and fertilization evolved analogously in different protoctists. The original selection pressure was probably toward the reduction of diploid sets of chromosomes that arose by fusion of two haploid sets (Cleveland, 1947). In animals and plants, mitosis and meiosis follow the typical textbook blueprint (see Figure 3-8); in fungi, however, mitosis and meiosis have often been extremely difficult to study, and various odd types of mitosis in protists have been known for years (Figure 8-1; Wilson, 1959). The presence of mitotic variations, which occur frequently in poorly known protoctists, may be independent of the type, color, and pigment composition of their plastids. Such idiosyncratic patterns of cell division are interpreted as legacies of the origins of mitosis. Information relevant to the evolution of mitosis and meiosis exists in plants, animals, and fungi; now that protoctists formerly regarded as plants have been united properly, a comprehensive interpretation will require the collaboration of many biologists. The problem is exacerbated because different nationalities and disciplines often use mutually unintelligible, specialized terminology for the same structures and processes. Examples include cryptomitosis (protoctist cell divisions differing from that of animals), archeoplasmic spheres (spherical bodies appearing around the spindles of certain mastigotes in division), blepharoplasts, kinetosomes or kinetosomal precursors), and NAOs (nucleolar-associated organelles, microtubule centers in fungi). To enhance understanding, we have prepared the *Illustrated Glossary of Protoctista* (Margulis, McKhann, and Olendzenski, 1992).

A direct relationship between undulipodia and mitotic centrioles (Figure 8-2) has been known for many years, although the meaning of the observations is debatable. All undulipodia grow forth from kinetosomes (unfortunately also called basal bodies) (see Figure 2-4). Kinetosomes develop from mitotic centrioles in many cases, but by no means all. In some ciliates short kinetosomes develop from preexisting kinetosomes: the offspring [9(3) + 0] kinetosome appears at a 90° angle, then a 45° angle, and then aligned parallel to the parent. A kinetosome reproductive-developmental cycle (Figure 8-3), a direct division of elongated centriole-kineto-

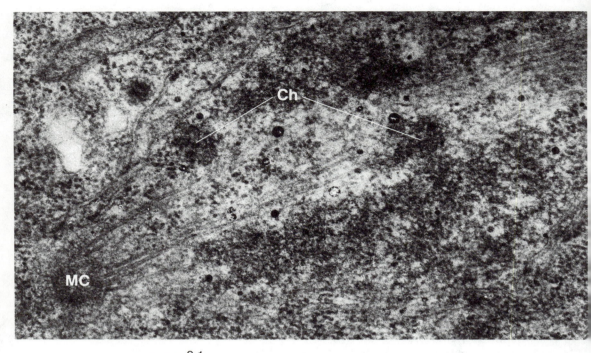

FIGURE 8-1

Mitosis in the basidiomycote fungus *Armillaria mellea.* Note the centriole-like but amorphous bodies at the poles. Presumably, these are MCs (microtubule-organizing centers). *Ch* = chromatin. Transmission electron micrograph, ×54,000. [Courtesy of J. Motta.]

somes, or the de novo appearance of the centriole-kinetosome all occur in different organisms. Electron microscopic observations unified much of this information: centrioles and kinetosomes are the same structure. They, the undulipodia, mitotic spindle fibers, and other related structures are all now known to be composed of microtubules, long hollow structures constant in diameter. Microtubules are composed of several closely related proteins called microtubule proteins, or tubulins. Microtubules compose not only the mitotic structure, but also an array of other organelles. They often develop from obvious centers, such as kinetosomes. Centrioles and kinetosomes seen with the electron microscope have the same [9(3) + 0] arrangement of microtubules in cross section. Centrioles are often at the poles of the mitotic spindle during mitosis in protists and animals, but

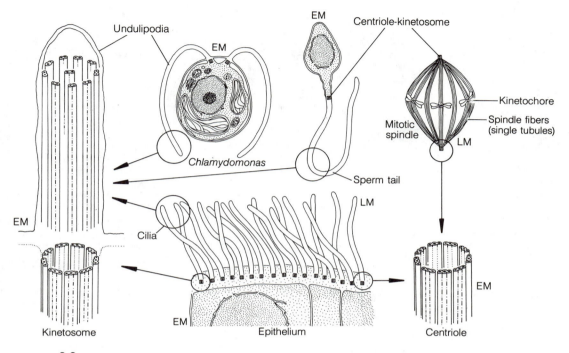

FIGURE 8-2

Homologies in microtubule systems: centriole-kinetosomes and undulipodia from algae sperm and ciliated epithelial tissue. Circles indicate sources of magnified images at arrows. LM (light-microscopic) and EM (electron-microscopic) images. [Drawing by Kathryn Delisle.]

centrioles *sensu stricto* are not required for normal mitotic division, since many organisms produce conventional spindles but lack [9(3) + 0] centrioles. Some organisms apparently replace centrioles with analogous structures, such as the "centrosomes" of some basidiomycotes and the centriolar plaques of yeast and red algae (Pickett-Heaps, 1974). Plants lack centriole-kinetosomes at the poles during mitosis. Apicomplexans have a polar cone structure resembling standard centrioles-kinetosomes but clearly unique to the group (Figure 8-4). To unify these observations, Pickett-Heaps (1974, 1975) developed the concept of the **microtubule-organizing center**, or **MC** (also often abbreviated MTOC); whether they are centrioles, their analogues, or amorphous granulofibrillar material, the structures or loci that

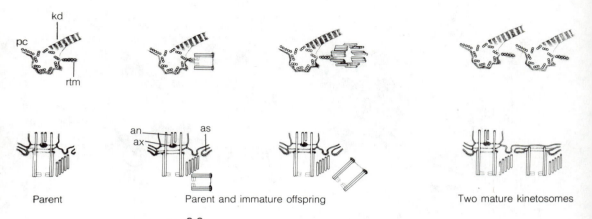

Parent   Parent and immature offspring   Two mature kinetosomes

FIGURE 8-3

Reproduction of kinetosomes as part of kinetids. *pc* = postciliary microtubules; *kd* = kinetodesmal fiber; *rtm* = radial transverse microtubules; *an* = axoneme; *ax* = axosome; *as* = alveolar sac. At left, parent kinetid only; after development, at right, both parent and mature offspring kinetids can be seen with kinetosomes parallel to each other. Top row: transverse sections; bottom row: corresponding longitudinal sections. [Drawing by Laszlo Meszoly.]

give rise to microtubular arrays are MCs. Kinetosomes, by definition, are centrioles that have shafts (axonemes of undulipodia) emerging from them. In no protoctist, animal, or plant cell are centriole-kinetosomes isolated structures. They are always associated with fibers, tubules, granules, axonemes or other species-specific features visible with the electron microscope. The centriole-kinetosome morphological complex, often useful as a taxo-

FIGURE 8-4

Mitosis in gregarines and microsporidians.
(**A**) Centrocone at pole in the mitosis of a gregarine (Phylum Apicomplexa). Gregarine mitoses are initiated by the duplication and differentiation of a peculiar structure called the centrocone (*c*), presumably an MC. It consists of a conical microtubular bundle, the top of which is a centriole made of nine singlet microtubules surrounding a single axial tubule (9 + 1). The distal portion of the microtubule bundle contacts the nuclear envelope (*ne*), whose disorganization is inversely related to the degree of development of microtubules. In most of the Apicomplexa the membrane remains nearly intact (closed mitosis) because of the poor development of microtubules. In some gregarines, the more extensive development of microtubules leads to a mitosis nearly the same as that of metazoa. Transmission electron micrograph, ×40,000. (**B**) Mitosis in the microsporidian (Phylum Microspora) *Stempellia mutabilis* showing plaques (*p*) at the poles of the spindle. *ch* = chromosomes; *mt* = microtubules of spindle. Transmission electron micrograph, ×15,000. [Courtesy of S. Molon-Noblot and I. Desportes.]

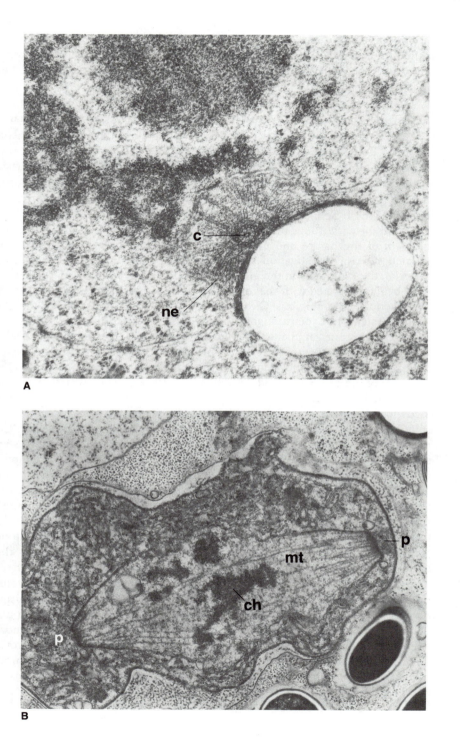

A

B

nomic or evolutionary indicator of relationships between organisms, is called the **kinetid**. Kinetids, by definition, contain at least one centriole-kinetosome; some contain two (dikinetids) and others more than two (polykinetids).

Structural and biochemical techniques now permit investigators to reinterpret the mitotic literature in an evolutionary context (Copeland, 1956; Cleveland, 1957; Grell, 1967; Ris and Kubai, 1974; Kubai, 1975; Ris, 1975). These investigators are academic descendants of the cytologists who, in the late nineteenth and early twentieth centuries, were concerned with the origin and evolution of mitosis but lacked the ultrastructural observations and biochemical techniques required to attack the problem (Belar, Dobell, Schaudinn; for review see Wilson, 1959).

From the early 1930s until his death in 1971, Cleveland struggled with the problem of the nature and evolution of mitotic and meiotic processes. His isolated style of investigation and his choice of unique and poorly known material, the symbionts in the guts of wood-eating insects, prevented his work from being integrated with that of other cytologists during his lifetime, yet his contribution was prodigious. Cleveland (1935, 1938, 1963) established that the mitotic systems of hypermastigote protists were functionally equivalent to the more orthodox ones of plants and animals. He devised experiments that distinguished nuclear chromosomes from the mitotic motility system that moves them. The latter, we know now, is a part of the microtubular–undulipodial structure of the cell. Cleveland described in *Trichonympha* a structure that he named the **long centriole**, showing how this structure replicates and forms the locus for the outgrowth of the mitotic-spindle fibers. We now recognize the spindle fibers as bundles of microtubules and the long centrioles as MCs. Although the structures function as centrioles, Grimstone and Gibbons (1966) showed that they are far larger than standard [9(3) + 0] centrioles. Complex bars of proteinaceous material at the anterior end of the cell, long centrioles are equivalent to the atractophores of several mastigotes described by Hollande and Carruette-Valentin (1970, 1971). In fact, the entire dipartite anterior (rostral) region in both *Trichonympha* and *Pseudotrichonympha* functions as mammalian centrioles seem to: the two apposed halves of the rostrum behave as mitotic poles, anchoring the mitotic spindle. In these and other hypermastigotes, the undulipodia are connected distally to the rostral halves, and they flail as division proceeds (Figure 8-5). The spindle, composed of typical microtubules, forms proximally between the rostral halves. There are few organisms besides Cleveland's hypermastigotes in which direct connections between undulipodia and spindle are seen. Most are obscure protists unfamiliar to those who work with animal and plant tissue. However, further work on microtubule structure and composition has confirmed earlier observations of a direct relationship among undulipodial, mitotic, and other microtubular systems in eukaryotes.

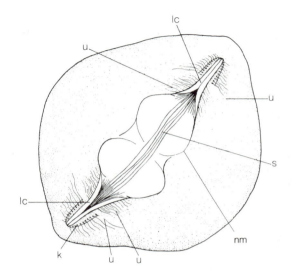

FIGURE 8-5

Spindle formation and cell division without chromosomes. The chromosomes of *Barbulanympha* have been destroyed by high concentrations of oxygen. *Barbulanympha*, a major symbiotic hypermastigote of wood-eating cockroaches, is similar to the termite hypermastigotes *Trichonympha* and *Pseudotrichonympha*. Even in the absence of the chromosomes, the cells produce the entire motile apparatus, "long centrioles" (*lc*) and spindles (*s*), which develop in a close relationship to the undulipodia (*u*), kinetosomes (*k*) and nuclear membranes (nm) (see Figure 3-3**B**). This cell division demonstrates the autonomy of the mitotic system. [After Cleveland (1956).]

CENTRIOLE-KINETOSOMES AND MICROTUBULES

In addition to the achromatic figure all the other extranuclear organelles, such as the flagella [undulipodia], parabasals, axostyles, etc., arise from the centrioles, so that the centriole is clearly an autonomous organelle, and the dynamic center of the cell since it reproduces itself and all the other organelles except the nucleus.

L. R. CLEVELAND, 1935

Microtubule research began long before microtubules were discovered.

D. MAZIA, 1975

Microtubules, hollow and composed of tubulin, are virtually universal constituents of eukaryotic cells (Little and Seehaus, 1988). They grow out from centrioles or other less obvious MCs to disappear and reappear at certain stages of cell development. Tubulin microtubules are conspicuously absent in prokaryotes (but see pages 294–303). Some microtubule-based structures are listed in Table 8-2. The homology of tubulin microtubules and their derivative structures is now accepted widely and has been reviewed well (Borgers and DeBrabander, 1975; Dustin, 1978).

TABLE 8-2

Some structures composed of or containing microtubules.

STRUCTURE	ORGANISMS
Undulipodia and their kinetosomes	Ciliates, vertebrate epithelium, mollusk gill, sperm tails (of ginkgo trees, mosses, male vertebrates, and many others), green algae, hypermastigotes, and zoospores of water molds
Axons and dendrites	Nerve cells of metazoans
Kinocilia	Ears, balance organs of vertebrates (see Figure 8-8)
Olfactory antennules	Crustaceans
Mechanoreceptors	Insects
Axopods (feeding organelles)	Heliozoa, Phaeodaria, Acantharia, Foraminifera (see Figures 8-17, 8-18C)
Oars (rowing organelles)	*Sticholonche*
Axostyles (motile organelles)	Pyrsonymphida (see Figure 8-18C)
Melanocyte processes	Pigment-bearing cells of metazoans
Haptonemes	Prymnesiophyes (coccolithophorids)
Mitotic spindle, asters	Nearly all eukaryotes (see Figures 8-1, 8-2, and 8-4)
Holdfasts (stalks) and tentacles	*Carchesium*, vorticellids and suctorians (ciliates)

Microtubules (Figure 8-6) are 24 nm in diameter and have walls 11 nm thick; they are made of helically arranged subunits, 13 per turn, composed of the dimeric tubulin protein. Approximately 60 tubulins have been sequenced, verifying that microtubules are, chemically, strikingly uniform. They are composed of two closely related proteins called α- and β-tubulin, each having a molecular weight of about 55 kilodaltons. Unique γ-tubulin is reported in *Aspergillus* fungi (Zheng, Jung, and Oakley, 1991). These tubulins are often associated with other proteins called **MAPs (microtubule-associated proteins)** — for example, dynein, the Mg^{2+}-sensitive ATPase protein. Microtubules are commonly sensitive to temperature and pressure change. For example, high hydrostatic pressures tend to dissolve them; restoration of normal pressure permits them to re-form. Microtubules generally are stable at 37° C and go into solution at about 4° C. They are also sensitive to calcium ion concentration. Not all of these features are observed in all microtubules, but enough work has been done to reveal homologies (Dustin, 1978).

Microtubule stability is enhanced by taxol. Many compounds inhibit the polymerization of tubulin into microtubules, including those in Figure 8-7 and many of their derivatives (Deysson, 1968; Malawista, Sato, and Bensch, 1968; Wilson, 1975), as well as certain carbamates, nocodazole, and others (Borgers and DeBrabander, 1975). Sensitivity to drugs and to physical conditions varies with concentration, species, developmental stage, and many other factors, but, in general, these microtubule-inhibiting compounds dissolve the mitotic spindle and cause cessation of chromo-

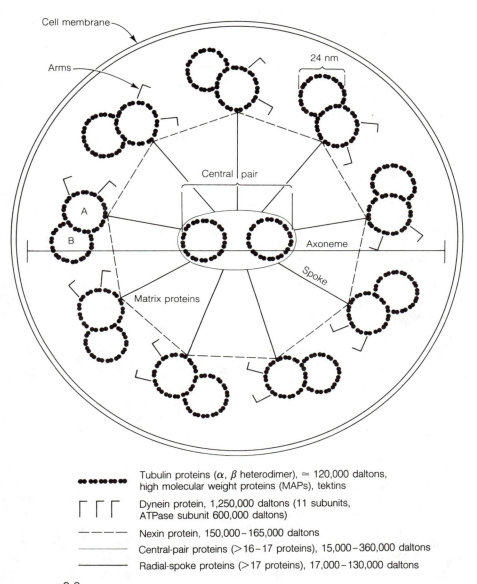

Tubulin proteins (α, β heterodimer), $\approx$ 120,000 daltons, high molecular weight proteins (MAPs), tektins

Dynein protein, 1,250,000 daltons (11 subunits, ATPase subunit 600,000 daltons)

Nexin protein, 150,000–165,000 daltons

Central-pair proteins (>16–17 proteins), 15,000–360,000 daltons

Radial-spoke proteins (>17 proteins), 17,000–130,000 daltons

FIGURE 8-6

An undulipodium and its microtubules. Diagrammatic transverse section through the [9(2) + 2] axoneme.

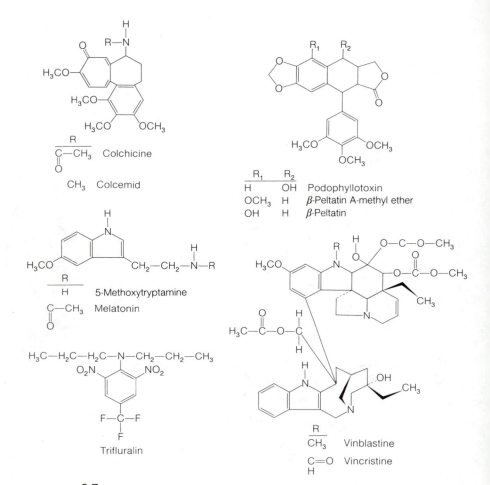

FIGURE 8-7

Structures of some microtubule inhibitors. These methoxylated compounds inhibit the polymerization of microtubule protein into microtubules. Colchicine and Colcemid affect plants and animals but generally not fungi. Melatonin and trifluralin affect plants and some protists but, apparently, not animals. Podophyllotoxin seems to inhibit microtubule polymerization in vitro in all systems studied. Vinblastine and vincristine precipitate microtubule proteins and other calcium-binding proteins. (See Dustin, 1978; Borgers and DeBrabander, 1975.)

some movement without affecting the chromosomes themselves. Other cell behavior correlated with the presence of microtubules (for example, maintenance of asymmetric structure and feeding in heliozoans) may be altered by these compounds at very low concentration.

The original use of microtubules in undulipodia and their subsequent development in mitosis preadapted them for many roles in protoctists, animals, plants, and fungi. The sensitivity of tubulin polymerization to inhibitors has been used to identify the roles of microtubules in activities of eukaryotes. Colchicine-sensitive microtubules form in protists during extension of the feeding axopods (Tilney, Hiramoto, and Marsland, 1966; Cachon and Cachon, 1974). Microtubules actually form "oars" that "row" the galleon-like radiolarian *Sticholonche* (Cachon et al., 1977). They are involved in the renodulation of the polyploid nucleus of ciliates (Margulis, Neviackas, and Banerjee, 1969) and in the pharyngeal basket of the ciliate *Nassula* (Tucker, 1968, 1971). They underlie the "feet" of ciliates that creep along blades of grass. Rows of microtubules compose the feeding tentacles of the predatory suctorians. [Other protoctist "mouths," such as the membranelles of heterotrichous ciliates, are made of bundles of undulipodia (Tartar, 1961).]

In animals, microtubules are important components of the nervous system; they form dendrites and axons of neurons and take part in axonal transport (Soifer, 1975). One of the most striking uses microtubules is in the cilium-based sensory cells of animals (Figure 8-8): the chemoreceptors of taste, olfactory receptors, balance organs, and, in insects, mechanoreceptors all contain modified undulipodia with microtubules. Microtubules in animals are part of intracellular transport systems (Soifer, 1975; Dustin, 1978).

In symbiotic hydras, microtubules help to transport ingested algae from the proximal, enteron side of the digestive cells to the distal side, ensuring the maximum amount of light for the symbiotic chlorellas (see Figure 7-13) (Cooper and Margulis, 1977; Fracek and Margulis, 1979). In fungi, microtubules probably have a role in nuclear migration, a mechanism that ensures perpetuation of the dikaryotic state (Raudaskoski, 1972; Ormerod, Francis, and Margulis, 1976). In legumes, the herbicide trifluralin, a microtubule inhibitor, retards severely the establishment of the nitrogen-fixing symbiosis by affecting mirotubule-based morphogenesis of plant cells (DeRosa et al., 1978).

In general, microtubules are important in at least these major functions: mitotic chromosome movements, the generation of asymmetrical cell structure, intracellular transport, undulipodial movement, and intracellular communication. They are intrinsic to the life of eukaryotic cells.

The genetic behavior of microtubule systems is known chiefly from studies of protists: ciliates and *Chlamydomonas*. Ciliates have a thick surface cortex made of complex, species-specific, stable structures composed of microtubules, membranes, and filaments (Figure 8-9; Corliss, 1979). From one to several micrometers deep, these cortices composed of kinetid arrangements are patterned elaborately and are distinctive enough to be studied in genetic experiments on crosses between stable cortical variants. The literature on the cortical genetics of ciliates establishes clearly the

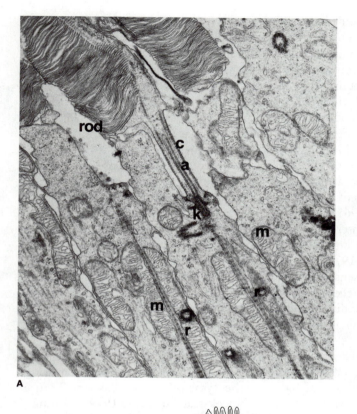

A

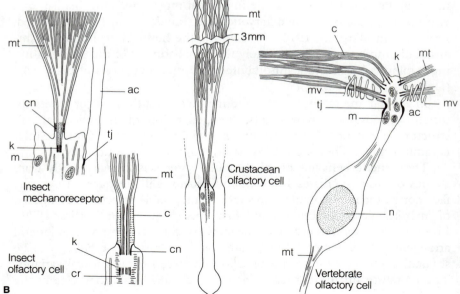

B

Insect mechanoreceptor

Insect olfactory cell

Crustacean olfactory cell

Vertebrate olfactory cell

FIGURE 8-8

Cilium-based sensory cells.
(A) Rod cell from the retina of a rat. The axoneme (*a*) of the immotile cilium (*c*) can be seen underlain by a kinetosome (*k*). A centriole lies at right angles to the ciliary kinetosome. Striated root fibers (*r*) underlie the kinetosomes (compare Figure 9-9B). Mitochondria (*m*) line the root fibers. The photosensitive rod is composed of infoldings of the ciliary membrane; it contains large quantities of photosensitive pigments bound to protein. Transmission electron micrograph, ×17,000. [Courtesy of David Chase.] (B) Cilium-based sensory cells. *ac* = accessory cell; *c* = basal portion of cilium; *cn* = ciliary necklace; *cr* = cilium rootlets; *k* = kinetosome; *m* = mitochondrion; *mt* = microtubules; *mv* = microvilli; *n* = nucleus of sensory cell; *tj* = tight junction. (For details see Atema, 1975.)

independence of the cortical pattern from the nucleus (Jinks, 1964; Beisson and Sonneborn, 1965; Tartar, 1967; Sonneborn, 1974). Conjugation, the ciliate mating process in which nuclei are exchanged but cytoplasm does not fuse, can place identical synkarya (zygote nuclei) in different paramecium cytoplasm. The cytoplasmic parent, never the nuclear parent, determines the transmission of certain cortical traits. The genetic determinants of these ciliate cortical patterns are in neither the soluble cytoplasm nor the

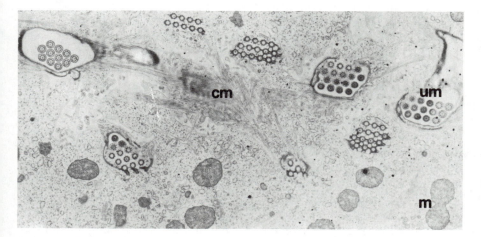

FIGURE 8-9

Oxytricha fallax cortex. A think section of the hypotrichous ciliate showing bundles of microtubules of the transverse cirri (undulipodia). The section cuts the axonemes at various levels. Microtubules in the cortex (*cm*) are at right angles to the microtubules of the undulipodia (*um*). A mitochondrion in division (*m*) is at the lower right. Transmission electron micrograph, ×8500. [Courtesy of G. W. Grimes and S. W. L'Hernault.]

mitochondria, but in the cortex itself. Undulipodia with both [9(3) + 0] and [9(2) + 2] patterns coexist within the same cell, presumably under the control of the same nucleus. In experimentally enucleated hypermastigotes, the undulipodial bands grow out, the number of undulipodia increases, and mitotic spindles form during the morphogenesis of gametes (Cleveland, 1956). Microtubule-based structures can be destroyed by radiation and by chemical treatment with agents that affect nucleic acid (Lederberg, 1952; Jinks, 1964; Simpson, 1968), but nuclear genes alone cannot re-form them. These observations support the concept of nucleic acid–based genetic autonomy of the undulipodial–mitotic apparatus.

Centriole-kinetosomes arise either from preexisting [9(3) + 0] structures or from the less definite MCs. Although kinetosomes were described at first as multiplying by division (Wilson, 1959; Lwoff, 1950), they seldom divide directly. They are products of a complex development that varies from species to species and from stage to stage (Figure 8-10). Centriole-kinetosomes may form from amorphous and small precursors that develop into immature **procentrioles**, most often in direct association with existing mature kinetosomes or centrioles (Dirksen and Crocker, 1965; Mizukami and Gall, 1966; Sorokin, 1968). The number of co-maturing procentrioles or prokinetosomes for each parent kinetosome may vary from one to about 250 or more; great numbers are found in plant sperm structures called "blepharoplasts" (Wilson, 1959; Mizukami and Gall, 1966). Procentrioles appear granulofibrillar under the electron microscope before they achieve their recognizable ninefold symmetrical morphology. Procentrioles themselves may be preceded by no recognizable microtubule-based structure at all. The absence of centriole-kinetosomes certainly does not imply the absence of the genetic capability of forming them. Some amoebomastigotes, such as *Naegleria* and *Tetramitus*, lack kinetosomes of any kind in their amoeboid stages, yet they retain the genetic potential for forming kinetosomes, undulipodia, and other related structures. The rapid production of kinetosomes is induced in a population of *Naegleria* amoebae when they are shaken in a nonnutrient solution, such as distilled water. From the kinetosomes, new undulipodia develop. The genetic determinants of kinetosomes and undulipodia are dedifferentiated to a size below the power of resolution of the electron microscope (Schuster, 1963; Outka and Kluss, 1967).

The **kinetochore** (centromere, or spindle-fiber attachment) is a differentiated portion of chromosomes in most animals and plants (Figure 8-11). Only after intensive studies of favorable material, such as some of Cleveland's protists from termites, was the relationship of the kinetochore to the rest of the mitotic spindle understood. Ontogenetically and phylogenetically at least, the microtubule-attachment portion of the kinetochore is part

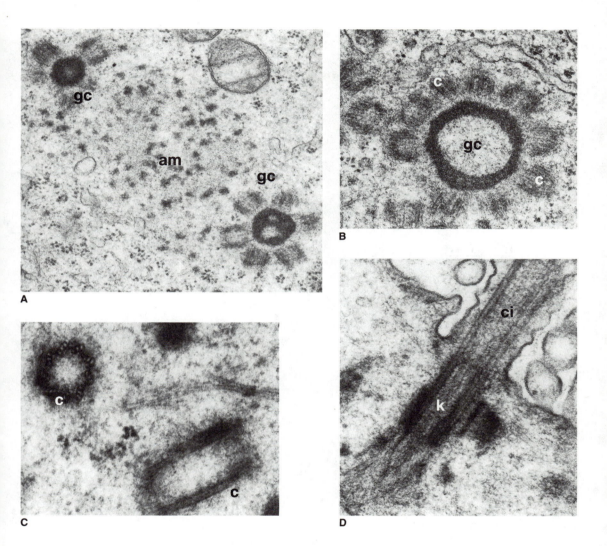

FIGURE 8-10

Kinetosome and undulipodium development in the ciliated epithelium of the oviduct of a mouse five days old.

(A) Two generative complexes (*gc*) lying in the cytoplasm. These MCs develop from the amorphous material (*am*) lying between them. (B) A generative complex (*gc*) surrounded by nine centrioles (*c*) during the maturation process. (C) Mature centrioles (*c*) [9(3) + 0] in transverse section (upper left) and in longitudinal section (lower right). (D) Longitudinal section of a kinetosome (*k*) and cilium (*c*) in final position facing the lumen of the oviduct. Transmission electron micrographs (A, ×30,000; B, ×41,000, C and D, ×60,000). [Courtesy of E. R. Dirksen.]

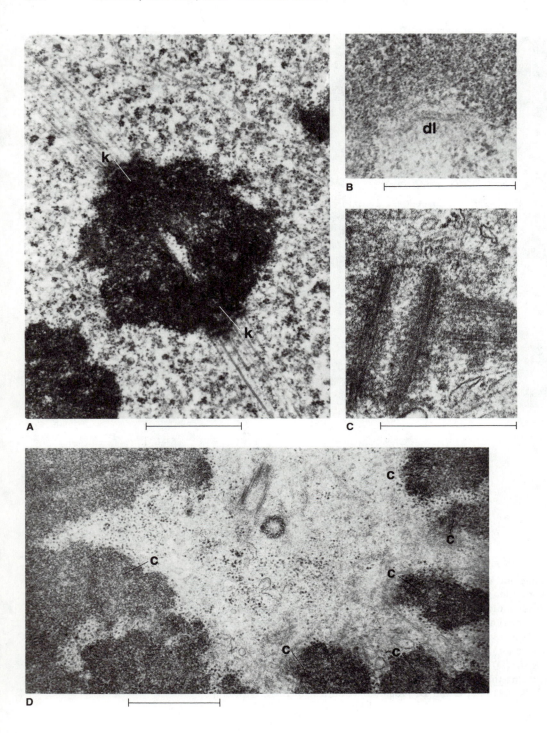

FIGURE 8-11

Kinetochores and centrioles.
(A) Kinetochores (k) in mitotic mammalian cell, showing the attachment of the chromosomes to the spindle microtubules. Transmission electron micrograph, bar = 1 μm. [Courtesy of J. R. McIntosh.] (B) A Chinese hamster fibroblast in mitosis, showing the double-layered (dl) structure of the kinetochore. Transmission electron micrograph, bar = 0.5 μm. (C) Centrioles at late interphase. The offspring centriole (right) is at the typical 90° angle to the parent in this longitudinal section. Transmission electron micrograph, bar = 0.5 μm. (D) Centrioles at right angles to each other, one in transverse section, in a Chinese hamster cell treated with demecolcine. Most mitotic microtubules have dissolved, but those of the centrioles have resisted demecolcine treatment. The centrioles are in the center of the metaphasic array of chromosomes (c) because of the failure of the normal spindle to form. Transmission electron micrograph, bar = 1 μm. [B–D courtesy of B. R. Brinkley and E. Stubblefield.]

of the mitotic apparatus and undulipodial system, and not part of the chromosomal system in which the kinetochores are usually embedded. In certain termite protists, Cleveland discovered that chromatin was destroyed preferentially by oxygen treatment, but that kinetochores nevertheless grew out from the bands of undulipodia: kinetochores were segregated by the spindle even after the chromosomes to which they are attached in normal cells were no longer present. Cleveland then dissociated spindle growth, kinetochore division, and cell division itself from chromosome replication, showing that chromatin had no direct role in control of the mitotic spindle and associated undulipodial structures (Cleveland, 1957). Rather,

the number of mitotic spindles is always a function of the number of centrioles and often, when there is a choice in these multicentriolar cells, chromosomes will move along other spindles than the central ones.

Because by "centriole" he meant a rostral region functioning as a mitotic center—rather than the much smaller centriole on the [9(3) + 0)] pattern—Cleveland was vulnerable to criticism. His point was that only growing spindle fibers (now known to be elongating microtubules that emanate from structures attached to undulipodial bands) determine the distribution of chromosomes to offspring cells. Chromatin itself, although capable of coiling and uncoiling, packing and unpacking, to form chromosomes, is incapable of intracellular deployment; that is, chromosomes are not responsible for their own movement into incipient offspring cells. Chromosome movement in plants, animals, fungi, and many protists is brought about only by structures related to the undulipodial system. For example, Cleveland (1963) says,

> Oxygen concentrations of 70–80 percent destroy all the chromosomes of the hypermastigote *Trichonympha*, provided the oxygen treatment was carried out during the early stages of gametogenesis when chromosomes are in process of duplicating themselves. This treatment does no damage to the cytoplasm and its organelles, following the loss of chromosomes. The centrioles function in the production of the achromatic figure [spindle and other parts of the mitotic apparatus], the flagella [undulipodia] and the parabasal [Golgi] bodies. Then the cytoplasm divides, thus producing two anucleate gametes which make some progress in the cytoplasmic differentiation characteristic of the normal male and female gametes of *Trichonympha* [Fig. 8-5].

On the other hand, Cleveland's observations (1963) of a binucleate, five-centriole cell showed that

> without centrioles no achromatic figure [spindle] is formed, there is no pole-ward movement of the chromosomes to form daughter [offspring] nuclei. The chromosomes reproduce themselves but the nucleus does not. However, two or more centrioles must be present and must be fairly close to the nucleus if the nucleus is to reproduce itself.

Thus, the generative and locomotory role of the spindle and undulipodial system in the segregation of chromatin was clear to Cleveland, but, unfortunately, he seldom communicated his findings clearly to his contemporaries.

Kinetochores function to attach the chromosomes to the mitotic spindle. Many cytogenetic studies have shown that chromosomes lacking kinetochores, because they fail to attach to the spindle, do not arrive at the poles and are thus not incorporated into offspring nuclei. Chromosomes proceed to the poles with kinetochores leading the way (Luykx, 1970). Certain abnormal chromosomes contain two kinetochores, tending to travel to opposite poles of the dividing cell. Such dicentric chromosomes usually break. Each chromosome fragment, attached to its kinetochore, is incorporated into one of the offspring cells. Kinetochores are not just DNA–spindle attachment sites. Only a small amount of DNA is required to attach chromatin to microtubules, either directly or indirectly, but that DNA must have one of many specific sequences. Minimally the DNA of kinetochores has fewer than 200 nucleotide base pairs (as in yeast artificial chromosomes). The kinetochore proteins themselves probably comprise the motor for poleward chromosome movement in anaphase (Nicklas, 1989).

Although there is a bewildering variety in the details, the function of centrioles, kinetochores, or MCs and their products during mitosis is to distribute chromosomes. The mitotic spindle even may be used to distribute mitochondria and plastids (see Table 10-6). Distribution of a large, single mitochondrion may be the reason for a striking phenomenon seen in trypanosomes, such as *Trypanoplasma* (Belar, 1915a). In each cell division, a second spindle of a rather standard appearance forms, an elongating structure at the base of the undulipodium (which in these cells is single and anterior). This structure (its division is called *"blephoplasteilung"*) divides simultaneous with and quite as conspicuously as the nuclear spindle. These protists apparently form a second mitotic apparatus attached to or at least oriented precisely toward the undulipodium.

In *Leishmania*, another trypanosome, the kinetoplast (the kinetosome-associated mitochondrion of protoctists belong to the class Kinetoplastida) divides each time the nucleus and the rest of the cell divides. On a cesium-chloride density gradient, a satellite band of DNA associated with the kinetoplast was found (DuBuy, Mattern, and Riley, 1965), which is consistent with the observation that labeled thymidine is incorporated into kinetoplast DNA (Gibor and Granick, 1964). This single special mitochondrion differentiates during the part of the life cycle in which respiratory oxidative phosphorylation occurs. The second spindle probably evolved to ensure proper distribution of the single mitochondrion in cell division. Strains of *Leishmania* lacking the kinetoplast can be produced by acriflavine; apparently, this drug selectively inhibits the synthesis of kinetoplast DNA. In acriflavine-treated cells, kinetoplast DNA was distributed equally to offspring cells for a few divisions, but then divisions occurred in which one product retained all the kinetoplast DNA and the other lacked any (becoming "dyskinetoplastidic"). From these studies, Simpson (1968) concluded that the dye interferes with the distribution of the kinetoplast DNA between offspring kinetoplasts, as well as with the replication of kinetoplast DNA.

No structure intermediate between undulipodia and bacterial flagella has been reported. The dramatic discontinuity between prokaryotes (which lack MCs and their products) and eukaryotes (which contain them) requires an evolutionary explanation. Certain eukaryotes, such as red algae, lack undulipodia at all stages of their life cycles even though they undergo fertilization and meiosis. Many biologists have suggested that, among the eukaryotes, the red algae are the most closely related to cyanobacterial prokaryotes. According to this view, the red algae are primitive because they never evolved undulipodia by compartmentalization. However, an elaborate MC composed of a ring of microtubules has been found at the poles in the mitosis of red algae (MacDonald, 1972), suggesting that rhodophytes descended from ancestors that renounced undulipodial motility. Red algae contain mitochondria and microtubular structures homologous to those of other eukaryotes. They have well-developed sexual systems, which cannot be considered intermediate between the cell-division systems of the nonmicrotubular bacteria and the microtubular eukaryotes.

If centriole-kinetosomes, kinetochores, and mitotic spindle originated from nuclei by direct filiation, these mitotic organelles should be more sensitive to treatments that affect the nucleus than to those that affect the undulipodia, but in fact the opposite is true (Weinrich, 1954; Cleveland, 1956). Centrioles, asters, spindles, kinetochores, and kinetosomes are all related by composition, behavior, and development to the microtubular–undulipodial system rather than to the chromatin system.

An important difference between mitosis and the distribution of the prokaryote genophore is the amount of DNA that is distributed to off-spring. If DNA were attached to an intracellular, self-replicating body whose copies could segregate at the time of cell division, this would ensure equal distribution of large quantities of genetic material independently of information that it carried. The hypothesis developed here is that adhering bacterial symbionts, namely spirochetes, providing replicating nucleic acid for producing their own attachment sites and that the attachment sites then evolved into centriole-kinetosomes. The spirochete bodies became undulipodia. This hypothesis, as bizarre as it may seem, is consistent with the principle that evolution is opportunistic, not foresighted.

A spirochetal origin of undulipodia helps to explain certain unusual phenomena, such as the direct morphological connections in protists between axopods or undulipodia and the mitotic spindle (see Figure 8-15) (Cleveland, 1938; Wilson, 1959; Bermudes, Margulis, and Tzertzinis, 1987). The presence of tubulin microtubules, the variation always in number rather than size of undulipodia, and the absence of any organisms linking prokaryotic and eukaryotic motility and cell division are all consistent with a spirochetal origin of undulipodia. By hypothesis, the genome of the spirochetes that became the original undulipodia is now the nucleic acid of the MCs. If one accepts the homology between undulipodia and the mitotic apparatus (as most biologists do) and the hypothesis that mitosis evolved by deployment of parts of symbiotic spirochetes (a contention with which most biologists do not agree), one can sketch a rough phylogeny of protoctists, based on the assumption that the evolution of mitosis determined the major patterns of variation in cell structure and life cycle.

TOWARD MITOSIS

A fundamental dualism exists in the phenomenon of
mitosis, the origin and transformation of the achromatic
figure [mitotic spindle microtubules] being in large
measure independent of those occurring in the chromatic
elements [DNA-containing chromosomes]. Mitosis
consists, in fact, of two closely correlated but separated
series of events.

E. B. WILSON, 1959

Observations of live mitosis date back to the beginning of this century. The elegant early work culminated in the chromosomal theory of heredity (Wilson, 1959). Fritsch (1935) summarized algal cell division. Recent ultra-structural studies based on electron microscopy have confirmed, extended, and sharpened the conclusions of the optical microscopists of the late nineteenth and early twentieth centuries. Three examples come to mind. The first is the demonstration by Mizukami and Gall (1966) that plant blepharoplasts (organelles long claimed to be homologous with centriole-kinetosomes, which give rise to plant sperm tails) indeed mature into large spherical structures composed of many oriented centrioles. The second is the ultrastructural confirmation of the light micrographic studies by Pickett-Heaps, Tippett, and Andreozzi, (1978, 1979) showing the specialized extranuclear spindle with interdigitating microtubules of a diatom described originally by Lauterborn (1896). The third is the electron-micrographic confirmation of the undulipodia – mitotic spindle structures in the *Dimorpha*-like mastigote, *Tetradimorpha*, by DeBrugerolle and Mignot (1984).

What kinds of selection pressures gave rise to mitosis? If a thermoplasma-like microbe was ancestral to the nucleocytoplasm, and if, as Searcy, Stein, and Green (1978) suggest, the thermoplasma contained actomyosin, some of the elements of the eukaryotic intracellular motility system — perhaps the filaments responsible for cyclosis — had already evolved prior to the acquisition of undulipodia, mitochondria, and plastids. The homologies of all the components of intracellular motility systems of eukaryotes have not yet been determined. However, the size of the cells and the diversity of proteins was limited by the total quantity of DNA available for performing and regulating protein synthesis. In the absence of efficient mechanisms for ensuring the equal distribution of newly synthesized DNA to progeny, the genetic complexity of the earliest eukaryotes must have been limited. Several examples survive of early nonmitotic strategies for distributing large quantities of DNA evenly to offspring cells. These include multinuclearity, gene duplication, and even entire genome duplication, as in the macronuclei of ciliates (Raikov, 1982) and asexual giant amoebae, such as *Pelomyxa palustris* (Daniels, Breyer, and Kudo, 1966; Daniels and Breyer, 1967). The presence of multiple copies of genes or genomes increases the probability that each offspring cell will receive at least one copy. However, the random segregation of multiple copies of genes and genomes is demonstrably inefficient (Gabriel, 1960). Even after the organization of chromatin into linkage groups, the problem of segregating these groups equally remained.

How did mitosis evolve? In some nucleated organisms it never did. In others, the behavior of the dividing nuclei themselves (called "division figures" in the early twentieth century) suggests a plausible sequence for the evolution of mitosis. Undoubtedly, the following account is hypothetical and oversimplified, but it has the virtue or organizing unruly observations.

The earliest eukaryotes had acquired surface spirochetes in becoming new holobionts: nonmitotic nucleocytoplasmic heterotrophs. As free-living organisms, these spirochetes already contained cytoplasmic tubules. With time, these complexes, the first protists, acquired the immense selective advantage of rapid motility, which allowed them to pursue their food. From different partnerships different motility symbioses evolved (Szathmary, 1987). As natural selection refined these systems, the spirochetes became undulipodia. With time, the nucleic acid of the spirochetes became the antecedent for much of the increased quantity of nuclear DNA, as well as for that of the MCs. A common spirochetal ancestry accounts for the characteristic and nearly universal structure of undulipodia. The interactions of populations of the surface spirochetes that became undulipodia underlie the species-specific patterns and autonomous genetic behavior of the ciliate cortex. The interactions of the genes of spirochetes with those of the nucleocytoplasm (to form pore-studded chromatin-filled nuclei), and later with those of the mitochondria (to form kinetoplasts), led to the differentiation of kinetochores, centriole-kinetosomes, and spindle. Natural selection was relentless; the advantage of a precise distribution of genes led to the final refinement of mitosis. A corollary of this hypothesis is that mitotic organisms maintain and replicate, generation after generation, genetic determinants derived originally from spirochetes.

How did the genome of symbionts differentiate to form the mitotic apparatus? Probably many series of mutations and gene transfers occurred. One set of mutations must have led to the development of an attachment between the nucleocytoplasm and remnant spirochetes. Permanent, precise connections were established; at first they were the rhizoplasts of the karyomastigont system.* A second series of mutations led to the segregation of spirochete and host nucleic acid to opposite poles of host cells. Tubulin protein, synthesized before mitosis begins, was put to use in protist and animal cells. Today, microtubules polymerized from previously synthesized tubulin form the spindle. Some mitotic movements are the direct consequence of the polymerization of tubulin into elongating microtubules (Mazia, 1967; McIntosh, Hepler, and Van Wie, 1969); others involve microtubule-associated "motility proteins," such as kinesin and dynein (Vallee, 1990). At each division, the progeny cells containing one euploid genome — that is, at least one copy of each gene — were selected for. Severe selection against an incomplete distribution of genes exerted continuing pressure for improved mechanisms of chromatin segregation. From the variety of present-day modes of protoctist cell division, one concludes that standard mitosis was achieved at various times in different lineages.

A phylogeny based on mitosis is sketched in Figure 8-12. Drawn to be consistent with the general biology of these organisms, it ignores plastids, which usually were acquired after mitosis had fully evolved. The following pages present a tentative scenario for the differentiation of centriole-kinetosomes from exogenous genomes, showing how certain protoctists can be considered representatives of adaptive radiations that resulted in the development and stabilization of mitosis. Although the scenario is oversimplified and speculative, I hope that it provides guidelines for the collection of new information. [I have seen many protoctists, but in only a few of those discussed here have I seen mitosis; I have relied greatly on other observers (Copeland, 1956; Pickett-Heaps, 1975; Ris, 1975; Page, 1976; Raikov, 1982; Heath, 1980).]

In this scenario, "MC" (microtubule-organizing center) refers to the genetic determinant of centriole-kinetosomes or equivalent division organelles. Such structures, including many **division centers**, are hypothesized to be repositories of nucleic acid descended from spirochete genomes. The light-microscopic term **division center** is undergoing reevaluation in the light of electron-microscopic findings. These centers are composed of centriole-kinetosomes or other microtubule structures. Thus the term "MC" is used here to mean a reproducing system that determines the presence of centriole-kinetosomes, undulipodia, division centers, or any other specialized examples of such organelles. (This is an extension of Pickett-Heaps's original meaning, which was simply to give a functional designation to the microtubule-organizing centers, without implying that they are self-replicating or of symbiotic origin. I hope he will forgive my taking the liberty of inventing a pedigree for his entity.)

---

*See footnote on page 220.

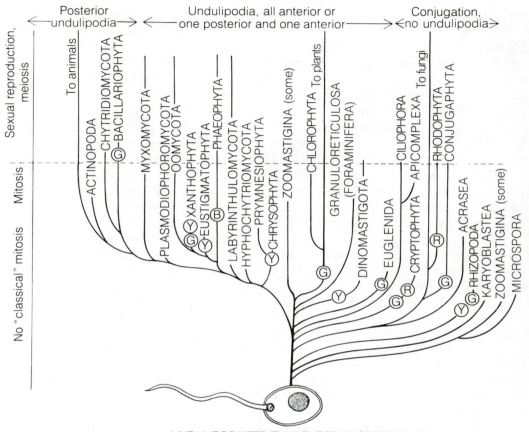

FIGURE 8-12

A phylogeny of mitosis in protoctists. The circled letters (B, G, R, and Y) refer to the acquisition of brown, green, red, and golden yellow plastids. Compare with Sogin's phylogeny based on 16S RNA sequence data (Sogin, 1993).

Figure 8-13 illustrates possible relations, during cell division, between the original nucleocytoplasmic genome and MC genomes at each step in the evolutionary scenario. The ratio of the number of functionally different MCs to the number of nucleocytoplasmic genomes in each generation is given. That is, although at least one dividing MC (that is, one kinetochore) is required on each linkage group to ensure euploidy, multiple chromosomes have been omitted in the schematic diagrams for clarity; the chromatin-complexed genome of the original nucleocytoplasmic host is taken as a unit and, for the moment, mitochondria and plastids have been ignored. In any case, very closely related species can differ in their total chromosome number (Chapter 12), so that the number per cell of any fundamentally replicative structures is a poor criterion for taxonomy (Dillon, 1962).

The following scenario "steps" are semes; they refer to sets of mutations that led to the suggested outcomes and left relics in the extant protoctists. They are included mainly for the plausibility and as a guide to organization of research results.

Step I: MCs were used as kinetosomes for undulipodia. In the earliest group of eukaryotes with undulipodia, the MC (the nucleic acid of the motile surface symbionts) was used only for its own synthesis and for the reproduction and synthesis of the undulipodia. The immediate selective advantage of acquiring these symbionts was motility. Microtubules were not used for chromatin segregation; thus, sexuality in the meiotic sense was precluded. A relic of this event may persist in some isolated groups of small mastigotes, such as *Prorocentrum* (Soyer-Gobillard, 1974). This stage is formally analogous to the spirochete motility symbioses of pyrsonymphids and other organisms discussed in Chapter 9.

Step II: MCs were incorporated into the nucleus for segregation of host chromatin; this resulted in the permanent loss of motility organelles. Spirochetal and nucleocytoplasmic nucleic acid became associated such that spirochetal nucleic acid was incorporated into or on the surface of the nucleus, where it was used as an **intranuclear division center**. Several types of intranuclear MCs evolved. Some are visible after staining with iron hematoxylin (Wilson, 1959) or as granulofibrillar material.

The incorporation of spirochetal nucleic acid into the host nucleus first produced amoebae that lacked undulipodia at all stages of their life history but contained intranuclear MCs and accompanying microtubules. Multinuclear and other asexual amoebae may be relics of this stage. For example, in *Amoeba lacertae*,

> No equatorial plate is formed and the "chromosomes," or chromatic granules wander irregularly toward the poles, the whole karyosome meanwhile drawing out into a spindle shape and finally dividing. It is doubtful whether we can here speak of chromosomes or even mitosis, but this type of division might well form the point of departure for the evolution of a true mitotic process (Dobell, 1912).

Eventually, different groups of mitotic organisms evolved from such amoeba: other amoebae, cellular slime molds, perhaps the microsporidian protists and the major phyla of amastigote fungi, the conjugating green algae, and the two great classes of red algae (Bangiales and Florideae). These organisms lack undulipodia at all stages in their life cycles, even though many are aquatic. When sexuality occurs, it is by conjugation.

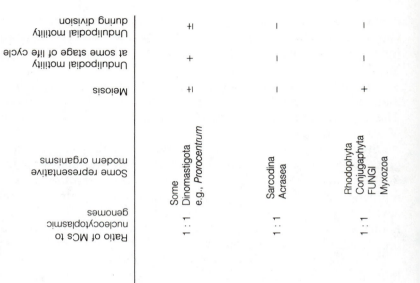

Ratio of MCs to nucleocytoplasmic genomes	Some representative modern organisms	Meiosis	Undulipodial motility at some stage of life cycle	Undulipodial motility during division
1 : 1	Some Dinomastigota e.g., *Prorocentrum*	±	+	±
1 : 1	Sarcodina Acrasea	−	−	−
1 : 1	Rhodophyta Conjugaphyta FUNGI Myxozoa	+	−	−

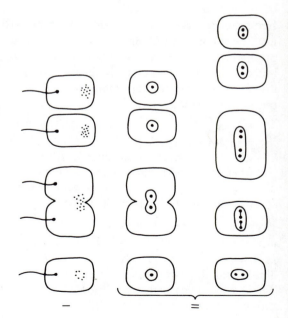

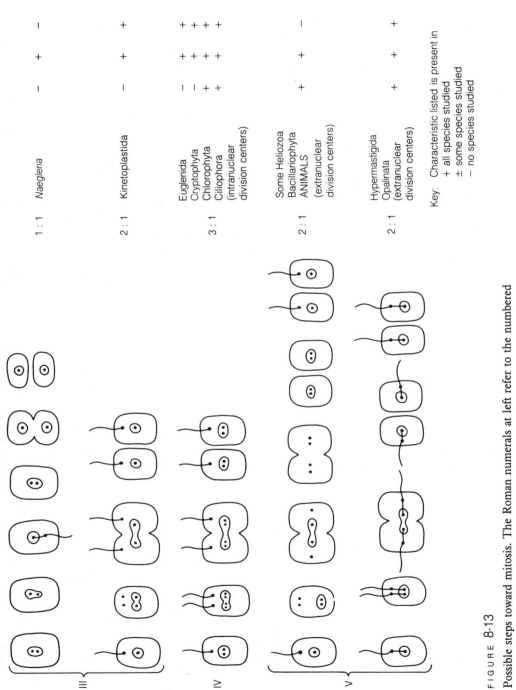

Key: Characteristic listed is present in
       + all species studied
       ± some species studied
       − no species studied

FIGURE 8-13

Possible steps toward mitosis. The Roman numerals at left refer to the numbered steps described in the text. The chart at the right lists some divisional and motile characteristics of modern organisms that represent each evolutionary step.

Whittaker (1969) raised the red algae (Rhodophyta) and the amastigote fungal groups Zygomycota, Ascomycota, and Basidiomycota to phylum status. The conjugating green algae (Conjugaphyta) are likewise different enough from other algae to deserve phylum status. The pit connections of the cell walls, the carbohydrate storage products, and the details of the conjugation process support the idea of a genealogical relationship between red algae and fungi (Kohlmeyer, 1975).

How did mitosis evolve in organisms without undulipodia? After their centers. A similar polar structure is observed in red algae (Bronchart and Demoulin, 1977). The centriole-like structures of *Penicillium* and *Catenaria* also lack a [9(3) + 0] substructure; they sit on the nuclear membrane attachments and eventually to MCs functioning as kinetochores. Other MCs, not attached to chromatin, functioned as morphogenetic centers, analogous to centrioles. A relic of this stage may be the intracellular centriole-like structure lacking the [9(3) + 0] substructure in the basidiomycote *Armillaria* (see Figure 8-1). The position of this body at the poles of the intranuclear spindle suggests a functional equivalence to other division centers. A similar polar structure is observed in red algae (Bronchart and Demoulin, 1977). The centriole-like structures of *Penicillium* and *Catenaria* also lack a [9(3) + 0] substructure; they sit on the nuclear membrane and have the appearance of a double parenthesis (Girbardt and Hadrich, 1975). The spirochete acquisition theory suggests that meiotic eukaryotes, including these that permanently lack undulipodia, are derived from motile spirochete symbioses in which the ancestors either absorbed their undulipodia or never reached the undulipodial stage (Szathmary, 1987).

Step III: MCs were used both as intranuclear division centers and as kinetosomes for undulipodia, but at different stages of the life cycle. Some protists evolved mitotic division in which the spirochete-derived genome was used as a division center, but they had to forfeit motility during cell division. In their motile stages, these organisms were incapable of division; during division, the MCs entered the nucleus and were "borrowed" as division centers for chromatin segregation. After replication of the MC, the replication product differentiated into the kinetosome from which grew standard undulipodia; motility was restored. Only the undifferentiated replicating form of the MC, presumably in the nucleus, was able to function as a division center. This led to amoebomastigotes and other protists still incapable of division during their motile stages. In some (for example, *Tetraselmis*), a morphological connection between undulipodium and nucleus may be seen in a calcium-sensitive contractile rhizoplast; in others, microtubule-based swimming or feeding behavior may be altered during division. The following groups may be relics of this line: *Naegleria* (Wilson, 1959; Schuster, 1963), *Paratetramitus* and other chromidia-forming "pro-mitotic" amoebomastigotes (Margulis, Enzien, and McKhann, 1990), and anisonemids or peranemids (Copeland, 1956). The use of the same structure to serve as centriole-kinetosome in *Chlamydomonas* (Hoops and Witman, 1983) is a remnant of this stage.

Step IV: Some MCs were used to form kinetosomes of undulipodia, while other MCs differentiated permanently as independent intranuclear division centers. Mutations in amoebomastigotes produced cells having two separate MC "clones." One produced offspring that became permanent intranuclear division centers; the other produced offspring that served only as kinetosomes. An intranuclear MC and, eventually, mitosis and meiosis evolved in a lineage in which MCs had been retained as kinetosomes of independent motile organelles. This series of mutations (which probably occurred more than once) led to many groups of protoctists, such as bodos, trypanosomes, chlorophytes, and chrysophytes. Some of them, such as the euglenoids, are idiosyncratic in the behavior of both their autonomous chromosomes and their massive endosome (a nucleolar homologue). The morphological and temporal connections between undulipodia and dividing nuclei are retained. In the earlier-evolved members of this line, meiosis is absent. The typical cell has anterior undulipodia, although they may be laterally directed or trailing.

One or perhaps several such protoctists were ancestors of mitotic chlorophytes and ultimately of plants, with their anteriorly undulipodiated motile stages and intranuclear MCs. Probably others gave rise to apicomplexans (Copeland, 1956). The failure to discover [9(3) + 0] centrioles in plant cells (except prior to sperm formation) implies that the MC homologue of plants generally lacks this phenotypic expression. However, the presence of (1) standard mitosis, (2) colchicine-sensitive microtubules during the formation of the phragmoplast or cell plate (Figure 8-14; Hepler and Jackson, 1968), and (3) in some groups, gametes with undulipodia suggests that an MC remnant of the former spirochete genome is present in plants and eventually will be identified.

With their two types of nuclei, ciliates evolved presumably by a Step IV series of mutations, probably from dinomastigotes, as suggested by Taylor (1978). Although ciliates have mitotic micronuclei with intranuclear spindles, the numerous kinetosomes of the ciliate cortex reproduce independently of nuclear division. The mitotic germ line of the micronuclei and premitotic "soma" of the macronuclei are strangely different. The chromosomes of macronuclei do not split lengthwise; the macronuclei contain a large number of copies of short pieces of DNA—a veritable bag of genes (Prescott, 1964; Prescott and Murti, 1973). Because colchicine-sensitive microtubules elongate during macronuclear division, macronuclear chromatin must be segregated by some idiosyncratic system that ultimately is

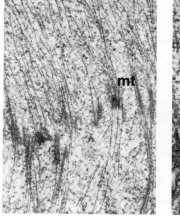
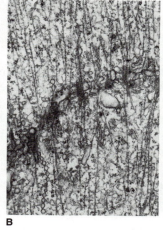

A   B   C

FIGURE 8-14

Relation between spindle microtubules and plant cell wall.
Early (**A**), intermediate (**B**), and late (**C**), stages in the development of the endosperm wall of *Haemanthus*, the African blood lily. Cellulosic walls in plants mature at right angles to the mitotic spindle of the previous cell division. *mt* = microtubules.
Transmission electron micrographs (**A** and **C**, ×12,000; **B**, ×16,500). [Courtesy of P. K. Hepler and The Rockefeller University Press.]

homologous with standard MCs. The series of mutations that was selected for in the evolution of micronuclear mitosis might have begun in a binucleate cell, in which many copies of the genome were produced regularly by one nucleus and reserved for the active administration of protein synthesis by the second, the macronucleus. Micronuclei gratuitous to all processes except meiosis and conjugation (Gorovsky and Woodard, 1968) lack cytochemically detectable RNA and do not synthesize messenger RNA. Neither mitosis nor meiosis ever evolved in the macronucleus. One can actually trace the history of nuclear differentiation in karyorelictid ciliates. Some ciliates must differentiate macronuclei from micronuclei after each cell division; in their descendants, the macronuclei can divide, but only by amitotic processes (Raikov, 1982; Corliss, 1979). Although the evolutionary route has been circuitous, the ultimate selective advantage is clear: ciliates and other representatives of this step retained both motility and the capacity to divide — most crucial was the retention of motility during division itself. Individual peculiar forms of meiotic sex appeared in this lineage.

Step V: Some products of MC reproduction differentiated terminally to produce centriole-kinetosomes or other microtubular structures; other products were reserved in undifferentiated forms capable of reproduction; cells containing differentiated MCs were no longer capable of division. In organisms already containing intranuclear division centers (that is, terminally differentiated organisms), mutations resulted in which other MCs became extranuclear division centers. These extranuclear MCs functioned as mitotic centers, either as centrioles or as other microtubular bodies, such as centriolar plaques. After an MC differentiates into a kinetosome that forms an undulipodium, it is permanently unable to reproduce. Hence, the extranuclear replicating form must be reproduced in each generation.

Light-microscopic studies reveal kinetosomes or kinetosome-containing structures functioning as extranuclear division centers during mitosis in many protists: *Dimorpha mutans* (Picken, 1960); *Clathrulina*, *Acanthocystis*, and *Wagnerella* (Wilson, 1959); and *Ochromonas* and *Centropyxis* (Doflein and Reichenow, 1929). During cell division, undulipodia may be attached at the poles of the mitotic spindle (Figure 8-15). One product of the MC division probably functions as the continuing reproductive form ("germ plasm"), while the other gives rise to kinetosomes that differentiate to undulipodia and therefore—because they cannot dedifferentiate (or because they lack kinetosomal DNA)—they are no longer capable of reproduction (Figure 8-16; Renaud and Swift, 1964).

Many protoctists have kinetosomes that function as centrioles. Some of these maintain a physical connection between the kinetid (undulipodial

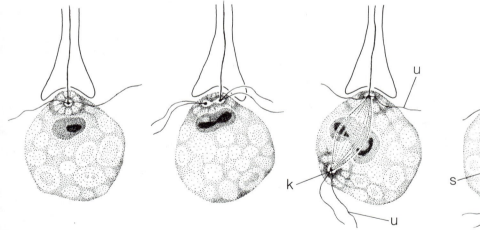

FIGURE 8-15

Early stages in the division of the mastigote *Dimorpha mutans*. Notice the relationship between the kinetosomes (*k*) of the undulipodia (*u*) and the mitotic spindle (*s*). [After Doflein and Reichenow (1929).]

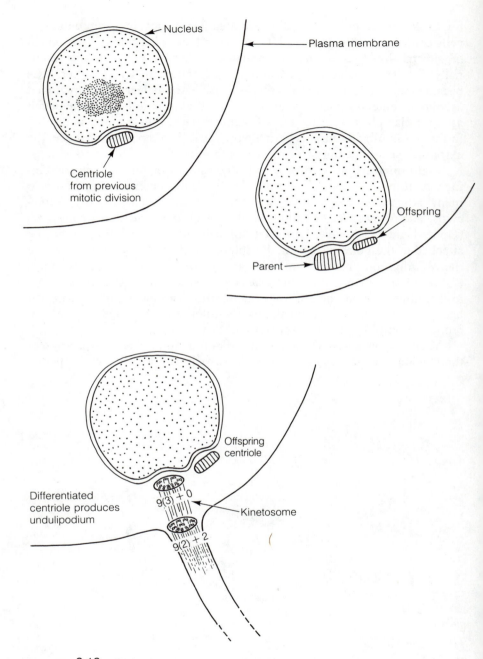

FIGURE 8-16

Kinetosome development in the chytrid *Allomyces*. [After Renaud and Swift (1964).]

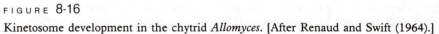

apparatus) and the mitotic structure throughout their life history—for example, *Trichomonas* and *Polymastix* (Wilson, 1959); *Heteromita* and *Eudorina* (Wenyon, 1926); and *Prowazekia* (Belar, 1915b). This connection may have been lost or never constructed in, for example, *Cryptobia* (Copeland, 1956); *Herpetomonas* and *Gurleya* (Wenyon, 1926); *Vaucheria* (Fritsch, 1935); *Dictyota* (Wilson, 1959); and *Paramoeba* (Goldschmidt and Popoff, 1907). Hypermastigotes, with their "long centrioles," are illustrative of the group retaining kinetid–mitotic spindle connections (Copeland, 1956; Cleveland, 1956, 1963). After studying the centrioles in the hypermastigote *Barbulanympha*, Cleveland (1956) concluded,

> The reorganization process [of the centrioles] just described shows several things clearly: a definite relationship between the hypermastigote centrioles and those of higher forms of life [animals and plants]; the ability of the centriole at certain times to function more than once in the formation of flagella [undulipodia], axostyles, and parabasals, just as it is able to do at all times in the formation of the achromatic figure [mitotic apparatus]; the ability of the centrioles to function in the production of extranuclear organelles without reproducing themselves, and also without accompanying nuclear or cytoplasmic reproduction; the inability of flagella [undulipodia], parabasals and axostyles to reproduce themselves; and most important of all, the fact that the anterior tip of these unusually large centrioles of flagellates [mastigotes] is their reproducing portion.

Heliozoans contain retractable axopods (Figure 8-17) composed of patterned microtubules sensitive to low temperatures, high pressure, and colchicine (Tilney, Hiramoto, and Marsland, 1966; Tilney and Porter, 1967). Homologous with other mirotubular structures (Figure 8-18), the axopods form from granular MC material. Early studies showed that in some cases axopods are organization centers for mitotic spindles during division (Figure 8-19). Thus, heliozoans and other actinopods belong to the "extranuclear" group of protoctists.

In animal cells, some clones of MCs may form centrioles functioning as division centers that, under proper physiological stimuli, grow axonemes to become kinetosomes. Once kinetosomes form undulipodia (after differentiation), these kinetosomes are no longer capable of producing (or reproducing) centrioles. A cell containing differentiated MCs (mature kinetosomes of undulipodia) cannot divide. Kinetosomes, with their undulipodia, are analogous to the body that can't reproduce, the "soma" [in Weissman's terms (1892)], whereas undifferentiated MCs are comparable to "germ plasm," tissue still capable of reproduction. In mammals, amorphous material (MCs) that gives rise to hundreds of kinetosomes in the formation of

FIGURE 8-17

Axopod (actinopodium) of the heliozoan *Echinosphaerium nucleofilum*. The axopods, which are retractable, are sensitive to mitotic-spindle inhibitors, such as colchicine. Each axopod is composed of hundreds of microtubules, each about 24 nm in diameter. Transmission electron micrograph, transverse section.

ciliated tracheal epithelium may have been generated ultimately from nucleic acid–based genomes that produced centrioles in previous mitotic divisions (Sorokin, 1968); the centrioles themselves arise from generative forms of the MCs, beneath the levels of resolution of the microscope.

Differentiated animal cells that have undulipodia or certain other microtubular systems (for example, neurotubules) are unable to divide. The

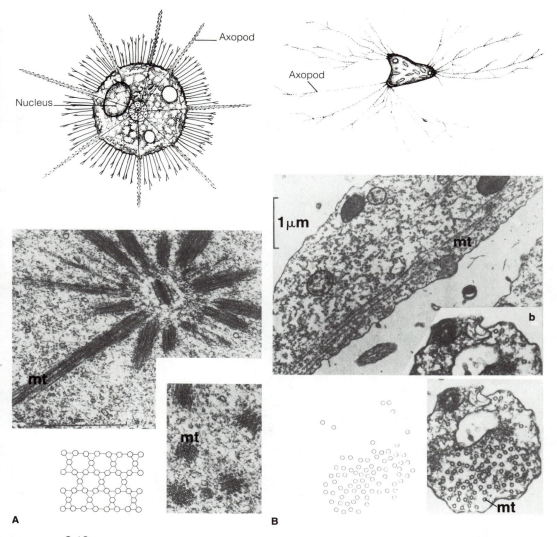

FIGURE 8-18

Structures composed of microtubules (*mt*). Although these organelles of intracellular motility have been specialized for different functions, they are all composed of microtubules in genetically determined arrays. In all photographs, the microtubules are about 24 nm in diameter. (**A, B** *above*; **C**, *following page*.)

(**A**) Structural and feeding axopods of the centrohelid heliozoan *Acanthocystis chaetophora*. Large transmission electron micrograph shows longitudinal section of organelles; inset shows cross section. (**B**) Feeding axopods of the foraminiferan *Allogromia*. Large transmission electron micrograph shows longitudinal section of an axopod; smaller micrograph shows cross section, enlarged somewhat in inset **b**. (**C**) The contractile axostyle of the polymastigote *Pyrsonympha*. Large electron micrograph shows longitudinal section of an axostyle (*a*); inset shows cross section.

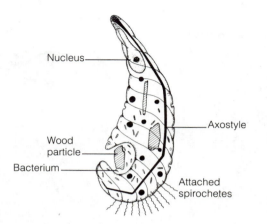

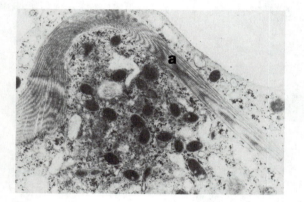

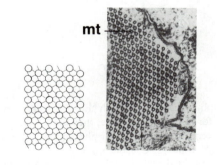

c

FIGURE 8-18 C

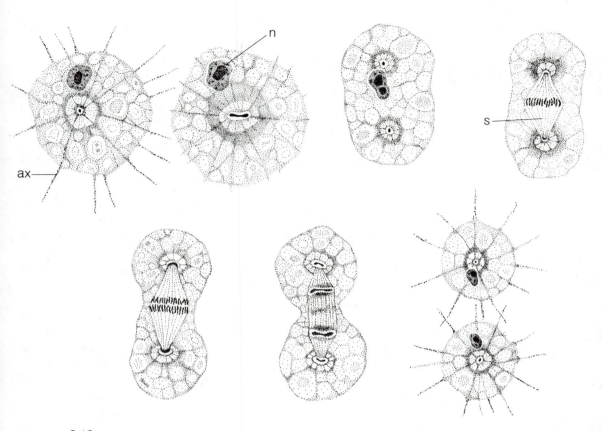

FIGURE 8-19

Mitosis in an actinopod. *Actinophrys sol*, a uninucleate heliozoan, withdraws its
feeding and locomotory axopods (*ax*) during mitosis. The microtubules apparently
dissolve and then are repolymerized immediately after cytokinesis. *n* = nucleus;
s = spindle. [After Belar (1923).]

inability to undergo cell division after kinetosomal–undulipodial differentiation probably had no selective advantage at all; rather, it solved a problem that plagued most protoctists: How can organelles of motility and
feeding, such as undulipodia, axopods, and so forth, be retained if their
components are required for cell division? In a unicellular organism, the
loss of the ability to divide is immediately fatal and hence is selected
against. Motility, lost when undulipodia were retracted, is subject to less
stringent selection pressure. Some protists solved the problem temporally,
by dividing at one stage and reserving motility for another stage in which

they were unable to divide. Other organisms solved the problem entirely and elegantly by differentiation of the two functions: undulipodia were retained, and division was accomplished by some means other than standard mitosis. In over 10,000 species of ciliates, for example, simultaneous cell division and motility are possible. However, the ciliate system, with its highly elaborate cortices and distinctive nuclear behavior, became so fixed that these organisms represent an evolutionary dead end. Mitotic organisms with extranuclear division centers never solved the problem at the cellular level; no animal cell that bears undulipodia ever divides. Animal cells, like other mitotic organisms, solved this problem only at the multicellular or organism level. Under pressure to retain undulipodial motility and the imperative to retain cell division, the ancestors to metazoa resolved the problem by multicellularity: one cell (the "soma") developed kinetosome and undulipodium, becoming terminally differentiated, and the other (the "germ plasm") retained the capacity for division. Mortality of the sperm cell, which can never grow and divide, is the price of sperm motility in the animal lineage.

If mitotic cell division and undulipodial motility are mutually exclusive, one solution is a two-celled organism, one cell maintaining the capacity for cell division while the other retains motility. The second cell is incapable of dividing after it has differentiated motile organelles, but these organelles can propel both cells. The failure to solve the problem of reproduction and motility on the single-cell level led, in several groups, to the origin of multicellularity. One of these groups is the choanomastigote–poriferan lineage; another is our own line of blastula-forming metazoans.

The potential for undifferentiated tissue cells to produce undulipodia is distributed widely, but undulipodiated metazoan cells never divide. Differentiated animal cells of many kinds contain elaborately arrayed 24-nm-wide microtubules — cilia, neurotubules of neurons, and the microtubules of pigment melanocytes, balance organ cells, rods, and cones. My interpretation of the prevalence of colchicine-binding, vinblastine-precipitable microtubule proteins in metazoan differentiated cells, such as those in the developing brain and in the sense organs, is that phenotypic products of the genome once belonging to symbiotic spirochetes became gratituous in nondividing animal cells and were used extensively for tissue differentiation. The formation of supernumerary kinetosomes and undulipodia in α-tocopherol-starved, oxygen-stressed cells (Hess and Menzel, 1968) and in the pituitary (Dustin, Hurbert, and Flament-Durand, 1975) indicates to me the lability of phenotypic expression of the original symbiont genome.

The diatoms, another phylum of mitotic–meiotic organisms, differentiate undulipodiated gametes that, like animal sperm, are incapable of division until after fertilization. Dividing diatoms lack undulipodia. They have an elaborate and conspicuous extranuclear centriolar plaque (an MC), the site of spindle formation. Mitotic heterotrophs, the common ancestors of animals, were perhaps ancestral also to diatoms. Because they have well-developed sexual systems, diatoms may not have acquired plastids until after the stabilization of their complex meiosis–fertilization pattern. This may account for the late arrival in the Cretaceous of traces of this great homogeneous group of algae on the fossil scene. If the production of fossilizable siliceous tests required contributions of plastid metabolism, diatoms may be protoctists of relatively recent origin.

Certain protoctists—for example, chytrids—are posteriorly undulipo-diated. These may share a more recent common ancestor with the meta-zoans than do the sponges. In working out relationships between protoctists and their conspicuous descendants, less attention must be paid to plastid color than to the mitotic cytology of the heterotrophic portions of cells. The number of different types of generative MCs per chromatin set in each group of organisms is in Figure 8-13. This number estimates the symbiont-genome/host-genome ratio that became fixed when these organisms di-verged from their common ancestors. The placement of any group of organisms in the scheme is, of course, tentative; it is based on interpretation of division figures from cytological drawings and descriptions more than half a century old. However, that eukaryotic microorganisms represent adaptive radiations leading to stabilization of mitoses and Mendelian ge-netic systems is correct in principle. By cooperation among protoctistologists—scientists who call themselves cytologists, botanists, protozoologists, mycologists, phycologists or molecular evolutionists—a phylogeny may be constructed that approximates past history. The impres-sive quantity of data collected by Heath (1980, 1981) on variations in mitotic morphology is an excellent beginning (Table 8-3).

CENTRIOLE-KINETOSOME DNA

Are centrioles autonomous? This question was first posed 100 years ago when Boveri and van Beneden, watching the course of mitosis in nematode embryos, concluded that centrioles . . . were "independent permanent cell-organs." . . . Meanwhile, several surprising findings in this paper [Hall et al., 1989] are intriguing to consider. The first surprise, of course, is the notion that DNA is present in the [centriole] at all.

U. W. GOODENOUGH, 1989

TABLE 8-3
Mitotic variation in Protoctista.[a]

PHYLUM[b]	POLAR STRUCTURES[c]	SPINDLE DEVELOPMENT[d]	NUCLEAR ENVELOPE[e]	KARYOKINESIS[f]	PERINUCLEAR ER[g]	SPINDLE VESICLES	METAPHASE PLATE	CHROMOSOME PATTERN[h]	MICROTUBULES/KINETOCHORE[i]	KINETOCHORE LOCATION[j]	CENTRAL SPINDLE[k]	ANAPHASE A[l]	ANAPHASE B[m]	EXTRANUCLEAR SPINDLE[n]	NUCLEOLUS BEHAVIOR[o]
1. Rhizopoda	none	1?	intact	ce	–	–	–	UC	–	?	?	?	?	?	?
2. Haplosporidia	cc, inuc plaque	?	intact	ce	–	–	?	UC	1	np	+	+	?	–	per
3. Paramyxea	?	?	cells divide inside cells	ce	?	?	?	?	Few?	?	?	?	?	?	?
4. Myxozoa	?	?	?	?	?	?	?	?	1–many	np	±	+	+	–	per
5. Microspora (C)	plaques inside and outside nuclear envelope	1	intact	S	–	–	–	chr	1–many	?	?	?	?	?	?
6. Acrasea	?	?	?	?	?	?	?	?	?	?	?	?	?	?	?
7. Dictyostelida	plaque, ring	2	polar fen	?	–	–	–	UC	1–3	np	+	+	+	–	per
8. Rhodophyta	sphere, granule, ring	2	intact, polar fen	ce, int ex	±	±	+	UC, chr, mass	1–many	np	±	+	+	–	ds
9. Conjugaphyta	none	2, 5	polar fen, di A, di P	O, int ex	–	±	+	chr	0, 1, many	np	±	+	±	–	as, ds
10. Xenophyophora	?	?	?	?	?	?	?	?	?	?	?	?	?	?	?
11. Cryptophyta	none	5	di P	O, int ex	–	–	+	mass	1	np	–	+	+	–	ds
12. Glaucocystophyta	?	varies with species	intact	?	?	?	?	?	varies	?	–	?	?	?	?
13. Karyoblastea (C)	none	?	?	ce, int ex	perinuclear bacteria	+	+	chr	1	np	–	?	?	–	ds

Taxon															
14-1. Amoebomastigota (C)	none	1?	intact	?	–	–	?	UC	?	?	?	?	+	?	–	per
14-8. Proteromonadida	none	?	intact	ce	–	–	?	chr	1	np	?	?	?	?	?	as
14-9. Parabasalia	attractophore (Fig. 32)	6	intact	ce	–	–	–	chr	many	nm	+	+	+	+	?	?
15. Euglenida	none, centrioles ‖	1	intact	ce, int ex	–	–	–	chr, mass	0–2	np	–	+	+	+	–	per
16. Chlorarachnida	?	?	?	?	?	?	?	?	?	?	?	?	+	?	?	?
17. Prymnesiophyta	centrioles ‖	?	di P	O	–	+	+	chr	?	?	–	+	+	+	–	ds
18. Raphidophyta	none	?	polar fen, di T	O, ?	–	+	+	chr	many	np	–	+	+	?	–	frag
19. Eustigmatophyta	?	?	?	?	?	?	?	?	?	?	?	?	?	?	?	?
20-1. Actinopoda: Polycystina	plaques, centrioles ‖	1	intact, di P	ce, O	–	+	–	chr	0–2	np	+	+	?	?	–	?
20-4. Actinopoda: Acantharia	plaques, centrioles ‖	1	intact	S	–	–	–	chr	1	np	+	+	+	+	–	ds
21. Hyphochytriomycota (C)	centrioles ‖	?	polar fen	int ex	?	+	+	?	1	np	?	?	?	?	–	ds
22. Labyrinthulomycota/ Thraustochytrids	centrioles ‖, procentrioles	2	polar fen, intact	int ex	–	+, –	+	mass	1	np	–	+	+	+	–	ds
23. Plasmodiophoromycota (C)	centrioles ==	2	polar fen	NlNM	+	+	+	mass	many	np	–	+	–	–	+	per
24. Dinomastigota (C)	spheres, striated plaques, none, centrioles ‖	6	intact	ce	–	–	–	chr, mass	1–many	nm	–	+	+	–	+	per
25. Chrysophyta	rhizoplast	5	di P	O	–	+	+	chr	?	?	–	+	+	+	–	ds
26. Chytridiomycota (C)	centrioles ‖	2	polar fen, intact	NlNM, int ex, ce	±	–	+	chr, mass	1	np	–	+	+	+	–	di, per, ds
27-1. Myxomycota (C)	none, centrioles ‖	1, 5?	intact, polar fen di P, di T	O, int ex	–	±	+	chr	1–2	np	–	±	+	+	–	ds
27-2. Protostelida (C)	none, centricoles ‖	?	di P	O	–	+	+	chr, mass	?	np	–	+	+	+	–	ds

TABLE 8 - 3 (continued)

PHYLUM[b]	POLAR STRUCTURES[c]	SPINDLE DEVELOPMENT[d]	NUCLEAR ENVELOPE[e]	KARYOKINESIS[f]	PERINUCLEAR ER	SPINDLE VESICLES	METAPHASE PLATE	CHROMOSOME PATTERN[h]	MICROTUBULES/KINETOCHORE[i]	KINETOCHORE LOCATION[j]	CENTRAL SPINDLE[k]	ANAPHASE A[l]	ANAPHASE B[m]	EXTRANUCLEAR SPINDLE[n]	NUCLEOL BEHAVIOR[o]
28. Ciliophora (C) Micronucleus	none	1	intact	ce, NINM, int ex	−	−	±	chr	0–10	np	−	+	+	−	abs
Macronucleus	none	4	intact	ce	−	−	−	mass	0	none	−	+	?	−	per
29. Granuloreticulosa (C)	spheres, centrioles ‖	1	intact	ce	±	−	+	chr	?	np	−	+	+	−	di, ds
30. Apicomplexa (C)	centrocones, rings, centrioles ‖, plaques, spheres	1, 2, 3, 6	intact, polar fen	ce, S	−	−	?	chr, UC	1 (many)	np	±	+	+	−	per, ds
31. Bacillariophyta	striated bar	5	di P	O	−	+	+	mass	0–many, C	np	+	+	+	+	ds
32-1. Prasinophyceae	centrioles ‖, rhizoplasts	1, 2	di ?, polar fen, intact	O, ?	−	−	+	mass	1	np	−	+	+	−	ds
32-2. Chlorophyceae (C)	centrioles ‖, none, sphere	1, 2, 3	polar fen, intact, di P	O, int ex, ce	±	±	±	mass, chr	0–many	np	−	+	±	−	per, ds, frag
32-3. Charophyceae	centrioles ‖, none	5	di P	O	−	+	+	chr	many	np	−	+	+	−	ds, as
32-4. Ulvophyceae	centrioles ‖, none, procentrioles	1, 2, 5	polar fen, intact, di P	O, int ex, ce	−	±	−	chr, mass	many	np	−	+	±	−	ds, per, frag, di

33. Oomycota (C)	centrioles ══	1	intact	ce	–	–	UC	1	np	–	+	+	–	per
34. Xanthophyta (C)	centrioles ‖══	1	intact	int ex	–	?	UC	?	?	+	+	–	–	frag
35. Phaeophyta	centrioles ‖══	2	polar fen	int ex	+	+	mass	?	?	–	+	–	–	ds

[a]Higher taxa from Margulis et al. (1990); data primarily from Heath (1980).

[b]Proctoctista; numbers refer to the phyla listed in the Appendix (number following a hyphen indicates a class). See Heath (1980) for genus and species names under each phylum. C = coenocytic (multinucleatic) organization in some genera.

[c]Morphology of organelles interpreted as individisial centers. "Centrioles ══" means that there are two centrioles at approximately right angles to each other; "centrioles ══ ══" means that the centrioles lie in approximately the same plane; procentrioles are immature or smaller centrioles (Figure 8-3); "cc" means that centrocone is present (see Figure 8-4); "inuc plaque" means that a centriolar plaque is found inside an intact nuclear membrane. Some organisms have kinetosomes with undulipodia retained during cell division that may lie close to the spindle poles; such organisms are listed as lacking polar structures. See Figure 4 in Heath (1980) for diagrams of spheres, attractophores, and rings. Fig. 32, Lophomonas, is in Margulis and Sagan (1986). Rhizoplasts are calcium-sensitive, striated, direct connections between kinetosomes and nuclei.

[d]Spindle forms inside nucleus from differentiated regions of nuclear envelope or plaques set into nuclear envelope (1); spindle pushes through polar fenestrae from extranuclear polar structures (2); spindle forms in nucleoplasm from permanently intranuclear centers (3) including numerous ill-defined sites (4); spindle forms in cytoplasm and sinks into nucleus as envelope disperses (5); spindle forms between extranuclear structures (6) and remains extranuclear throughout division cycle (7).

[e]Polar fen = polar fenestrae, large openings at the poles of the nuclear membrane through which the spindle forms; di = dispersal of the nuclear membrane during prophase (P), metaphase (M), anaphase (A), or telophase (T); intact = nuclear membrane that is present throughout karyokinesis does not dissolve or form perforations detectable in live organisms.

[f]During telophase, the nuclear membrane may become constructed at the equator (ce) or constricted in two places so that the interzone is excluded from the offspring nuclei (int ex); a new and independent nuclear membrane may form within the old nuclear membrane (NiNM); or a constricting septum may form centripetally (S). If the nuclear membrane disperses before telophase, none of the above events occurs and the division is open (O).

[g]ER = endoplasmic reticulum (ribosome-studded membranes surround nucleus). Perinuclear bacteria are Gram-negative rod-shaped symbionts surrounding each of the many nuclei.

[h]Mass = chromatin in fused mass; chr = separate chromosomes; karyotyping feasible; UC = uncondensed, insufficient condensation of chromosomes to detect a karyotype pattern.

[i]— = no structures detected at microtubule-chromatin boundary; C = collar-shaped structure on fused chromosome mass (Surirella); many = more than 20 microtubules per kinetochore.

[j]np = in the nucleoplasm; nm = inserted in the nuclear membrane; none = no kinetochores.

[k]Spindle comprised of a bundle of nonkinetochore microtubules forms before telophase.

[l]Movement of chromosomes to poles.

[m]Spindle elongation.

[n]Permanent extranuclear spindle outside nucleus, which permanently remains closed.

[o]The fate of the nucleolus during mitosis: per = persistent; ds = dispersed; di = discarded; frag = fragmentary; abs = nucleolus absent; as = nucleolar fragments associate with chromatin to ensure their distribution.

More than 15 motility mutants of the haploid green algae *Chlamydomonas reinhardtii* were studied by Hall, Huang, Adams, Dutcher, Luck, and their colleagues in the 1980s. Some mutations are correlated with abnormal kinetosome assembly and undulipodial structure, and other have no discernible effect other than limiting normal phototactic motility of these autotrophs. To the surprise of all the investigators, all of these motility mutants, when mapped by standard genetic techniques, were found to be clustered together on their own linkage group (linkage group XIX, also called the **uni linkage group**). Uni can be distinguished from the other eighteen linkage groups, which correspond to well-known nuclear chromosomes.

Peculiarities of the uni linkage group were revealed by further study: the motility markers mapped in a circle (rather than linearly like markers on a "normal" chromosome) and showed a preponderance of second-division segregation (unlike the first-division segregation of ordinary Mendelian markers that map near the centromere). Their linkage relations showed a strange sensitivity to slightly elevated temperatures applied at a specific time prior to meiosis. Using pulsed-field gel electrophoresis to resolve chromosome-sized DNA fragments, Luck and his colleagues were able to identify the DNA corresponding to the "motility chromosome" (Hall, Ramanis, and Luck, 1989). Restriction-fragment-length polymorphisms detected with the subclones of this "uni DNA" showed standard $2:2$ segregation patterns in crosses of *C. reinhardtii* (mutant) with *C. smithii* (wildtype for motility). In situ hybridization using cloned segments from two regions of the uni DNA as probes localized the uni DNA with fluorescence microscopy. The localized uni DNA in situ showed two bright spots, one at the base of each undulipodium per cell: "centriole-kinetosome DNA" fluoresced brightly, shocking its discoverers. Using chromosomal DNA from yeast and *Neurospora* as standards, they estimated the total quantity of the centriole-kinetosome DNA to be very large: about 6 megabases (which is the amount of the total genomic DNA of a hefty prokaryote).

As discussed in *Microcosmos* (Margulis and Sagan, 1991), *Origins of Sex* (Margulis and Sagan, 1986), here, and elsewhere (Margulis and McMenamin, 1990), the discovery of centriole-kinetosome DNA (or, if not DNA, at least RNA remnant nucleic acid) was anticipated. We were gratified to see that Luck and his colleagues performed such elegant experiments with *Chlamydomonas*. This experimentally ideal protist, familiar to geneticists, has a well-marked nucleus and a known pattern of inheritance of organelles (several mitochondria and one chloroplast per cell).

During each mitotic division, *Chlamydomonas* resorbs its axonemes, and each naked kinetosome becomes a mitotic centriole. Unlike those of many protoctists and animals, each centriole-kinetosome in *Chlamydomonas* is capable of genetic continuity. We do not expect all mature kinetosomes to harbor DNA — quite the contrary: since so many are incapable of further reproduction, it is predicted that they lack DNA. Kinetosomes incapable of reproduction [for example, membranellar kinetosomes of *Stentor* (Younger et al., 1972); all animal kinetosomes] probably lack DNA at all times. Luck's DNA probes, highlighted with the fluorescent label, allowed unequivocal visualization of two 0.25-μm-diameter centriole-kinetosomes per *Chlamydomonas* cell. The photographs decorate the cover of the journal *Cell* (see Hall, Ramanis, and Luck, 1989). Some of the *Chlamydomonas* centriole-kinetosome DNA probes must cross-react with centriole-kinetosomes of ciliates, of mammalian basal epithelium, and of the many other cells in the biological world where these generative $[9(3) + 0]$ structures exist.

The findings of Hall et al. (Hall, Ramanis, and Luck, 1989) have not been corroborated. On the contrary: after studying *Chlamydomonas*, using immunological techniques for the general visualization of DNA, Johnson and Rosenbaum (1990, 1991) denied the presence of centriole-kinetosome DNA even though a large amount of fibrous material associated with centrioles and interpretable as kinetosome DNA is beautifully visible in William Dentler's micrograph that illustrates their own article (Johnson and Rosenbaum, 1991)! Although plastid, nuclear, and mitochondrial DNA were evident in abundance with DAPI (diamidinophenylindone) staining, Kuroiwa et al. (1990), using fluorimetry with a video-intensified photon-counting system, also could not detect any fluorescence associated with centriole-kinetosome DNA in *C. reinhardtii*. In genetic studies, Johnson and Dutcher (1991) found linkage group XIX (the uni linkage group) to be present in the same number of copies per cell as the other nuclear linkage groups. They reported that, in mutants lacking kinetosomes, the copy number of linkage group XIX is unchanged. These investigators concluded that the uni linkage group has a nuclear, rather than a kinetosome, location.

Since the size of the motility DNA's linkage group is so large, argue many (including Goodenough, 1989), it seems unlikely that it would have gone unnoticed for nearly a century. Given the observations collected in this book, the most parsimonious hypothesis to me is that Luck and his colleagues have discovered the "spirochetal secret agent" genome predicted by Margulis and Sagan (1986, 1991) and that all potentially undulipodiated eukaryotes will have DNA homologous to that in the *Chlamydomonas* motility linkage group (even though that DNA may often take up nuclear residence). The molecular virtuosity of centriole-kinetosome DNA is its association with rapid cell motility systems that are its phenotypic

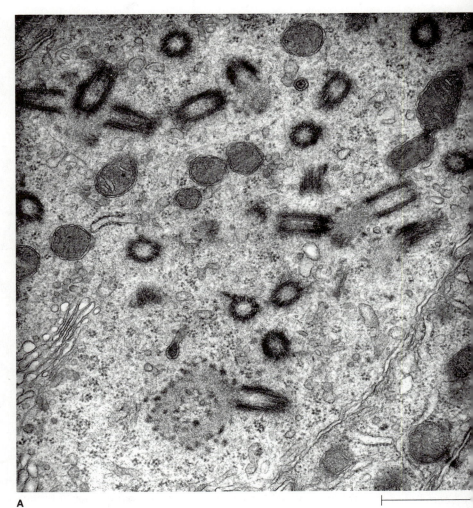

A

FIGURE 8-20

Centriole-kinetosome DNA in a mouse.
(**A**) More than 15 newly developing oviduct kinetosomes from a five-day-old mouse; fibrous cytoplasmic material associated with ribosomes is visible. Transmission electron micrograph. (**B**) A single, newly-developing kinetosome (*k*) showing the fibrous material (f) in two clumps, which may correspond to duplicated kinetosome DNA. Earlier these comprised a single clump. Transmission electron micrograph. Bars = 0.5 μm. [Courtesy of E. R. Dirksen.]

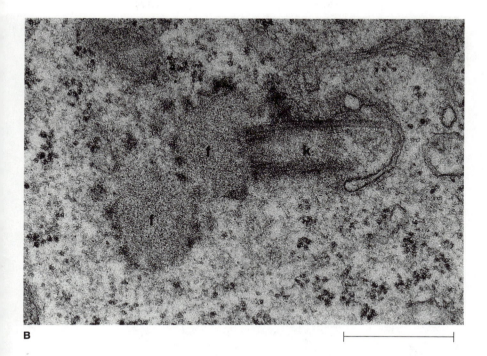

B

products and the integration of these products in the evolution of mitosis. Is not a periodic nuclear residence of this ancient genome not only possible, but highly likely?

The centriole-kinetosome itself generally does not reproduce; reproduction of this structure is determined by the presence of granulofibrillar material — that is, DNA — that, it can be claimed, has already been seen in the cytoplasm even of mammalian cells (Figure 8-20). If Maul (1991, personal communication) is correct, this DNA (like DNA in general) requires at least a membrane vesicle to ensure its distribution upon replication. When such a membrane vesicle (seen here inside the centriole, Figure 8-21) is available, the centriole retains its capacity for reproduction. When the membrane is lost with maturation, so too is the capacity for further reproduction.

I hypothesize that the incorporation and redeployment of spirochete DNA and its motility protein products was the major event of eukaryosis. If spirochete remnant DNA corresponds to the linkage group discovered by Hall and his coworkers (Hall, Ramanis, and Luck, 1989), we are still able to detect it 1500 million years after integration of the original bacterial ge-

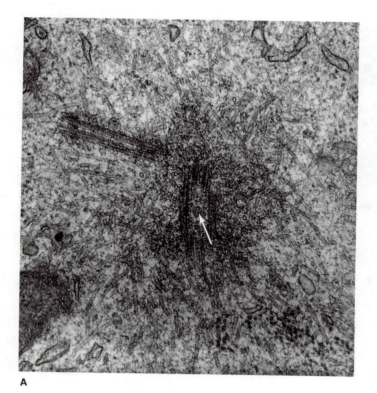

A

FIGURE 8-21

Centriolar membrane vesicle in human cells.

(**A**) Developing centriole-kinetosome containing a membrane vesicle; melanoma tissue culture cells in early G_1 phase. The unit membrane of the vesicle is apparent in the lumen of the centriole (at arrow) but absent in the second centriole. Transmission electron micrograph. (**B**) The membrane vesicle, seen here at higher magnification, may be required for DNA segregation in that its absence may preclude centriolar reproduction. Transmission electron micrograph, bar = 0.5 μm. [Courtesy of G. Maul.]

nomes occurred. We should not be naive, though, and expect that in 1500 million years of coevolution inside a symbiotic complex, this moving, reproducing entity has ever failed for long to move and reproduce.

The first evaluation of symbiogenesis by English-speaking scientists was published as a series of essays (Margulis and Fester, 1991). The history of symbiogenesis (the Russian work reviewed by Khakhina) and symbionticism (Wallin's concept reviewed by Mehos) is available (Margulis and

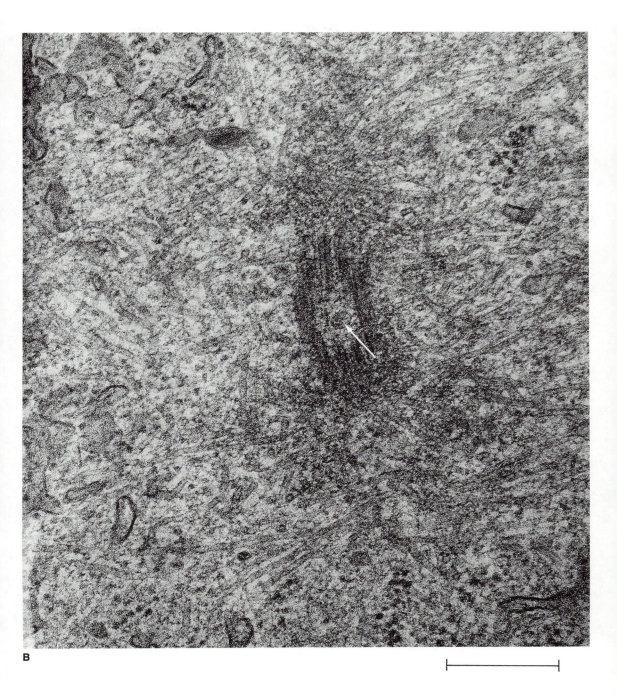

B

McMenamin, 1992). Together, these works make it clear that symbiogenesis is a principle of evolution of far wider application than merely the origin of the eukaryotic cell. Sulfide-oxidizing chemoautotrophic animals at deep-sea vents, families of luminous bacteria- and nematode-ridden lepidopterans ("glow worms") owe their existence to symbiogenesis. Entire communities (for example, mycorrhizae of the northern boreal forests and tropical coral reefs) depend unequivocally on microbial symbionts for their existence. As aptly noted by Goodenough (1989), the story of centriole-kinetosome DNA and the symbiotic origins of this organelle in particular (and of symbiogenesis as a major mechanism of evolutionary innovation in general) is far from over.

CHAPTER 9

UNDULIPODIA FROM SPIROCHETES

SYMBIOSIS AND MICROTUBULE SYSTEMS

To my knowledge, B. M. Kozo-Polyansky (1924) tried to answer this question for the first time in his now almost forgotten book, *New Principles of Biology*. Concerning the origin of centrioles and blepharoplasts, which he justifiably considered as completely homologous structures, Kozo-Polyansky arrived at the conclusion that they represent "flagellated cytodes." (Kozo-Polyansky, after Haeckel, understood cytodes to be non-nucleated organisms, bacteria and blue greens.) He concluded a short exposition of his hypothesis with these remarkable words: "At least the suspicion of the bacterial nature of these kinetoplasmatic (mobile) organoids, without a doubt is legitimate (p. 57)." This daring hypothesis for the time did not find supporters and went by unnoticed. It was so discrepant with the established point of view that it could not be discussed at respectable scientific meetings. But after 43 years this same idea was proposed by a young biologist from Boston University (USA).

A. L. TAKHTAJAN, 1973 (in Russian; see 1992 translation of Khakhina, 1979)

Two theories of the origin of microtubule systems (compared in Table 9-1) are in the literature of cell biology. In the theory of endogenous compartmentalization or direct filiation, microtubules differentiated within

TABLE 9-1

Exogenous and endogenous theories of the origin of microtubular systems.

EXOGENOUS (SYMBIOTIC)	ENDOGENOUS (DIRECT FILIATION)
Undulipodia and other microtubule systems in eukaryotes began as ectosymbiotic spirochetes.	Microtubules and their organizing centers preceded the origin of undulipodia, which arose endogenously, perhaps from nuclear membrane.
Microtubules will be found in spirochetes. Spirochetes with cytoplasmic tubules (in pairs or otherwise) related to the [9(2) + 2] array may be found.	Microtubules will be found in the cytoplasm of the prokaryotes that are ancestral to the eukaryotes. There is no reason to expect them in spirochetes.
MCs[a] replicates; DNA replication accompanies the "reproduction" of centriole-kinetosomes and their undulipodia.	MCs[a] are systems for the self-assembly of structures from proteins; they do not require the presence of nucleic acid.
All meiotic organisms derive from ancestral motility symbioses.	Some meiotic organisms are primitively nonmotile (red algae, fungi); thus, [9(2) + 2] homologue organelles will not be found in them.
The genes that code for tubulin and other motility proteins are homologous with the code for such proteins in certain spirochetes.	The genes that code for tubulin and associated proteins are in the nucleus, where they originated; such genes are expected only among ancestral prokaryotes on the direct line to eukaryotes.
Undulipodia are monophyletic; mitosis and meiosis are polyphyletic and evolved in several lines of undulipodiated protoctists and spirochete motility bionts.	Mitosis is monophyletic; undulipodia evolved in mitotic organisms and were retained for their selective advantage of motility.
Once evolved, mitosis was not lost; mastigotes and amoeboids that lack mitosis are primitive. The earliest microtubule-containing eukaryotes were undulipodiated heterotrophs.	Mitosis was lost in several lines of undulipodiated protists. The earliest microtubule-containing eukaryotes were nonmotile photoautotrophs.
The earliest eukaryotes were anaerobic mastigotes lacking mitochondria but having fully-developed centriole-kinetosomes and undulipodia.	The earliest eukaryotes were mitotic organisms that lacked centriole-kinetosomes and undulipodia.

[a]MC = microtubule organizing center (also called MTOC).

photosynthetic organisms in the evolution of the mitotic apparatus of red algae from cyanobacteria. They were later organized into undulipodia (Bisalputra et al., 1975; Pickett-Heaps, 1974; Taylor, 1978; Jensen, 1989). Thus, the nucleus and the evolution of mitosis preceded the appearance of undulipodia. This chapter explores in detail only the alternative, exogenous theory: the origin of undulipodia from spirochetes. Free-living spirochetes that already had microtubules within their protoplasmic cylinders became surface symbionts on the predecessors to the nucleocytoplasm, as discussed in Chapter 1 (pages 6-7, 16-17), 7 (pages 194-200), and 8 (pages 244-245). The spirochetes eventually integrated into host bacteria that became the first nucleated cells (protoctists). The former spirochete genome produced undulipodia, centrioles, kinetochores, and other microtubular manifestations of the original spirochete genome.

Spirochetes live together with other organisms, often with protists and animal tissue cells. For example, the family to which *Dienympha* and

Pyrsonympha belong (Pyrsonymphidae), has been described as having "elongate flagellates [mastigotes], the four or eight anterior flagella [undulipodia] adherent to the body and spirally twisted with it, free at their distal ends. Often they are beset with spirochetes which have been mistaken for additional flagella [undulipodia]; the family has been misplaced in the order Hypermastigina" (Copeland, 1956). The same is true of another family of protists, the Devescoviidae: "Spirochetes which share the habitat of these organisms are commonly found adhering to their cell membranes, and were mistaken for additional flagella [mastigotes] in the original descriptions of some of the genera" (Copeland, 1956). Recently, work on spirochete–protist associations has resumed, using modern techniques. Here we ask: What are spirochetes? May some be related to ancestors of undulipodia? What is the evidence that undulipodia originated from spirochetes and that microtubules evolved in prokaryotes? How might this theory of the spirochete origin of undulipodia be tested?

SPIROCHETES

I have also seen a sort of animalcule that had the figure of the river eels: These were in very great plenty, and so small withal that I deemed 500 or 600 of 'em laid out end to end would not reach to the length of the full grown eel such as there are in vinegar. These had a very nimble motion, and bent their bodies serpentwise, and shot through the stuff as quick as a pike does through the water.

A. VAN LEEUWENHOEK, 1681
(in Dobell, 1958)

Accordingly, I took (with the help of a magnifying mirror) the stuff off and from between my teeth further back in my mouth where the heat of the coffee couldn't get at it. This stuff I mixt with a little spit out of my mouth (in which there were no air bubbles) . . . then I saw with as great a wonderment as ever before, an inconceivably great number of little animalcules . . . the whole stuff seemed alive and a-moving . . . these moved their bodies in great bends so swift a motion, in swimming first forwards and then backwards, and particularly with rolling around on their long axis, that I couldn't but behold them again with great wonder and delight.

A. VAN LEEUWENHOEK, 1692
(in Dobell, 1958)

Spirochetes are Gram-negative bacteria — highly motile, slender, flexuous, helical, unicellular prokaryotes that divide across their short axis; that is, by transverse fission. Van Leeuwenhoek first saw these microbes in a sample of his own diarrhea, his "excrement being so thin" and in his saliva. The drawings that he sent in 1681 to the Royal Society in London show spirochete-like forms. Spirochetes were studied later by Dobell (1912), Noguchi (1921), Kirby (1941), and other investigators. They have been identified as causative agents in Lyme disease, yaws, and syphilis (Balows, et al., 1992). Their distinctive structure consists of a protoplasmic cylinder bounded by a plasma membrane and an elaborate Gram-negative wall. Unlike those of other motile bacteria, the flagella of spirochetes lie in the periplasmic space and therefore are surrounded by other layers of cell wall, called the sheath (Figures 9-1 and 9-2). Spirochetes are chemoorganoheterotrophs. In suboptimal media, they grow in length without forming cross-walls. For this reason, their diameters are more constant, reliable, and strain-specific than their lengths. Twelve genera have been described:

Members of the genus *Borrelia*, often carried by arthropods, are pathogenic or symbiotic in mammals. They range in diameter from 0.2 to 0.5 μm and have from 7 to 20 flagella. Tick-borne borrelias are the causative agents of Lyme disease.

Spirosymplokos deltaeiberi is a large (0.7 to 2 μm by 50 to 150 μm), free-living mud spirochete from beneath microbial mats at the mouth of the Ebro river in northeastern Spain (Guerrero, Ashen, and Margulis, 1992; Margulis et al., 1993). This complex spirochete (Figure 9-3) bears smaller (0.2 to 0.5 μm diameter) but similar spirochetes with conspicuously granulated cytoplasm. The smaller spirochetes probably form desiccation-resistant spores in response to aeration. They are 5:10:5 spirochetes (see footnote page 279).

Members of the genus *Clevelandina* are symbiotic inhabitants of the hindguts of dry wood – eating termites (North American reticulitermitids).* They range in diameter from 0.4 to 0.8 μm and have a distinct chambered inner coat of the outer membrane, a sillon, and 30 to 45 periplasmic flagella.

Members of the genus *Cristipira* are commensal in mollusks and have hundreds of flagella that form a bundle, the crista. These large cells range from 0.5 to 3 μm in diameter. They live in the gelatinous crystalline style, an organ of the digestive system of clams, oysters, and other bivalve mollusks.

---

*Informal names of termite families: subterranean termites are reticulitermitids, dry wood – eating termites are kalotermitids, etc. *Kalotermes schwartzi* = *Incisitermes schwartzi*.

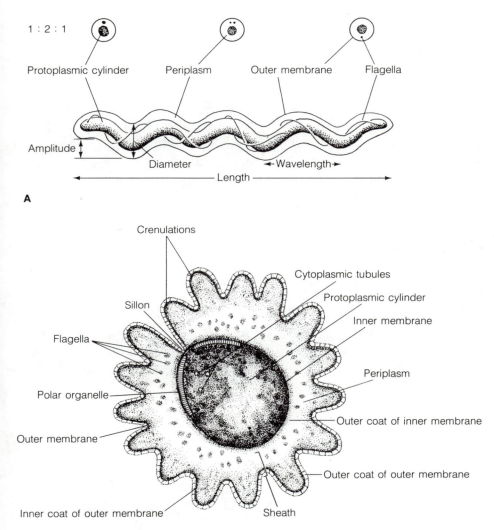

1 : 2 : 1

Protoplasmic cylinder   Periplasm   Outer membrane   Flagella

Amplitude

Diameter

Wavelength

Length

A

Crenulations

Cytoplasmic tubules

Protoplasmic cylinder

Sillon

Inner membrane

Flagella

Polar organelle

Periplasm

Outer membrane

Outer coat of inner membrane

Outer coat of outer membrane

Inner coat of outer membrane

Sheath

B

FIGURE 9-1

Spirochetes.
(A) Generalized *n:2n:n* spirochete in which *n* = 1. (See footnote on page 279.) (B) Diagram of *Pillotina*. For details, see Margulis and Hinkle (1992).

Members of the genus *Diplocalyx* are described as symbionts of the hindguts of kalotermitids. The outer coat of their inner membrane is thickened (and in the type case, split; hence "diplo"). They have 40 to 60 periplasmic flagella and are 0.7 to 0.9 μm in diameter. Both polar organelles and cytoplasmic tubules have been found in *Diplocalyx*.

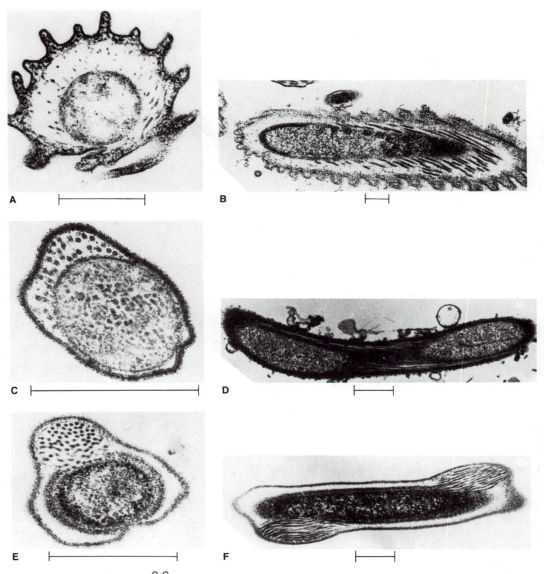

FIGURE 9-2

Large spirochetes symbiotic in termites; family Pillotinaceae. Transmission electron micrographs, bars = 0.5 μm.

(A, B) Transverse and longitudinal sections of *Pillotina*. (C, D) Transverse and longitudinal sections of *Hollandina*. (E, F) Transverse and longitudinal sections of *Diplocalyx*. [A, B, and D courtesy of David Chase. C courtesy of L. P. To. E courtesy of I. D. Gharagozlou.]

Members of the genus *Hollandina* are symbiotic in a wide range of wood-eating insects (cockroaches, hodotermitids, kalotermitids, mastotermitids, and rhinotermitids). Their diameter ranges from 0.4 to 1 μm, and they have 30 to 60 periplasmic flagella. *Hollandina* tends to have a conspicuous outer coat on the outer membrane, polar organelles, and cytoplasmic tubules.

Members of the genera *Leptonema* and *Leptospira*, about 0.1 μm in diameter, with curved ends and few flagella, are the only strictly aerobic spirochetes reported. Differing in ultrastructure, they are often found in mammals and can be pathogenic.

Members of the genus *Mobilifilum*, found free-living in anaerobic, sulfurous muds below laminated microbial mats in Baja California, Mexico, have a conspicuous outer coat of the inner membrane that extends to surround the 10 periplasmic flagella in their bundled distribution. They are about 30 μm in length and their diameter is 0.25 μm (Margulis, Hinkle, Stolz, Craft, Esteve, and Guerrero, 1990).

Members of the genus *Pillotina*, symbiotic in the hindguts of kalotermitid and rhinotermitid wood-eating termites, are large spirochetes (0.6 to 1.5 μm in diameter) distinguished easily by their crenulated (ruffled) surfaces (Figure 9-4) and the extensive inner coats of their outer membranes. Both cytoplasmic tubules and polar organelles have been detected in their protoplasmic cylinders (diagrammed in Figure 9-1; micrograph in Figure 9-2A).

Members of the genus *Spirochaeta* are free-living marine or pond-water microbes from 0.25 to 0.75 μm in diameter. They have few flagella arranged in a 1-2-1 or 2-4-2 pattern (Canale-Parola, 1992).*

Members of the genus *Treponema*, slender spirochetes that range from 0.18 to 0.5 μm in diameter, are anaerobic and, by definition, live with other organisms; in many mammals, they are pathogens or symbionts. Certain strains cause syphilis and yaws. Although they often have a greater number of flagella, they generally have flagellar patterns like those of *Spirochaeta*, from which they probably evolved.

Detailed information on all of these except *Spirosymplokos* is in Balows et al. (1991, vol. IV).

---

*The *n*:2*n*:*n* numbers refer to three consecutive transverse sections—end, center, end. The basal insertions of spirochete flagella are at the ends of the cell. Spirochetes having a single flagellum at each end, overlapping at the center, show a 1:2:1 pattern; those having two at each end, overlapping at the center, show a 2:4:2 pattern. When the two flagella are short and therefore do not overlap (e.g., in some *Leptospira*) the expression becomes 1:0:1.

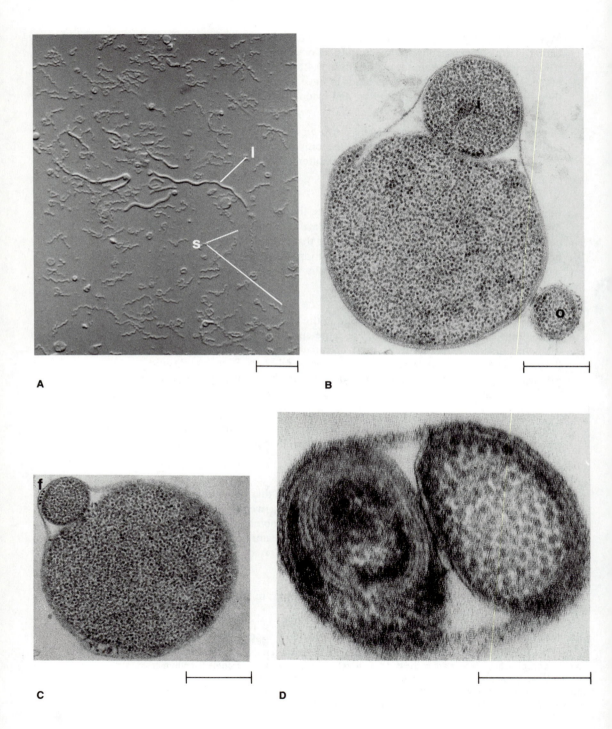

FIGURE 9-3

Spirosymplokos deltaeiberi, composite spirochete with granular cytoplasm from an enrichment culture collected from microbial-mat mud at the Ebro river delta, Catalonia, Spain.
(**A**) Large (*l*) and small (*s*) spirochetes enriched under anoxic conditions. Live, phase-contrast light micrograph, bar = 10 μm. (**B**) Small to medium-sized spirochetes seen inside (*i*) and outside (*o*) the outer membrane of an organism interpreted in (**C**) to be a 5:10:5 spirochete (see footnote on page 279). Transverse electron micrograph sections, bars = 1.0 μm. [**B** courtesy of F. Craft.] (**D**) Sporelike structure in a small granulated organism. Transverse electron micrograph section, bar = 0.5 μm.

The large spirochetes have been observed for many years by microscopists who study the microbial community of the termite hindgut (Leidy, 1881; Dobell, 1912; Dubosq and Grassé, 1927). The spirochetes symbiotic in termite hindguts and wood-eating cockroaches, referred to collectively as **pillotinas,** are in the family Pillotaceae (Hollande and Gharagozlou, 1967), now corrected to Pillotinaceae (Balows et al., 1991). Members of the family Pillotinaceae are recognized in live termites by their large size and flexuous swimming; electron microscopy is required for recognition of genera. Some contain within their protoplasmic cylinders longitudinally aligned cytoplasmic tubules about 21 nm in diameter (Margulis, To, and Chase, 1981; Bermudes, Chase, and Margulis, 1988). Peculiar rosette structures, but no cytoplasmic tubules, are seen in *Cristispira* (Margulis, Nault, and Sieburth, 1991). Because of their microtubules, distribution, and behavior, free-living relatives of pillotina spirochetes are candidates for ancestors of undulipodia.

Pillotina calotermitidis, discovered by Hollande and Gharagozlou (1967) in the hindgut of an obscure termite, *Calotermes praecox* (now reclassified as *Proelectrotermes praecox*), was known only from the forested western end of the island of Madeira. For many years, pillotina spirochetes were not available for study. While experimenting with the microtubular axostyle of the protist *Pyrsonympha,* symbiotic in the hindgut of *Reticulitermes flavipes,* the common subterranean termite of the northeastern United States, Bloodgood et al. (Bloodgood and Miller, 1974; Bloodgood et al., 1974) noticed large spirochetes. Perusing Bloodgood's micrographs, I saw spirochetes extremely similar, if not identical, to the Madeiran *Pillotina* of Hollande and Gharagozlou. A systematic search revealed pillotina spirochetes in many North American and European termites (To, Margulis, and Cheung, 1978; To et al., 1980).

Pillotinas are found with the diverse and abundant microbiota in insects with diets restricted to wood. In more than 30 species of dry wood–eating, damp wood–eating, and subterranean termites examined, pillotina spirochetes were abundant components of the hindgut microbiota. They were

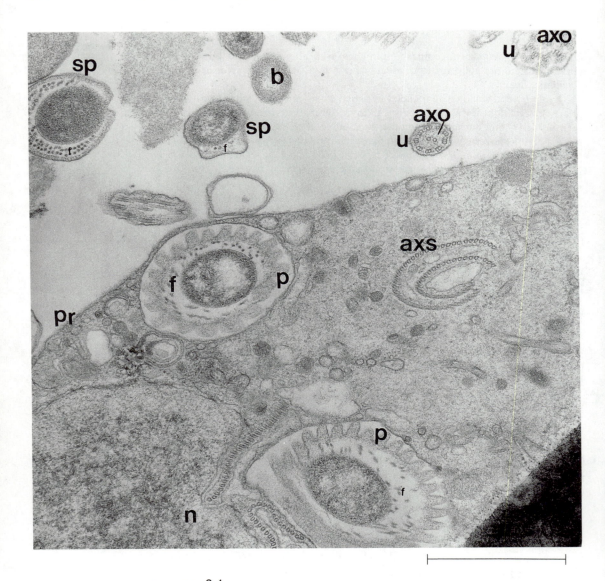

FIGURE 9-4

Spirochetes (*sp*) and protist axonemes (*axo*) from hindgut preparation from
Reticulitermes hesperus, a subterranean termite. Two crenulated large *Pillotina*
organisms (*p*) are inside the protist (*pr*), whose microtubule axostyle (*axs*) is also
seen. *b* = unidentified bacterium; *f* = flagella; *n* = nucleus; *u* = undulipodium.
Electron micrograph, bar = 1 μm.

found in 21 out of 21 species of kalotermitids and in 5 out of 5 rhinotermitids; approximately 323 species of dry wood–eating and 180 species of
subterranean termites remain to be investigated. Although also common in
the wood-eating cockroach *Cryptocercus punctulatus*, no pillotinas (or very
few) were found in termites with eclectic diets that include rotted wood (for
example, hodotermitids and nasutitermitids). The hindgut communities
of wood-eating insects usually include — in addition to pillotinas —
pyrsonymphids, hypermastigote protists (especially trichonymphids), trichomonads (especially devescovinids), and many prokaryotes: nitrogen-fixing bacteria, large peritrichously motile rod bacteria, Gram-positive cocci,
the filamentous spore-forming bacteria of the genus *Arthromitus* (Margulis,
Olendzenski, and Afzelius, 1990), and many treponeme-like spirochetes
(Breznak, 1973; To, Margulis, and Cheung, 1978; To et al., 1980).

Both mastotermitids and kalotermitids are judged entomologically to
have evolved early; they are related to wood-eating cockroaches —
Cryptocercidae, Order Blattaria — from which the termite order Isoptera is
thought to have originated. Wood-eating larvae, nymphs, and pseudergates
of kalotermitids consistently harbor enormous populations of large spirochetes. Soldiers and alates contain them as well. Termites from the same
population contain the same genera and similar population densities of
spirochetes, although seasonal variation and differences between termites
collected from disjoint geographical locations have been noted. Pillotina
spirochetes are found between the undulipodia of hypermastigotes, such as
Trichonympha agilis. They even enter protists (Margulis, Chase, and To,
1979) (Figure 9-5).

Pillotinas probably have not emerged from their hindgut ecosystems for
more than 200 million years. Because they cannot survive even a few hours
outside of their hosts without elaborate precautions, it is likely that they
coevolved in association with the termites. Although uncultivable, pillotinas can be enriched by heat and starvation of the termite (Grosovsky and
Margulis, 1982). No genetic or metabolic information about these microbes
is available; they are distinguished by morphological criteria: length, diameter, amplitude, degree of crenulation of the sheath, and number and
distribution of the flagella (Bermudes, Chase, and Margulis, 1988).

With corkscrew-type movements, pillotina spirochetes can penetrate
bundles of *Arthromitus* trichomes, just as they "nose up" between mastigote undulipodia. An extreme example was observed in Alicante (Spain):
kalotermitids in which there were very few protists but huge numbers of
Arthromitus bundles. The spirochetes penetrated these bundles longitudinally, first moving forward and then reversing, but remaining inside the
bundles.

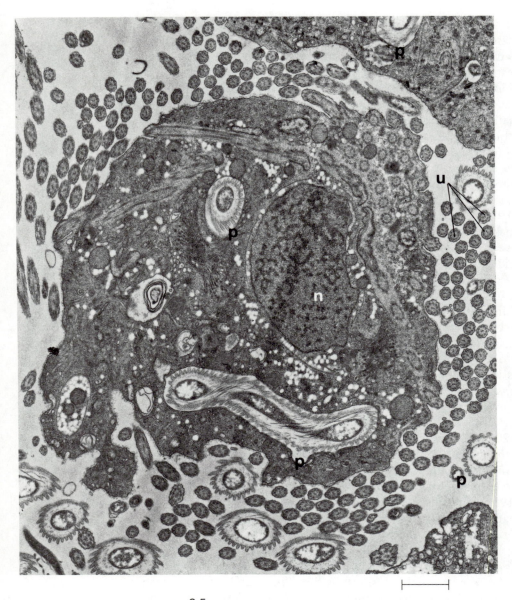

FIGURE 9-5

Pillotina spirochetes (*p*) live in the cytoplasm of an unidentified protist in the hindgut of the dry wood–eating termite *Reticulitermes hesperus*, San Diego, California. Pillotinas also swim freely in the lumen of the termite intestine. The nature of the association between spirochetes and protists is unknown, although the association often can be observed in live material. *n* = nucleus of mastigote; *u* = undulipodia of mastigote. Transmission electron micrograph bar = 1 μm [Courtesy of David Chase.]

Spirochetes that contain cytoplasmic tubules include treponemes (Hovind-Hougen, 1976; Bermudes, Chase, and Margulis, 1988) and members of the family Pillotinaceae. Cytoplasmic tubules (24 nm diameter) are especially conspicuous in *Diplocalyx* from *Cryptotermes cavifrons*, where tubules are seen in nearly every thin section (Ashen, 1992). Spirochetes about 0.4 μm in diameter have been found in the anaerobic portions of microbial mats at Laguna Figueroa, Baja California (Horodyski and Vonder Haar, 1975). The spirochetes (*Spirochaeta bajacaliforniensis*; Fracek and Stolz, 1985) react with fluorescein-labeled antibody made against guinea pig brain tubulin because of a nonflagellar, periplasmic, 65-kilodalton protein (S1) that runs the length of the spirochete. Peptide sequencing and other analyses show S1 to have little amino acid sequence in common with tubulin but several tubulin-like properties: temperature-dependent polymerization, nucleotide binding, and a shared epitope (Hinkle, 1991). Since S1 protein is unambiguously a heat-shock or chaperonin-type protein, homologous to GroEL of *E. coli*, its relation to tubulin, if any, is not clear. Although 24-nm cytoplasmic tubules are indisputably present in some spirochetes, the question of whether or not they are composed of tubulin is unresolved. Microbial mats, unlike termite intestine, are locally depleted in eukaryotes; if cytoplasmic tubules and tubulin protein are present in mat spirochetes, they are not likely to have come from contaminating eukaryotes.

PREADAPTIVE SPIROCHETE BEHAVIOR

An important question . . . is whether flagella [undulipodia], cilia, cirri, membranelles, and other units of structure are always proportionate in size to that of the individual protozoan; do they grow with the cell, or are they the same size in normal animals, tiny regenerants, and giant forms? It has been assumed that cilia and cirri are always proportionate to cell size. . . . There is in fact good evidence that ciliary organelles of any one type are constant in size through varying in number. For example in the largest stentors [$\approx$750 μm, nearly a millimeter] and the smallest regenerated fragments [$\approx$7 μm] the length and width of the membranelle structures composed of rows of cilia [see Figure 9-6], and hence the lengths of the peristome structure composed of membranelles at the oral opening of the single-celled organism vary enormously.

V. TARTAR, 1961

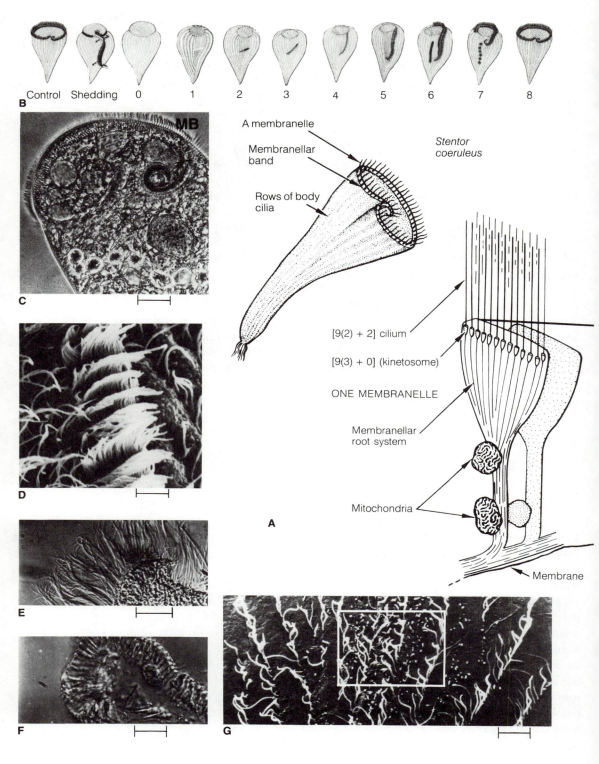

Control Shedding 0 1 2 3 4 5 6 7 8

B

MB

C

D

E

F

G

A membranelle

Membranellar
band

Rows of body
cilia

*Stentor
coeruleus*

[9(2) + 2] cilium

[9(3) + 0] (kinetosome)

ONE MEMBRANELLE

Membranellar
root system

Mitochondria

Membrane

A

FIGURE 9-6

Oral membranelles of *Stentor coeruleus*.
(**A**) Membranelles, patterned rows of cilia, make up the oral membranellar band of
the heterotrichous ciliate *Stentor coeruleus*. The membranelles direct particles of
food into the gullet of this "huge" protist. (**B**) When exposed to certain substances,
such as sucrose, glycerin, urea, or sea water, stentors shed their oral membranelles
in synchrony. If removed into a fresh medium, the proceed synchronously to
regenerate their oral kinetosomes and axonemes—some 20,000 in each cell. The
morphogenesis of the oral band takes about 8 hours at 22° C. The regenerating
band always appears at the stripe-contrast zone, where the wide pigmented cortical
stripes meet the narrow ones. The numbers signify hours after shedding (Tartar,
1961). (**C**) Live stentor, showing macronuclear nodes and membranellar band (*MB*).
Phase-contrast light micrograph, bar = 50 μm. (**D**) A regenerating oral band. Sparse
cilia on the left are body cilia, which are not shed and regenerated in this process.
Scanning electron micrograph, bar = 10 μm [Courtesy of J. Paulin.] (**E**) Isolated
membranellar band shed by *S. coeruleus*. Membranellar bands and their kinetosomes
may be collected for cytological and biochemical study. If a 12.5 percent solution of
sucrose is used to induce shedding of the band, the cilia remain attached to the
band. Phase-contrast light micrograph, bar = 20 μm [Courtesy of C. van Wie.] (**F**)
Membranellar band shed by *S. coeruleus*. A 2 percent solution of glucose induces
the shedding of the band, but the ciliary axonemes are also sloughed off separately.
Phase-contrast light micrograph, bar = 20 μm. [Courtesy of C. van Wie.] (**G**) The
stripe-contrast zone, the wide stripes on the right and the narrow ones on the left.
About 1.5 hours after a stentor has shed newly emerged cilia, the membranellar-band
analgen (within white square) can be seen at the stripe-contrast zone. The anlagen is
also shown in part **B** of this figure as an oblique white line to the left of the
stripe-contrast zone at 1 hour. Scanning electron micrograph, bar = 10 μm. [Courtesy
of J. Paulin.]

Termite spirochetes swim freely in the lumen of the intestine, or they are
modified for attachment to specific protists. Stages intermediate between
these states also exist.

Motility is indispensable to survival of unattached microbes in the
termite gut, and the types of possible patterns of organization of motility
organelles in these protoctists are limited. Zoomastigina show "karyomastigont" cell structure: undulipodia are attached to nuclei by proteinaceous
material involved in peculiar mitoses (see Table 8-3). Although most are 24
nm, tubulin microtubules in eukaryotes range from 15 to 30 nm in diameter (Cachon and Cachon, 1974; Margulis, To, and Chase, 1978). The
diameter of the [9(2) + 2] undulipodium composed of such 24-nm-diame-
ter microtubules is even more uniform (0.25 μm). In some cases (Phylum
Zoomastigina, Class Parabasalia), cell organization and division limit the
number of undulipodia to four per nucleus. To prevent being defecated,

mastigotes may remain small and motile by virtue of their few undulipodia, as do *Foania* or *Tricercomitus*. Alternatively, the number of undulipodia may increase with the number of nuclei, as in polymonads (Cleveland, 1956). However, the multinucleate condition leads to problems of gene dosage and distribution, precluding a diploid–haploid life cycle. Thus, to increase in size but remain motile, mastigotes in microbe-rich environments tend to develop motility symbioses, convergently and, in some cases, spectacularly. This phenomenon is particularly conspicuous in devescovinids, some of which developed motility symbioses with spirochetes or peritrichously flagellated bacteria (see Figure 7-11; Tamm, 1982; Copeland quotes, page 275).

Preadaptations to the formation of motility symbioses are observed in termite hindguts, where the population density of bacteria, including spirochetes, can exceed 10^{11} per milliliter of gut fluid. Spirochetes temporarily form a living covering on moribund trichomonads or devescovinids. Packed and entrained spirochetes form rows attached loosely to the dying carcasses of the protists; after some time, they beat together in synchrony within a row, but in metachrony from row to row. That attached spirochetes in these symbioses beat in synchrony is not mysterious. When highly motile microbes or sperm are sufficiently close together, they exert mechanical forces upon each other that compel synchronous behavior (Machin, 1963; Lighthill, 1976). When the protists die, the number of attached spirochetes increases enormously, so that the carcasses become entirely covered with bundles of synchronously beating spirochetes. At low magnification these immense populations of coordinated spirochetes resemble diffraction patterns (Figure 9-7). Dense spirochete populations attach to wood, grease globules, and other debris as well. That they are not permanently attached in bundles is evident because, from time to time, individuals leave the pack and swim away. In some cases, spirochetes moving in greater or lesser synchrony and attached only to each other will move in a bundle. Spirochetes have "sticky" surfaces; often, other bacteria are transported by them temporarily. Bundles of pillotinas and smaller spirochetes may generate forces large enough to rotate protists in their vicinity. The tendency of motile cells to beat together is understood on fluid-hydrodynamic principles (Brennen and Winet, 1977). Thus, pillotinas are preadapted for coordinated movement.

MOTILITY SYMBIOSES

Mixotricha is propelled not by its flagella [undulipodia] but by the undulation of its attached spirochetes.

L. R. Cleveland and A. V. Grimstone, 1964

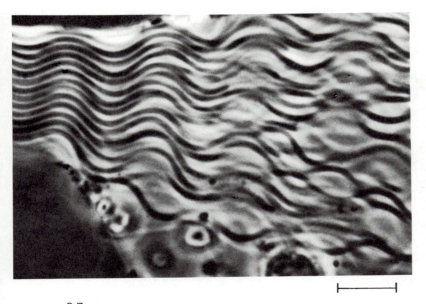

FIGURE 9-7

Spirochetes from hindgut of *Pterotermes occidentis*. Even when not attached to protists, pillotina spirochetes beat synchronously, apparently because of physical proximity. Nomarski phase-contrast light micrograph, bar = 5 μm.

Motility symbiosis, in which spirochetes attach to the surface of protists and contribute to or even cause the host's motility, has been described for *Mixotricha paradoxa*, an intestinal symbiont in *Mastotermes darwinensis* (see Figure 7-9; Cleveland and Grimstone, 1964). Motility symbioses are common and involve different bacteria and protists (Margulis, Chase, and To, 1979; Tamm, 1982).

Many termite spirochetes are attached permanently to larger protists, often in tufts but occasionally in fringes that cover the host's entire surface. Well-developed spirochete attachment sites appear in electron micrographs (Figure 9-8); Smith and Arnott, 1974; Smith, Buhse, and Stamler, 1975; Smith, Stamler, and Buhse, 1975; Bloodgood and Fitzharris, 1977). Attached spirochetes are "polarized": one end is highly differentiated; protists produce corresponding cytoplasmic "docking sites" (Margulis, Chase, and To, 1979). Some spirochetes have flattened proximal ends with fibrous connections to their hosts. Because spirochetes, their protist hosts, and their attachment sites differ, the associations must have evolved polyphyletically.

Do attached spirochetes confer additional motility on their mastigote hosts? It is obvious from observing the huge mastigote *Mixotricha paradoxa*

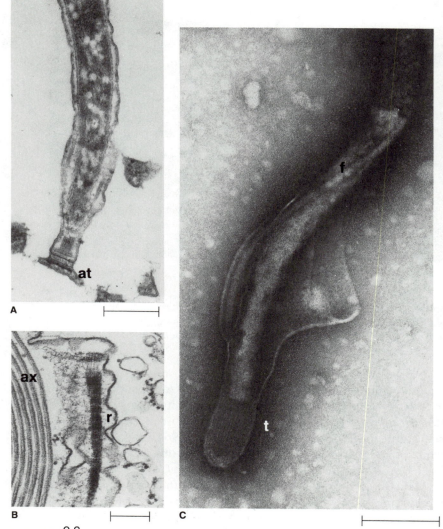

FIGURE 9-8

Modifications of spirochetes for attachment to other cells.
(**A**) Spirochete modified for attachment (*at*) to *Pyrsonympha* from *Reticulitermes flavipes*. Transmission electron micrograph, bar = 0.5 μm. (**B**) Like a root fiber, an attachment organelle (*r*) extends deep into the cytoplasm of the host. In the live state, the organisms are attached as in Figure 9-9**B**. *ax* = axostyle of host. Transmission electron micrograph, bar = 0.5μm. (**C**) Tip (*t*) of unidentified spirochete from *Reticulitermes hesperus* showing modification for attachment. *f* = flagellum. Transmission electron micrograph, negative stain; bar = 1μm. [**A** and **B** courtesy of H. Smith and H. J. Arnott; **C** courtesy of David Chase.]

that it is propelled by its half-million, synchronously beating surface spiro-chetes and not by its four undulipodia (Cleveland and Grimstone, 1964). Photomicrographs show regular connections between raised portions of the protist, the spirochetes, and other ectosymbionts (surface bacteria that may anchor the spirochetes or provide nutrient for energy generation) (see Figure 7-9).

The thrust generated by undulipodia of a given length is limited by the length of the load; bodies moved by one or very few undulipodia are generally far less than 50 μm long (Lighthill, 1976). From the lengths of the four or fewer undulipodia of *Mixotricha*, compared to the length of the entire cell, one concludes that the spirochetes are responsible for the motility of the complex. In protists more than 100 μm long, the spirochetes have been "borrowed" as organelles of motility; many cases are analogous to *Mixotricha*. Cleveland's films (Cleveland, 1956) show that the helical movements of attached spirochetes are correlated with the forward move-ment of *Mixotricha*.

Kirby (1936) described a motility symbiosis inadvertently in his de-scription of a devescovinid (*Pseudodevescovina uniflagellata* Sutherland) from the Australian dry wood–eating termite *Kalotermes (Neotermes) in-sularis*. He remarked that this protist is normally covered with

> a dense investment of short spirochetes. . . . The length is about 8–10 microns, and is quite uniform in the great majority. . . . Often the coat is especially thick on the papilla. Movement of the spirochetes is extremely rapid, but becomes slower under unfavourable conditions, which also slow down the activity of the flagellates. . . . The attached ends of the spiro-chetes seem actually to be embedded in the surface of the body although for a very short distance only. In Schaudinn-iron hematoxylin material the embedded part [Kirby's Figure 3] may appear thicker than the rest and heavily stained, while the free part of the microorganism is pale. The rather evenly distributed coat of spirochetes then in stained material is almost indistinguishable from a coat of cilia with basal granules [ki-netosomes].

The average size of the devescovinid was recorded as 73 by 41 μm. The undulipodia, one trailing and three anterior, are about 30 μm long, less than half the length of the body. The anterior undulipodia are used like whips, not for forward motility. The movement of the organism is ex-tremely vigorous. Hydrodynamically, Kirby's conclusion that movement is caused by the single, short, trailing undulipodium is highly unlikely (Lighthill, 1976). Kirby was probably describing a motility symbiosis be-tween spirochetes and unidentified microbes. We have observed, and in some cases filmed, motility symbioses involving other unidentified mi-crobes: free-living spirochete–protist associations from microbial mats at the Ebro river delta (Catalonia, Spain), and from *Incisitermes schwartzi* (Flordia; see Figure 7-10), *Pterotermes occidentis* (Arizona), and *Marginitermes hubbardi* (Arizona). The motility symbioses described by Cleveland and Grimstone (1964) are thus not an isolated phenomenon.

How did motility symbioses originate? In termite guts, spirochetes often congregate around the sensitive, wood-ingesting posterior end of a protist. Remaining in this nutrient-rich location gives the spirochetes access to leftovers. The wood-ingesting regions of many termite protists are covered densely with temporarily adhering motile spirochetes, without evidence, in most cases, of motility symbiosis. Since contact of wood particles with the posterior ingestive zone of these protists prompts a unique rapid ingestion response, the moving spirochetes may first have been retained by protists because their movements encouraged more efficient wood ingestion (Lavette, 1967). With time, many spirochete–protist associations became regular and, eventually, permanent.

Spirochetes tend to adhere to many surfaces and form associations (Margulis, Chase, and To, 1979). Complex attachments of spirochetes to pyrsonymphids have been described by Bloodgood and Fitzharris (1977); Smith, Buhse, and Stamler (1975); and Smith, Stamler, and Buhse (1975). Raikov (1971) described spirochete attachments to ciliates. Treponemes attach to intestinal and testicular epithelia of mammals. In microbial mats, living spirochetes attach to filamentous cyanobacteria. *Treponema*-like spirochetes attach to the mesogleal tissue of freshwater hydras. The fatty acid and other complex nutritional requirements of spirochetes may have preadapted them for association. How is the continuity of their partnership through cell divisions ensured? Some spirochetes remain attached during their own transverse division. Some replicate or, at least, redifferentiate their attachment sites. This is analogous to the behavior of kinetosomes in the development, for example, of clam gill cilia; both cilia and attached spirochetes have conspicuous root-fiber systems (Figure 9-9).

The motility of spirochetes and undulipodia can be compared. Pillotina spirochetes are visible even at fairly low magnifications. Attached bacteria and other debris are visual markers for observing their movements. Free-

FIGURE 9-9

Comparison of spirochete attachment sites with kinetids of clam gill cilia.
(A) Clam gill cilium. ax = axostyle; k = [9(3) + 0] kinetosome; r = root fiber. Transmission electron micrograph, bar = 0.5 μm. (B) Attachment of a spirochete to a *Pyrsonympha*-like protist from the hindgut of the California subterranean termite *Reticulitermes hesperus*. at = attachment site; r = root-fiber-like structure; s = protoplasmic cylinder of spirochete. Transmission electron micrograph, bar = 0.5 μm. (C) A row of clam gill cilia. ax = axostyle; r = root-fiber-like structure. Transmission electron micrograph, bar = 1 μm. [A and C courtesy of Fred Warner; B courtesy of David Chase.]

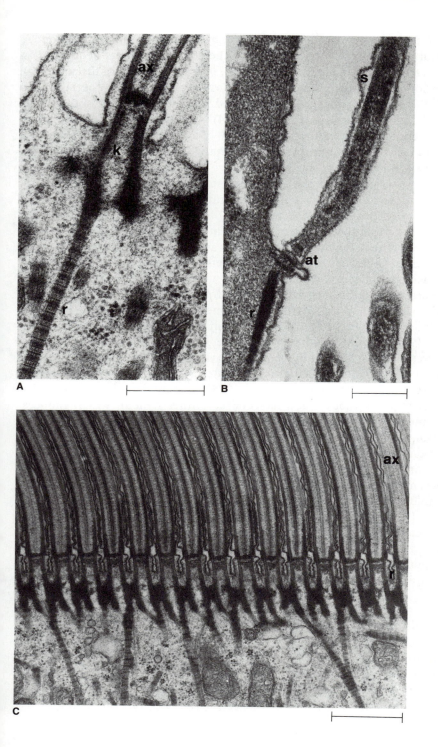

swimming spirochetes translate forward by rotating in either direction. A helical wave passes down the body of the active swimmers. The wave is generated subterminally from either end, probably because of the subapical insertion of the periplasmic flagella. Most undulipodia, by contrast, beat with the effective stroke in the plane and recovery stroke out of the plane (Sleigh, 1962; Brokaw and Gibbons, 1974)—but not always; helical motion has been observed in the undulipodia of opalinid protists parasitic to amphibians (Cheung and Jahn, 1975), in the tails of moribund hyphochytrid zoospores, and in the body cilia of moribund *Stentor coeruleus*. A helical wave occurs in the propulsion of the motile (yet $[9(2) + 0]$ in transverse section) sperm tails of gregarines (Desportes, 1970). The beat frequency of most undulipodia (10 to 20 hertz), the longitudinal propagation of the wave, their uniform size and morphology, the tendency toward synchrony, the potential helicity of the wave, and the nature of the attachment sites (underlain by striated rootlets) suggest an analogy between attached spirochetes and undulipodia.

Bacterial motility (including that of *Spirochaeta*) depends on the conversion of an electromotive force into mechanical movement where the basal disk of the flagellum is embedded in the membrane (Mitchell, 1966; Berg, 1975). In undulipodial motility, the generation of mechanical energy requires dephosphorylation of ATP. At least two axonemal proteins— tubulin and dynein (a magnesium-activated ATPase)—interact as microtubules slide past one another (Satir, 1974). The minimal isolated motility system requires tubules and an ATPase (either dynein, which induces movement toward the "minus" end of a tubule, or—in brain tissue— kinesin, which leads to movement toward the "plus" end). No comparable work on the motility mechanism of the large, complex pillotinas exists.

If undulipodia originated from spirochetes, the sequence homologies between the proteins and DNAs of the appropriate spirochetes and those of the microtubule organizing centers (MCs) of mitotic eukaryotes will be found. Furthermore, the mechanisms that generate helical waves ought to be homologous. If the microtubular system originated endogenously by compartmentalization in eukaryotes, then the genome and components of the motility system of these obscure spirochetes are expected not to be homologous with those of the undulipodia.

CYTOPLASMIC TUBULES IN PROKARYOTES

In absolute symbiosis, the adjustment responses of the
two symbionts have been completed, making it difficult to
recognize their exact nature.

I. E. WALLIN, 1927

Evidence for cytoplasmic tubules in spirochetes is based on three different techniques. The first is direct observation of glutaraldehyde-fixed *Pillotina*, *Diplocalyx*, and related spirochetes by transmission electron microscopy (Hollande and Gharagozlou, 1967; Gharagozlou, 1968; Margulis, To, and Chase, 1978; To and Margulis, 1978). In the second, polyclonal fluorescent antibodies made to various tubulins are tested for cross-reaction in cytological preparations; the immunofluorescent, diluted antitubulin serum is applied directly to the fixed spirochetes and observed by fluorescent microscopy. The third technique entails the isolation of a spirochete protein that comigrates on acrylamide gels with tubulin from mammalian brain. The tubulin-like protein was obtained from a termite hindgut fraction containing only prokaryotes (primarily large spirochetes in which cytoplasmic tubules have been seen) and from the small spirochete that has fibers, not tubules: *Spirochaeta bajacaliforniensis*.

Electron micrographs of three termite spirochete genera are shown in Figure 9-2. Sections containing tubules have been seen occasionally in these pillotinas, taken from the subterranean termites *Reticulitermes flavipes* and *R. hesperus*, and from the dry-wood termites *Incisitermes schwartzi*, *I. minor*, and *Pterotermes occidentis* (Figure 9-10). Often the tubules, longitudinally aligned, look as though they follow the contours of the spirochetes' helical bodies; by no means are they seen in all sections. Except in *Diplocalyx* from *Cryptotermes*, they are not necessarily abundant. None is arranged in ninefold symmetry pattern. Similar cytoplasmic tubules are seen in unidentified long, thin, nonflagellated termite bacteria. Cytoplasmic tubules as large as 20 to 24 nm in diameter are limited to large spirochetes. Only smaller tubules have been described in *Treponema* and *Leptonema* (Hovind-Hougen, 1976).

Antibodies to microtubule protein have been prepared by many investigators. Evidence is strong for the presence — in unidentified spirochetes from *I. schwartzi* and *P. occidentis*, and in *Spirochaeta bajacaliforniensis*, *S. halophila*, *S. litoralis*, and *Borrelia burgdorferi* — of protein that cross-reacts with anti-tubulin antibody (Margulis, To, and Chase, 1978). The cross-reacting proteins from *S. halophila* and *S. bajacaliforniensis* both have a molecular mass of 65 kilodaltons. Significant and regular staining by anti-tubulin antibody bound to fluorescein was seen along the entire length of large spirochetes. The fluorescent antibody in the spirochetes gave a signal comparable in intensity to that of the undulipodia attached to mastigotes unavoidably present in termite hindgut preparations. The intensity of the spirochete fluorescence, even in spirochetes lacking tubules, suggests that pillotinas contain homologous proteins that are not necessarily organized into tubules (Figure 9-11). However, although many controls were performed to verify that the fluorescence was due to tubulin, in the absence of complete sequence data the results of immunofluorescence experiments

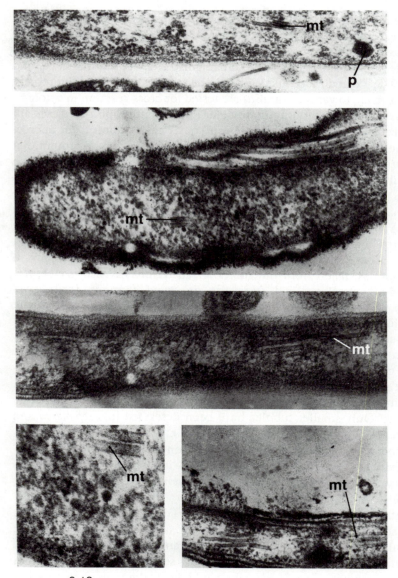

FIGURE 9-10

Longitudinally aligned microtubules (*mt*) about 24 nm in diameter, in spirochetes and unidentified bacteria from the hindguts of dry wood–eating termites. *p* = phage. [Courtesy of David Chase and L. P. To. From Margulis, To, and Chase (1978).]

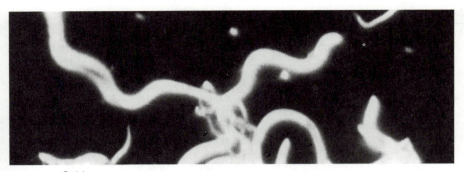

FIGURE 9-11

Fluorescing *Pillotina* sp. from *Incisitermes schwartzi*. The hindgut preparation stained with fluorescein-labeled antibody made against brain tubulin presumably reflects staining of S1 protein (as discussed on pages 285 and 295). Light micrograph.

are not conclusive. Recent DNA sequence studies lead us to conclude that the "tubulin-like protein" is a chaperonin, and hence the question of prokaryotic tubulin is still unanswered.

ORIGINS OF UNDULIPODIA

The colonization of host cells by modern symbionts is surveyed. The morphological distinction between extracellular and intracellular symbionts is not sharp, and the various kinds of association can be arranged in a graded series of increasing morphological integration of the symbiont into the host cell. Apart from some aggressive parasitic infections, the great majority of symbionts are enclosed by a host membrane in a vacuole. Those not enclosed in a host vacuole usually cannot be cultivated outside the cell. It is therefore surmised that the encirclement by a vacuolar membrane would only disappear, if at all, in the later stages of the evolution of intracellular symbioses.

D. C. Smith, 1979

If undulipodia indeed began as symbiotic spirochetes, what became of their DNA, their protein-synthesizing machinery, their external and plasma membranes, and the rest? Clearly, the spirochete hypothesis leaves many

TABLE 9-2

Symbiotic origin of microtubule organizing centers (MCs) and their products.

EVOLUTIONARY QUESTION	HYPOTHESIS
What was the free-living form of the protoundulipodium?	Anaerobic or microaerophilic motile prokaryotes (spirochetes) exhibiting ATPase-sensitive motility and containing microtubules and other proteins homologous to those of axonemes. Tubules arranged in pairs or triplets (Figure 9-12).
In what host cells did the protoundulipodia first become established?	In heterotrophic archaebacteria capable of glucose fermentation by the Embden-Meyerhof pathway. Host most likely had histone-like nucleosome structure rendering its DNA acid- and heat-resistant.
What environmental agent selected for and maintained the symbioses?	Scarcity of food; made motile by acquisition of spirochetes and thus better able to seek nutrients, protists evolved from the symbiotic complex.
When did these symbioses become established?	In the Proterozoic eon, during and after transition to an oxic atmosphere and as a prerequisite to the evolution of mitosis (see Table 5-5).
Are these symbioses obligate?	Yes, in all mitotic organisms, because of the utilization of "motility proteins" from the free-living spirochete for mitosis and other intracellular motility processes.
What genetic legacies of free-living prokaryotes do MCs and undulipodia retain?	Evidence for DNA (Hall et al., 1989) and kinetosomal RNA, most likely in the lumen of centrioles and kinetosomes (Stubblefield and Brinkley, 1967; Dippell, 1976); new RNA synthesis accompanies new kinetosome production (Younger et al., 1972); and kinetosomal RNA may function in MC formation (Heidemann et al., 1977).
What did the protoundulipodia lose after becoming symbionts?	Peptidoglycan walls and some membranes; most biosynthetic functions, including synthesis of DNA and ribosomal proteins, probably were integrated to form the nuclear gene–controlled nucleocytoplasmic system.
What changes the ratio between the number of original host genomes (now nucleocytoplasm) and the number of MCs?	Subjection of cells to low concentration of vitamin E and high concentration of oxygen (Hess and Menzel, 1968); treatment with colcemid (Stubblefield and Brinkley, 1966); with pargyline, a monoamineoxidase inhibitor (Milhaud and Pappas, 1968); or with pituitary hormone (Dustin, 1978); normal tissue differentiation (Sorokin, 1968); life cycle stages (Dirksen, 1991); and irreversible mutations (speciation in ciliates) (Lynn and Small, 1990).
Can permanent loss of the former symbiont from the nucleocytoplasm be induced?	No. Because integration led to mitotic eukaryotes, loss, in principle, would be lethal. Cleveland (1956) treated developing hypermastigote gametes with low concentrations of oxygen; this caused preferential destruction of the undulipodial system, producing a nucleated cell the cytoplasm of which was unable to divide (although, because nuclear chromatin had been produced, DNA presumably had been synthesized so chromatids did separate but did not move apart, i.e. karyokinesis occurred).
How much may MCs and their [9(2) + 2] homologues dedifferentiate?	To below the limit of resolution by the electron microscope (Schuster, 1963; Stubblefield and Brinkley, 1966; Dirksen, 1991; MCs in plants and fungi, presumably composesd only of nucleoprotein, apparently are normally dedifferentiated to below the [9(3) + 0] level (see Figures 8-1 and 8-4).
How are offspring MCs produced?	Usually they form from parent MCs; for example, new kinetosomes form at right angles to the parent in a complicated sequence with many variations (Mizukami and Gall, 1966; Dirksen and Crocker, 1965; Dirksen, 1991; see Figure 8-3). Centrioles and associated organelles that duplicate in mitotic organisms before becoming visible must have undergone nucleic acid replication, by hypothesis (see Figure 8-20).
What organisms that became eukaryotes evolved from spirochete symbioses?	Amitochondrial undulipodiated protoctists and, after mitochondrial acquisition, nearly all other protoctists. All plants, animals, and fungi.
What do the genomes of the original spirochete symbionts code for now?	Not for tubulin, but otherwise unknown; they may code for a tubulin-based herbicide resistance (James et al., 1988) or RNAs or proteins of the 80S ribosomes. Membrane-associated RNA or DNA synthesis may be related to kinetosome reproduction (Dippell, 1976; see Figure 8-20).

TABLE 9 - 2 (*continued*)

EVOLUTIONARY QUESTION	HYPOTHESIS
What is the evidence for genetic autonomy of kinetosomes?	Inheritance of stentor undulipodia independent of the macronucleus (Tartar, 1961, 1968); inheritance of cortical structures in paramecia independent of the macronucleus, nucleus, mitochondria, and soluble cytoplasm (Beisson and Sonneborn, 1965). Growth and development of undulipodiated bands and related microtubular structures of hypermastigote gametes in absence of the nucleus (Cleveland, 1956). See Jinks (1964), Bermudes et al. (1987), To (1987) for reviews.
How are MCs transmitted from parent to offspring generations through complex eukaryotic life cycles?	Centrioles are carried in sperm in many animals (Wilson, 1959; Satir, 1980; Turner, 1968; Renaud and Swift, 1964, for the chytrid *Allomyces*). The parental source of MCs in sexually dimorphic eukaryotes has not been studied with modern methods. The kinetochore motor (Nicklas, 1989), which includes a dynein ATPase (Vallee, 1990), actively transmits MCs to offspring cells; it should be examined for protein homologies to spirochete motility systems.
Why are most undulipodial functions under nuclear genetic control?	Mitosis involves coordination between nuclear chromatin and the MCs responsible for its segregation; selection against redundancy. Eukaryosis was equivalent to integration of archaebacteria (host) with spirochete symbiont.
Why do the numbers of centriole-kinetosomes and undulipodia per cell vary, yet the unit sizes of microtubules, kinetosomes, and axonemes remain constant?	At first, they varied as a consequence of the fact that the symbiont/host ratio is seldom 1/1; as control evolved, they varied because of differentiation for motility and the evolution of mitosis. Size ($0.25 \times 1 - 100\ \mu m$) is determined by the original spirochete genome; like the diameter of prokaryotic cells in general, individuals vary very little.

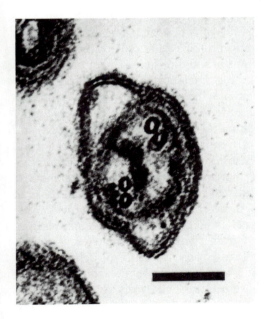

FIGURE 9-12

Cross section of microtubular structures in rod-shaped bacteria from the hindgut of *Incisitermes schwartzi*. Note the paired cytoplasmic tubules. Electron micrograph, bar = $0.2\mu m$. [From Bermudes (1987).]

questions unanswered, but the present state of the evidence deserves examination (Table 9-2). The early observation of ciliate kinetosome reproduction (Lwoff, 1950) prompted the search for kinetosomal DNA. This search ended with the realization that kinetosomes do not divide: they develop from MCs. Even in *Stentor*, whose 20,000 oral kinetosomes can be experimentally induced to form in less than two hours, DNA synthesis does not accompany the production of this enormous number of [9(3) + 0] structures. No evidence for kinetosome DNA synthesis accompanied kinetosome reproduction in *Stentor*. In some early reports, kinetosomal DNA was probably confused with mitochondrial DNA (Younger et al., 1972). Centrioles and kinetosomes do contain RNA organized as ribonucleoprotein (Dippell, 1976), and RNA synthesis must consistently accompany the production of new oral kinetosomes (Younger et al., 1972). The discovery of centriole-kinetosome DNA by Hall, Ramanis, and Luck, (1989) raises again the issue of the chemical basis of the helical structure seen in the lumen of centriole-kinetosome from protoctists and animals (see discussion on pages 266–270).

How — or even whether — DNA or RNA functions in centriole-kinetosome formation and other MC differentiation is not known. MC duplication may begin when membrane-bound cortical polymerases catalyze the replication of kinetosomal DNA; during the maturation of kinetosomes, protein would assemble on DNA–RNA templates at these sites. This hypothesis, consistent with the idea that the MC nucleoprotein of undulipodia and other microtubule organelles is derived from the original genome of former spirochetes, could explain why ciliate cortical patterns are inherited independently of nuclear and mitochondrial genes. The synthesis of the nucleotides, amino acids, and proteins acquired for production of new undulipodia — even the synthesis of tubulin itself — takes place, under the direction of nuclear genes, in the nucleocytoplasmic system. Many consider this to support the endogenous theory of the origin of undulipodia. I believe, however, that kinetosome reproduction and patterning independent of the nucleus and mitochondria must occur at specific MC sites and that this is the last vestige of free-living existence. If correct, Hall, Ramanis, and Luck's centriole-kinetosome DNA will prove to be homologous with that from spirochetes that move using homologous motility proteins. Undulipodial nucleic acids and proteins must resemble those in appropriately chosen spirochetes more than those in randomly chosen bacteria or in less-related cell organelles such as mitochondria and plastids. Of course, if undulipodia evolved from symbiotic spirochetes, they certainly have changed immensely. Nearly every aspect of spirochete growth and synthesis was relegated to the host genome in the complex act of integration, which, in retrospect, we see was eukaryosis.

The spirochete hypothesis for the origin of undulipodia predicts homologies of the motility system, which we continue to explore. Undulipodia have approximately 600 proteins; kinetosomes contain some 350 different proteins, and axonemes may have up to 250 more (Hall, Ramanis, and Luck, 1989). Of these, very few—primarily tubulins and microtubule-associated proteins (MAPS) such as dynein ATPase—are known in any detail. We have begun a systematic search for spirochete–undulipodia homology, which, of course, depends absolutely on studies of centriole-kinetosome–axoneme nucleic acid and protein performed by others. Our results are summarized here.

The 65-kilodalton protein having properties in common with tubulin has been isolated and characterized from pure cultures of spirochetes. Absent in *E. coli*, *Spiroplasma*, and *Treponema* at least, it has been found in *Borrelia* and in members of the genus *Spirochaeta* grown in axenic culture: *S. bajacaliforniensis*, *S. littoralis*, and *S. halophila*. In the alternating warm–cold treatment employed routinely for tubulin purification, the anti-tubulin antiserum–reactive, 65-kilodalton protein copurifies with a 45-kilodalton protein. These proteins, definitely not components of the flagellum or its basal rotary motor, have been localized in fibers that extend the length of the periplasm (Bermudes, 1987). Whether the 45- and 65-kilodalton proteins are also in the protoplasmic cylinder (cytoplasm) must be determined. In many experiments the 65-kilodalton protein has demonstrated its immunoreactivity with polyclonal antibodies made to mammalian brain tubulin, including antibodies affinity-purified with tubulin (Hinkle, 1992). Nearly 80 percent of the 65-kilodalton protein has been sequenced (R. Obar, G. Tzertzinis, and D. Munson, unpublished data); from both DNA and peptide analyses this protein, which has nucleotide binding sites, shows little primary sequence homology with tubulin but is indisputably homologous with one class of chaperonins (Hemmingsen et al., 1988). The spirochete protein has some 70 percent homology with, for example, the 70-kilodalton heat-shock protein GroEL from *E. coli*. (This class of proteins is absolutely required for importation of nuclear gene products into plastids and mitochondria; see page 315.)

Tektin proteins, which may coat the shared wall between the A- and B-tubules (see Figure 2–4), certainly extend down the axonemes of sperm tails (Linck and Stephens, 1987). Using antibodies affinity-purified with three axonemal tektin proteins, we have showed that the most cross-reactive of the three, an antibody to a C-tektin, cross-reacted with a 30-kilodalton protein in *S. halophila*; the protein is unidentified (Barth, Stricker, and Margulis, 1991). Clearly, far more detailed work comparing undulipodial to bacterial proteins must be done before evolutionary conclusions are drawn.

Our search for cultivable motile bacteria that might still show homology with undulipodia led us to discover an entirely new group of large, free-living spirochetes more than 100 μm long and up to 2 μm in diameter. Enriched from muds near the commercial saltworks from below microbial mats at the Ebro river delta (Catalonia, Spain) these Gram-negative bacteria

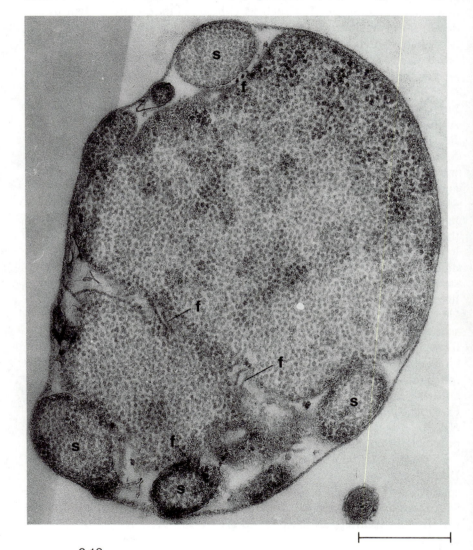

FIGURE 9-13

Spirosymplokos deltaeiberi, a large, free-living, composite spirochete with granulated cytoplasm (as in Figure 9-3). It appears to be developing at least four smaller, similar spirochetes inside (*s*). In transverse section. *f* = flagella. Electron micrograph, bar = 1.0 μm. [Courtesy of F. Craft.]

display a remarkable ultrastructure: smaller spirochetes reside inside the larger ones. The entire membrane-bounded complex, which may harbor up to eight or nine spirochetes, behaves as a single unit. It is likely that conditions of oxygen exposure induce the formation of desiccation-resistant endospores, up to four in each spirochete (see Figure 9-3 **D**). The periplasmic flagella (described in terms of end-center-end) are in a 5:10:5 arrangement, at least in the smaller internal spirochetes (Figure 9-3 **C**).*
No cytoplasmic tubules are seen in the large, complex spirochetes, but the cytoplasm of both large and small is replete with large granules (32 nm) of unknown composition (Margulis et al., 1993). The paucity of flagella and abundance of granules in these large, highly motile Ebro delta spirochetes lead us to question the role in motility, if any, of the granulated cytoplasm.

The discovery of morphologically complex, aerotolerant spirochetes that survive the rigors of rapidly changing intertidal environments by forming spores upon exposure to air widens our view of the diversity in the group. Rapid dedifferentiation and morphogenesis in response to environmental change occur in spirochetes that live in a membrane-bounded population of their own relatives. These properties argue that at least some spirochetes were already capable of transformation to spore-like structures and of regeneration from these membranous, small, nonmotile forms. That is, some free-living spirochetes could already undergo rapid morphological change before they became motility symbionts. Undulipodia — remarkable for their rapid development from centrioles that become kinetosomes and other specialized structures — are far more comprehensible if we realize that they were once free-living bacteria with similar transforming capabilities, which only later were insinuated into cells that became eukaryotes.

---

*See footnote p. 279.

CHAPTER 10

MITOCHONDRIA

MITOCHONDRIA FROM PROTEOBACTERIA

More recently, Wallin (1922) has maintained that
chondriosomes [mitochondria] may be regarded as
symbiotic bacteria whose associations with other
cytoplasmic components may have arisen in the earliest
stages of evolution. . . . To many, no doubt, such
speculations may appear too fantastic for present
mention in polite biological society; nevertheless, it is in
the range of possibility that they may some day call
for more serious consideration.

E. B. WILSON, 1959

Comparative macromolecular sequence studies of the DNA coding for
16S ribosomal RNA from many cultured strains of respiring and phototrophic bacteria tell us that a great group of microorganisms — proteobacteria
— are more closely related to each other than they are to archaebacteria or
other bacteria (Stackebrandt, Murray, and Trüper, 1988). Mitochondria are
closer relatives to proteobacteria than they are to the rest of the cell in
which they reside. Eukaryotes may lack mitochondria genetically, may
have permanently altered (nonrespiratory) mitochondria, may dedifferentiate mitochondria in response to environmental change, or — as most
do — may have an obligate requirement for respiratory-competent mitochondria throughout their life history. Mitochondria are organelles that
originated as bacteria (Gray 1983; Table 10-1). Indeed, even "polite"
biological society now concludes that the bacterial ancestry of the origin of

TABLE 10-1
The symbiotic origin of mitochondria.

EVOLUTIONARY QUESTION	HYPOTHESIS
What free-living bacteria were ancestral to mitochondria?	Gram-negative aerobic eubacteria with the Krebs-cycle enzymes and the cytochrome system for total oxidation of carbohydrates to carbon dioxide and water. Members of the "purple group" or "proteobacteria," eubacteria of Woese (1981). *Paracoccus denitrificans* (John and Whatley, 1977a, b), *Bdellovibrio* (Starr, 1975), and *Daptobacter* (Guerrero, 1991).
In what free-living host did protomitochondria become established?	Mastigotes capable of glucose fermentation to pyruvate by the Embden-Meyerhof pathway. Microaerophilic organisms with fermentative metabolism, characteristic of the nucleocytoplasm, like *Thermoplasma acidophilum* (Searcy et al., 1978; Searcy and Hixon, 1991) after integration of microtubule systems (Szathmary, 1987).
What environmental agents selected for and maintained the symbioses?	Atmospheric oxygen, depletion of nutrients, acidity, elevated temperature.
When and how did these symbioses become established?	In the Proterozoic eon, during or after the transition to an oxidizing atmosphere by genetic integration of symbionts.
Is the symbiosis obligate?	Yes, in all descendants who inhabit oxic environments.
What traits of free-living prokaryotic cells do mitochondria retain?	Circular DNA, not histone-bound (Nass 1969; Kroon et al., 1976; Gray, 1989; Gillham, 1978). DNA synthesized throughout life cycle and distributed equally to offspring mitochondria (Reich and Luck, 1966). Non-Mendelian inheritance of mitochondria in meiotic eukaryotes (Ephrussi, 1953; Jinks, 1964; Gillham, 1978).
What new syntheses were made possible by the acquisition of mitochondia?	Synthesis of steroid derivatives and some polyunsaturated fatty acids (see Table 6-8), ubiquinone, and especially in plants, probably many secondary metabolites.
Why are mitochondria packaged into sperm cells, mold spores, seeds, and other propagules?	They, or at least their genetic material, must be retained throughout life cycles because host nuclei lack the full genetic potential for forming mitochondria; mitochondria are required by the nucleocytoplasm for ATP synthesis, calcium control, and other functions.
What mechanisms maintain mitochondria in cells throughout the life history of the host?	See Table 10-6.
How can loss of mitochondria be induced?	Because the symbiosis is obligate, loss of mitochondria is lethal; but in certain facultatively anaerobic eukaryotes mitochondrial DNA and structures can be diminished permanently by mutagens, such as acriflavine (Ephrussi, 1953; Roodyn and Wilkie, 1968), agents to which mitochondrial nucleic acid is more sensitive than is the host (Kusel et al., 1967).
How much may mitochondria dedifferentiate?	Very little or not at all in most eukaryotes. In facultative anaerobes, such as yeast and some protists, they may dedifferentiate beyond the resolution of the electron microscope. In yeast the mitochondrial differentiation system is inducible by, and sensitive to, concentration of glucose and oxygen (Roodyn and Wilkie, 1968).
What kinds of RNA do mitochondria synthesize?	Ribosomal and transfer RNAs. Ribosomal RNA sequences of mitochondria are more similar to those of bacteria than to those of the cytoplasm in the same cell.
What proteins do mitochondria synthesize?	Ribosomal and enzyme proteins: protein synthesis sensitive to inhibitors of prokaryotic, but not of eukaryotic, protein synthesis (Table 10-5) (Gillham, 1978).
Do mitochondria synthesize lipids?	Yes (Parsons, 1967; Clark-Walker and Linnane, 1967). The lipid fractions of offspring mitochondria receive equal amouns of labeled choline fed to the parent cell (Reich and Luck, 1966; Roodyn and Wilkie, 1968).

TABLE 10-1 (*continued*)

EVOLUTIONARY QUESTION	HYPOTHESIS
Why are mitochondrial nucleic acids and enzymes "packaged" in all fungi, plants, and animals?	They were acquired together by heterotrophic ancestors as an intracellular symbiont.
Why are some mitochondrial functions under nuclear control?	Obligate symbionts relegate redundant or dispensable metabolic functions to the host. Production of mitochondrial RNA polymerase by the nuclear genome (Barath and Kuntzel, 1972) and the integration of nuclear and mitochondrial functions (Gillham, 1978) imply gene transfer from mitochondria to nucleocytoplasm.
Why do numbers and sizes of mitochondria per cell vary?	As in all host–symbiont relationships, the ratio of host to symbiont is altered by nutritional and other environmental factors.
Why are the mitochondria of plants, animals, and fungi similar?	Similar protomitochondria were acquired by the common protist ancestors of eukaryotes, fermenting mastigotes, before the ancestors of algae acquired photosynthetic plastids. Yet the variety and frequency of protist–bacterial associations suggest polyphyly of mitochondria.
Why are mitochondrial genes in pieces?	Intron acquisition occurred during hundreds of millions of years of endosymbiosis.
Why do mitochondria have cristae?	Cristae are adaptations that increase the surface area of oxidative enzymes, analogous to the internal membranes of many prokaryotes.
Why are yeast and regenerating rat liver mitochondria sensitive to streptomycin, chloramphenicol, spectinomycin, paromomycin, and so forth, but not to cycloheximide?	Because the proteobacterial ancestors of mitochondria were sensitive to drugs that block prokaryote protein synthesis; cycloheximide blocks eukaryotic ribosomes (see Table 10-5).
If they originated as symbionts, why do mitochondria lack cell walls?	Walls are dispensable in the controlled, osmotically regulated, buffered conditions of the cytoplasm in which the mitochondria live; therefore, they were selected against.

mitochondria has been proven (Gray, 1983). Mitochondrial ancestors came, probably polyphyletically, from the proteobacteria (or "purple bacteria"), a large group of eubacteria with many respiring and phototrophic members (Stackebrandt, Murray, and Trüper, 1988). The data on nucleotide and protein sequence that provide overwhelming evidence of symbiotic bacterial origins, now found in comprehensive textbooks of cell biology such as Alberts et al. (1989), cannot concern us here. Yet the review of current insights into mitochondria–nucleocytoplasm relationships permits us to ask what the mitochondrial example reveals, in principle, about cell evolution.

Viewing mitochondria as highly integrated endosymbionts raises new questions. What hosts acquired **protomitochondria**, that is, the respiring bacteria that, after integration, became mitochondria? Was acquisition polyphyletic? What genetic changes accompanied the protomitochondria-to-mitochondria transition? How does mitochondrial DNA content and genetic organization vary (Table 10-2)?

TABLE 10-2

Some characteristics of mitochondrial DNA.

	GENOME SIZE
SOURCE OF DNA	NUCLEOTIDE BASE PAIRS $\times 10^3$
Animals (e.g., flatworms, insects, mammals)	16–19
Plants	150–2500
Fungi (e.g., yeasts, molds)	17–78
Protoctists (e.g., *Chlamydomonas*[a], *Paramecium*[a], *Trypanosoma*)	16–40

	DISTRIBUTION OF DNA		
CELLULAR SOURCE	NUMBER OF MITOCHONDRIA/ CELL	DNA MOLECULES/ MITOCHONDRION	PERCENT OF TOTAL CELL DNA
Rat liver	10^3	5–10	1
Frog egg	10 million	5–10	99
Yeast	1–50	1–10	15

[a]Linear molecules; other mitochondrial genomes circular.

No experimental manipulation can cause eukaryotes to lose mitochondria permanently, yet some yeasts that obtain energy by fermentation may reversibly dedifferentiate mitochondria to **promitochondria**—small, membranous spheres capable of redifferentiation into mitochondria. (The **pro**mitochondrion is distinguishable from the mitochondrial evolutionary ancestor, referred to as a **proto**mitochondrion (Figure 10-1). Promitochondria lack the convoluted membranes (cristae) typical of mature mitochondria. Retaining the potential to redifferentiate their mitochondria, some yeasts grow without respiratory metabolism. Both the absence of oxygen and high concentrations of fermentable substrates such as glucose repress mitochondrial development from promitochondria, although in different ways. This unusual eukaryotic style of facultative aerobiosis permits the transmission of lethal or severely deleterious mitochondrial genes under fermenting (anoxic or high glucose) but nonrespiring conditions. Other eukaryotes, obligate aerobes, carrying such genes would not survive. Consequently, nearly all definitive information about mitochondrial genetics and development comes from studies of yeast. Plants, animals, and most fungi other than certain yeasts require mitochondrial function at all times. The few eukaryotes that lack mitochondria are protoctists—many mastigote groups (diplomonads such as *Giardia*; retortamonads; *Psalteriomonas*; and all para-

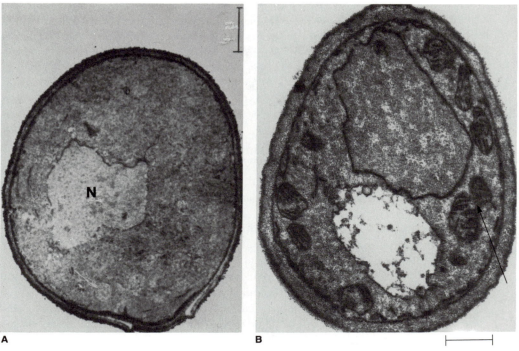

FIGURE 10-1

Yeast with and without mitochondria.

(A) Mitochondrial development is repressed in yeast grown under anoxic conditions or in the presence of high concentrations of glucose. The capacity for redifferentiation of the mitochondria is retained, as is the mitochondrial DNA, even though there is no morphological evidence for the presence of mitochondria. N = nucleus. Transmission electron micrograph, bar = 1 μm. (B) When oxic conditions are restored and glucose is depleted, the mitochondria reappear. A mitochondrion apparently in division can be seen in the lower right part of the cell (arrow). Transmission electron micrograph, bar = 1 μm/ [Courtesy of A. W. Linnane.]

basalia, for example). Some ciliates, amoebae, and other obscure mastigotes, primarily from the anoxic or micro-oxic hindguts of wood-eating cockroaches or from other animal tissues seem never to have mitochondria. Few, besides the chytrid *Neocallimastix* and some cellulolytic forms (Yamin, 1980), have been cultured axenically. No genetic information about any of them is available.

The earliest protoeukaryotes, bacteria that evolved into nucleocytoplasm that lacked mitochondria, would also have lacked mitosis and undulipodia. In the absence of mitochondria or other respiring symbionts, they would be microaerophils or anaerobes. Could such microbes be distinguished, even in principle, from prokaryotes or from protists that lost mitochondria when they reinvaded anoxic environments? Perhaps not. Dennis Searcy and his co-workers have argued that *Thermoplasma acidophilum*, a heat- and acid-tolerant archaebacterial mycoplasm, is an excellent candidate for a living descendant of the protoeukaryotes (Searcy and Stein, 1980; Searcy, Stein, and Green, 1978). A combination of traits supports Searcy's idea — lack of cell walls, presence of histone-like and actin-like proteins, Embden-Meyerhof fermentation pathway, ease with which associations are entered, sterol requirements, and sensitivity to cytochalasin B (Ghosh, Maniloff, and Gerling, 1978), as well as acid and heat tolerance (Darland et al., 1970). *Sulfolobus* — a sulfur-oxidizing heat- and acid-resistant prokaryote related to *Thermoplasma* — unlike other prokaryotes, may even be capable of particle ingestion and therefore could easily acquire intracellular symbionts. However, the 16S RNA of *Thermoplasma* and *Sulfolobus* indicates that both are related more to the other archaebacteria (methanogens and halobacters) than to eukaryotic nucleocytoplasm or other mycoplasmas (Woese et al., 1975).

Perhaps the large anaerobic amoeba *Pelomyxa palustris* is the closest living descendant of the protoeukaryotes. *Pelomyxa* lacks mitosis and mitochondria but, unfortunately, cannot be cultured (Daniels, Breyer, and Kudo, 1966; Whatley and Chapman-Andresen, 1990). Because it has nuclear membranes, *Pelomyxa* is classified as a eukaryote, although lack of standard eukaryotic features makes it anomalous. Since it may never have acquired mitochondria, *Pelomyxa* may be a direct link to the kind of organism that acquired protomitochondria as endosymbionts. Alternatively, *Pelomyxa*, with its atypical, stiff, undulipodia-like structures, may have lost its mitotic spindle, mitochondria, and other eukaryotic features. On the principle that semes tend not to be lost without a trace (see Table 5-2), however, this is unlikely.

Comparing the microaerophilic metabolism common to the cytoplasm of eukaryotes with the metabolism of candidate organisms such as *Thermoplasma* would aid the search for protoeukaryotes. Especially intriguing is DeDuve's (1991) suggestion that peroxisomes, cell organelles of animals, structures that have been seen to divide by binary fission, were also acquired as microbial symbionts. By DeDuve's reckoning, peroxisomes were present in protoeukaryotes before establishment of the mitochondrial symbiosis.

The outer layer of the double membrane of mitochondria is more similar to membranes of the endoplasmic reticulum (ER) than it is to inner mitochondrial membranes (Table 10-3). How do the ER and inner mitochondrial membranes compare to the plasma membranes of *Bdellovibrio*, *Paracoccus*, and *Thermoplasma*? The inner membranes should be homolo-

TABLE 10-3

Mitochondrial, bacterial, and endoplasmic reticulum (ER) membranes compared.[a]

PROPERTY	MITOCHONDRIAL MEMBRANE		SMOOTH ER	BACTERIAL MEMBRANE[b]
	INNER	OUTER		
Thickness	5.5 nm	5.5 nm	5.5 nm	—
Fine structure	"Globules," 9.0-nm membrane subunits	"Globules," 2.8-nm pits, 6.0-nm membrane subunits	"Globules," polysaccharide fringe	—
Density (g/cm³)	1.21	1.13	1.13	—
Protein:lipid ratio	1:0.275	1:0.829	1:0.385	—
Cardiolipin (% of total phospholipids)	21.5	3.2	0.5	High
Phosphatidyl inositol (% of total phospholipids)	4.2	13.5	13.4	—
Phosphatidyl serine (% of total phospholipids)	Not detected	Not detected	4.5	—
Phosphatidyl ethanolamine	—	—	—	High
Cytochrome $a + a_3$ (μmol/g)	0.24	<0.02	0.0	—
Cytochrome b_5 (μmol/g)	0.17	0.51	0.79	—
Permeability	Small molecules	Probably large and small molecules	Small molecules	Small molecules
Osmotic response	Responsive	Not responsive	Slightly responsive	Responsive
Effect of ATP	Contraction	No effect	No effect	—
Cholesterol	Very little	Present	Large amounts	None[c]

[a]After Parsons (1967).
[b]From Sokatch (1969). Bacterium studied was *Serratia*.
[c]With the exception of the mycoplasm as (pleuropneumonia-like organisms; PPLOs).

gous to those of the protomitochondria, and the outer and ER membranes to those of *Thermoplasma*, if indeed it is a good candidate for codescendant of the protoeukaryote.

Recognition of the polyheterogenomic nature of the eukaryotic cell relative to the single genomic system of all prokaryotes aids interpretation of certain facts of physiology and genetics, such as steroid synthesis. The aerobic eukaryote is homologous not to a single bacterial ancestor but to a community of microorganisms—archaebacteria, proteobacteria, and perhaps even spirochetes—that have co-evolved since the Proterozoic eon.

Steroid synthesis is a nearly unique metabolic capability of eukaryotes. Eukaryote membranes regularly contain large quantities of these cyclic lipids. Steroid synthesis may be the product of interaction at the metabolite level (see Figure 7-6) between more than one genome. Steroids are apparently indispensable for the formation of the flexible dynamic membranes of eukaryotes. In spite of bewildering diversity, steroids can be traced back to their common biosynthetic precursos, lanosterol. This compound is converted to cholesterol in animals, cycloartenol in plants, and ergosterol in fungi. Lanosterol is formed from the universal isoprenoid precursor squalene, which is the product of a pathway from acetate through isopentenyl pyrophosphate. Free gaseous oxygen is required to cyclize the ring in the formation of lanosterol from squalene. Atmospheric or dissolved oxygen, used in the final step of aerobic respiration, is available at mitochondrial sites.

Complex terpenoid syntheses are talents belonging almost exclusively to plants (Table 10-4). Biosynthesis of oils, diterpenes, and triterpenes may require the presence of more than one type of genome: the genetic control of cyclic and oxygenated terpenes, so characteristic of plants, may be due to their trigenomic nature—nucleocytoplasm, mitochondrion, and plastid. To test this concept, biosynthesis must be studied during the suppression (either physiologically or by mutation) of mitochondrial metabolism, plastid metabolism, or both. In yeast, mitochondrial metabolism can be turned off entirely and reversibly, and in some euglenids, photosynthetic plastid metabolism can be turned of both irreversibly and reversibly. A correlation of steroid and terpenoid synthesis with organellar functions would help to determine both the division of organellar genomic control and the origin of

TABLE 10-4

Distribution of terpenoid synthetic capability.

COMPOUNDS[a]	SYNTHESIZING ORGANISMS
Diterpenes Phytyl, oleic acid, vitamin K	All organisms except a few bacteria
Acyclic terpenes and oxygenated derivatives Bayberry wax, oil of citronella, oil of rose, "essential oils," lycopene, rubber	Certain angiosperms
Squalene	All eukaryotes; universal precursor of steroids, through cycloartenol in plants, ergosterol in animals, and lanosterol in fungi
Monocyclic terpenes Oil of lemon, peppermint, ginger	} Certain angiosperms
Bicyclic terpenes Oil of turpentine, oil of ginger, camphor	
Sesquiterpene alcohol Farnesol	

[a]Terpene (by custom), $C_{10}H_{16}$; sesquiterpenes, $C_{15}H_{24}$; diterpenes, $C_{20}H_{32}$; triterpenes, $C_{30}H_{48}$; polyterpenes $(C_5H_8)_n$.

these metabolic pathways. (See Table 11-2 for a summary of analogous work with carotenoids.)

For all organelles that began as free-living organisms, primary metabolic pathways were present in the partners at the onset of the association. As natural selection reduced inherent redundancy, partners became progressively more interdependent. Any nutrient essential for the development and replication of a symbiont must be supplied by the environment (either the aqueous medium or the host cell), or it must be synthesized by the symbiont. A primary metabolite or enzyme made and required by both the host and symbiont tends with time to be supplied by only one partner, usually the host.

Selection pressure against cyclical association and for permanence is strong, especially under relatively stable environmental conditions. Metabolic redundancy is selected against as long as the associates remain together, so that gene products and nutrients are exchanged. If the host is a sexually reproducing eukaryote, it must contain precise mechanisms for the segregation of symbiont–organellar genomes to offspring cells. In such cases, symbionts use the convenient metabolic facilities of hosts. If the host is a normal Mendelian diploid cell and the supply of the nutrient in question depends on the presence of dominant genes, recessive mutations may then result in lower reproduction rates and impaired development or function of symbionts. The phenotype of such chromosomal recessives can be expressed only if the symbiont is present in the cytoplasm; the restoration of a dominant gene that controls a metabolite required for symbiont development (for example, a required amino acid) can never restore the genome of a lost symbiont. This analysis can be applied to the observation that metabolites, as products of nuclear genomes, have crucial effects on the expression of cytoplasmically transmitted organelles. The transmission of certain lysine auxotrophs of yeast is interpreted in this way (Mounolou, Jakob, and Slonimski, 1967). For the genetic details of such effects on mitochondria and plastids, see Gillham (1978).

Mitochondrial Legacy and Cytoplasmic Heredity

The evidence for calling mitochondria bacteria, rests upon the following attributes: Their general behavior in the cell is similar to that of known microorganisms which live symbiotically in the cells of higher organisms; for example the root-nodule bacteria of legumes. When grown independently in artificial culture media, they behave in all

observed particulars like bacteria. They divide like
bacteria. They are similar to bacteria in structure and
shape. They exhibit no cultural characteristics foreign to
bacteria.

I. E. WALLIN, 1927

All mitochondria contain DNA, RNA, and ribosomes. Mitochondrial DNA is double-stranded, generally forming a closed circle (Figure 10-2). The molecular mass of 10,000 to 100,000 kilodaltons is not enough to code for all the proteins found in animal mitochondria. Fully developed mitochondria are products of the interaction of at least two genomes, mitochondrial and nuclear. These facts about mitochondrial physiology and genetics are unintelligible except in the light of the peculiar evolutionary history of these organelles. We concur with Gray (1989) that the "endosymbiont hypothesis is generally regarded as the best explanation of the origin of the mitochondria genome, as well as of the structural and functional complexity of the mitochondria itself."

About 50 chromosomal genes, called *pet* genes, affect respiration in yeasts. Yeasts that carry these genes (designated *pet* 1, *pet* 2, and so on) exhibit the **petite** phenotype—small colonies when the yeast is grown on plates. The genes are unlinked—located on different chromosomes. Yeast cells containing *pet* recessives are not able to grow on nonfermentable substrates, such as ethanol, lactate, and glycerol. They depend on the fermentation of six-carbon carbohydrates and related compounds, which they metabolize anaerobically to nonfermentable two- or three-carbon compounds. They cannot respire, even though genetic experiments show

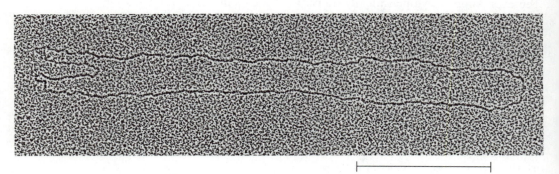

FIGURE 10-2

Circular mitochondrial DNA, isolated from chicken liver mitochondria by the Kleinschmidt technique. Transmission electron micrograph, negative stain, bar = 0.2 μm. [Courtesy of E. F. J. van Bruggen.] See Palmer (1985) for details of organellar DNAs.

that they contain the cytoplasmic mitochondrial genetic determinant (ρ^+) found in normal wild-type yeast. The symbiotic theory explains this phenomenon by the reduction of the metabolic redundancy inherent in symbioses.

It is curious that a nuclear mutation, of a *pet* gene, results in both the petite phenotype and the appearance of a lysine requirement (Mounolou, Jakob, and Slonimski, 1967). The nuclear product lysine is apparently required for normal mitochondrial development; lysine is produced in low yield or not at all by cells carrying the recessive alleles. If lysine is added to the medium, phenocopies of wild type may be made, restoring functional mitochondrial activity. An analogy can be drawn to Jeon's amoebae (see Figure 10-3). Mitochondrial multiplication is possible only if the essential metabolite is provided. Besides lysine, mitochondrial requirements under nuclear control include those for polypeptides and proteins. Chloramphenicol and other potent protein-synthesis inhibitors active against prokaryotic ribosomes selectively inhibit the synthesis of enzymes in redifferentiating yeast and in regenerating animal mitochondria (Table 10-5). Yet the chloramphenicol-binding peptide of the mitochondrial ribosomal protein is coded for by the nucleus. These observations imply (1) that mitochondrial ribosomes are more closely related to those of free-living prokaryotes than to the ribosomes of the cytoplasm in which the mitochondria reside, and (2) that the information coding for the chloramphenicol-binding protein has been transferred from the mitochondria to the nucleus.

In plant and animal cells, mitochondria participate in the synthesis of major classes of metabolites: for example, hemes, pyrimidines, and amino acids. Although mitochondria contain their own genomes (see Table 10-2) and bacteria-style protein-synthesizing systems for making a dozen or more peptides, most of the hundreds of proteins that comprise the mitochondria are coded for by nuclear DNA. The peptides of several large proteins (for example, F_0F_1-ATPase and cytochrome oxidase) of respiratory chains are made jointly by nucleocytoplasmic and mitochondrial protein synthetic systems. Only five proteins, identified recently, are absolutely essential for viability, at least in yeast. All five involve the import of nucleocytoplasmic gene products into the developing mitochondrion. They are: (1 and 2) the two subunits of the water-soluble matrix protease; (3) a 60-kilodalton chaperonin that mediates the refolding of newly imported mitochondrial proteins (homologous, like the spirochete protein described on page 301, to the *E. coli* GroEL heat-shock protein); (4) an ATP-sensitive mitochondrial heat-shock protein (mhsp70), that may provide the driving force for the transmembrane movement of polypeptides (and is similar to the *E. coli* dnaK gene product); and (5) a 42-kilodalton protein, located in the mitochondrial outer membrane, that may be a subunit of a transport channel for peptide movement (Baker and Schatz, 1991).

TABLE 10-5

Antibiotic sensitivities of bacterial, mitochondrial, and cytoplasmic ribosomes.

	RIBOSOMES		
AGENT[a]	MITOCHONDRIAL	BACTERIAL	CYTOPLASMIC
Puromycin (a)	+	+	+
Thiostreptin [siomycin] (d,h)	+	+	−
Cycloheximide (b) actidione	−	−	+
Anisomycin (c)	−	−	+
Chloramphenicol (d,f) [Chloromycetin]	+	+	−
Macrolides such as erythromycin (b)	+	+	−
Lincomycin (a)	+	+	−
Aminoglycosides such as neomycin, paromomycin, streptomycin, kanamycin	+	+	−
Tetracyclines (e,f)	+	+	−

[a]Probable mechanisms of action: (a) prematurely terminates polypeptide synthesis and breaks down polysome; (b) blocks peptidyl tRNA transfer; (c) inhibits formation of peptide bond; (d) binds 50S ribosomal subunit; (e) binds 30S ribosomal subunit; (f) blocks attachment of amino acid to tRNA; (g) inhibits by forming an abortive complex; (h) stops translation. See Gillham (1978) and Bücher et al. (1976) for discussion.

Mitochondria are incessantly active; their numbers per cell and their shapes are not constant. Mitochondria divide by fission and then fuse and fission again differently, which probably ensures the opportunity for combination and recombination of mitochondrial DNA (Bereiter-Hahn, 1990). From data on numerous mitochondrial properties—genetic recombination, uniparental inheritance of the mitochondrial genome, the presence of introns, circular or linear DNA configuration, deviations from the universal set of codons, diminished numbers of transfer RNAs, large differences in mitochondrial features in congenic (that is, belonging to the same genus and therefore closely related) organisms, the paucity of DNA repair mechanisms, the capacity for differentiation and differential proliferation in tissues, and expression of only one mitochondrial genotype in cells with multiple heterogeneous copies of the mitochondrial genome and other peculiarities—we deduce that mitochondria in all eukaryotic lineages have co-evolved vigorously with their respective nucleocytoplasm. Such genetic and molecular studies emphasize that mitochondrial evolution did not cease with the ancestral bacterial acquisition; on the contrary, the differences between extant, free-living, mitochondria-like bacteria and mitochondria themselves alert us to the rapidity of mitochondrial evolution in some lineages and the lack of correlation among evolutionary rates in the different (nucleocytoplasmic, mitochondrial, and plastid) eukaryotic organellar systems. The flexibility of molecular systems and the great divergence of molecular detail after acquisition of the original symbionts warn us that we are dealing with the integration of community members to form individuals at greater levels of complexity. We are alerted to the dangers of precipitous conclusions about evolutionary history made from limited molecular comparison. Evidence that mitochondrial genomes are genetic mosaics, novel products of recombination after cell fusion, is reviewed by Gray (1989).

PATHOGEN TO ORGANELLE: A MITOCHONDRIAL ANALOGY

Pathogenicity is not the rule. Indeed, it occurs so infrequently and involves such a relatively small number of species, considering the huge population of bacteria on the earth, that it has a freakish aspect.

L. THOMAS, 1974

The transition from nefarious pathogen to new organellar system, in some ways analogous to the origin of mitochondria, was documented by cell biologist Jeon (1991). In the late 1960s and early 1970s, Jeon studied nucleocytoplasmic relations in *Amoeba proteus*. Healthy amoebae, in cultivation for years, suddenly became ill with a bacterial infection that probably was introduced into the laboratory inadvertently with a new batch of amoebae. Electron-microscopic observations of moribund cells showed about 150,000 bacteria inside each infected amoeba. The growth rate of the infected amoebae declined precipitously. Some survived, however, and after a number of years recovered their original growth rate. The survivors, finally as healthy as uninfected amoebae, still harbored the bacterial symbionts—about 42,000 per cell (Figure 10-3).

In an ingenious set of nuclear-transplant experiments, Jeon proved that the host nucleus, which by then had been living in association with the new bacterial symbionts for about five years, had become dependent on these symbionts. The transplanted host nucleus no longer supported the functioning and division of cytoplasm that lacked the once-pathogenic bacteria. In the presence of the symbionts, however, the transplanted nuclei functioned indefinitely. Since the bacteria still have not been grown in pure culture, the function that they perform for the amoeba nuclei has not been ascertained. With the stabilization of the symbiosis, an increased sensitivity

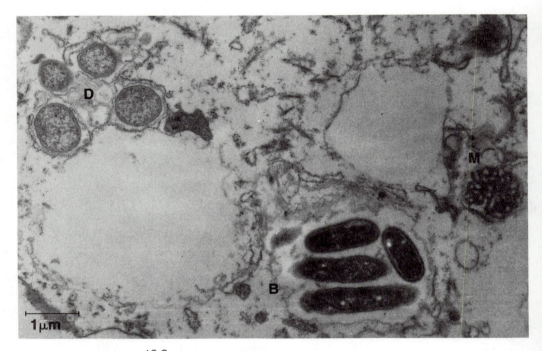

FIGURE 10-3

Infected amoebae with membranous vesicles (symbiosomes) containing bacteria (*B*) that once were rampant pathogens. These amoebae also normally contain "D" bodies (*D*) and mitochondria (*M*). Both D bodies and mitochondria contain DNA and probably are relics of earlier intracellular bacterial symbioses. Transmission electron micrograph. [Courtesy of K. W. Jeon.]

of the partnership to elevated temperatures, starvation, crowding, and certain antibiotic treatments evolved (Jeon and Ahn, 1978).

No defined medium for axenic growth of these amoebae has ever been developed, and sexuality in amoebae is unknown, which precludes genetic and metabolic analysis of the new symbiosis. However, much is known about the only symbiosis whose establishment from pathogenesis has been captured in the laboratory (Jeon, 1991).

The unidentified Gram-negative rod symbionts (x-bacteria) of the amoeba harbor at least two kinds of plasmid DNA (21- and 59-kilobase). These former pathogens are remarkably resistant to digestion by the amoebae: not only are their walls tough, but somehow they prevent lysosomal fusion with membranes of the vacuoles in which they are contained (these vacuoles are called **symbiosomes**). Although the x-bacteria can be transferred and grown in symbiosomes of previously uninfected amoebae, even after many attempts the investigators have not discerned the trophic needs of the bacteria. Like mitochondria, these symbiotic bacteria are entirely and permanently dependent on the nucleocytoplasm of the cells in which they reside. What formerly was the "disease agent" is now an intrinsic part of the self (Sagan and Margulis, 1991).

The symbiotic bacteria produce and release a large quantity of trimeric protein (3×29 kilodaltons $= 87$ kilodaltons), composed disproportionately of hydrophobic amino acid residues. This protein crosses the symbiosome and cytoplasmic membranes and is detected in the amoeba cytoplasm. Furthermore, as a result of the laboratory-induced symbiosis, at least two new proteins are synthesized by the amoeba. Within two weeks or so after being infected, the bacteria-ridden amoebae produce a new 200-kilodalton protein which, in nuclear transplantation experiments, has been shown to have a lethal effect on the nuclei of uninfected amoebae. The second protein, a 150-kilodalton glycoprotein, is limited to symbiosome membranes; this restriction may be related to prevention of lysosomal fusion.

Another consequence of the association is the accumulation of a certain amoeba protein in the symbiosomes. A 43-kilodalton fibrous protein synthesized on amoeba ribosomes finds its way into the symbiosome vesicles and attaches to the surface of the bacteria. This protein, also in the cytoplasm of uninfected amoebae, is identified as amoeba actin. The mysterious accumulation of amoeba actin on the surface of the once pathogenic x-bacteria becomes comprehensible as an aborted behavior when the analogous actions of *Listeria*, a pathogen of humans and other mammals, are compared.

Listeria monocytogenes, found in contaminated milk products, is known to cause illness and even death — especially in pregnant women, newborn infants, and others with compromised immune systems. The bacterium, a Gram-positive rod, is sensitive to antibiotics until it enters the cytoplasm of macrophages, fibroblasts, and other mammalian cells. *Listeria* is attacked actively by phagocytotic macrophages as part of the immune reaction. The listeria are brought into phagocytic vesicles (analogous to the symbiosomes of Jeon's amoebae), where, although typical fusion of lysosomal and phagosomal membrane occurs, the resulting phagolysosomal vesicle is swollen more than usual and filled with debris. *Listeria* escapes destruction by lysosomes not by preventing lysosomal–vesicle membrane fusion, but by escaping from the vesicles by dissolving the vesicle membrane with hemolysin. (If the bacteria are mutated such that the toxic protein hemolysin can not be made, *Listeria* fails to escape into the cytoplasm.)

Unlike Jeon's x-bacteria, the listeria flee the membrane-bounded vesicles and come to lie free in the cytoplasm of the human white blood cells. Accumulation of intracellular *Listeria* stimulates the synthesis of fibrous actin by the cells. The listeria become coated with huge numbers of actin filaments that have been likened to a toboggan or a comet's tail; up to 5 μm long, the actin bundles move through the mammalian cell, propelling each *Listeria* that "rides" the toboggan (Figure 10-4). The mammalian actin bundles, thus co-opted by the wily *Listeria*, are put to use as the populations of bacteria propagate themselves, spreading to neighboring blood cells: The bacterial aloft the actin toboggan protrude from the cell in a membrane projection that is phagocytosed by a neighboring white blood cell.

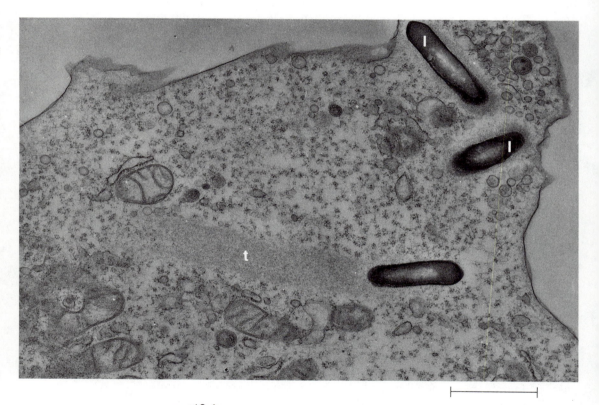

FIGURE 10-4

Listeria (*l*), eubacteria that are transported from one white blood cell to another by actin fibers ("toboggan"-shaped bundle, *t*) such that the pathogen is never in direct contact with the serum; thus, antibodies against it are not made. Electron micrograph, bar = 1 μm. [Courtesy of Lewis and Molly Tilney.]

The basis of systemic *Listeria* infection, then, is bacterial control of protein hypertrophy: the use of mamallian actin by *Listeria* so that they can be transported from one cell's cytoplasm to another's without entering the lymph or blood (Portnoy et al., 1989; Tilney and Portnoy, 1989). Since we know that Jeon's x-bacteria can travel from amoeba to amoeba, x-bacterial strategy in crowded natural conditions may involve an actin-mediated attack comparable to the drama of *Listeria* but one in which the investigators have missed direct observation of the tragic last act. Pathogenesis, like urban blight, is acutely sensitive to population density. As numbers increase, bacteria spread rapidly and lethally through a population which, when less dense, can keep the potential pathogen in check. The irony is that evolutionary innovation which proceeds by enemy absorption and by conflict resolution through population control preceded cities by 1500 million years!

REFINING AND INTERFACING: THE MIX-MATCH PRINCIPLE

There is a tendency for living things to join up, establish
linkages, live inside each other, return to earlier
arrangements, get along, whenever possible. This is the
way of the world.

                    L. THOMAS, 1974

The evolution of mitochondria did not stop with their acquisition by early eukaryotes. Residence in a cell capable of chromosomal recombination and many complex biosyntheses has led mitochondria to relegate most of their protein-synthesizing activity to the nucleocytoplasm, and, with respect to introns, for example, to evolve a genomic organization that resembles, in some aspects, that of the original host.

As builders of great civilizations and of large computers know, innovation does not mean reinventing the wheel. Rapid change is the result of mixing, matching, and "interfacing" previously evolved modules. The mix-match principle is a pervasive preadaptive mechanism in evolution — from horizontal transfer of single genes or plasmids to wholesale inheritance of acquired genomes. In each case, refined and well-functioning systems with high selective advantage in certain environments at a particular time are brought together into new combinations. This principle is invoked to account for yeast mitochondrial genetic organization. It can also be seen in the following:

1. The origin of linkage groups (Gabriel, 1960), in which cistrons are brought together on chromosomes and retained by the selection of interactive gene complexes (Todd, 1970; Todd, 1992).
2. The origin of transducing and lysogenic bacteriophages and plasmids, which transfer fragments of useful genomes between bacteria (Sonea, 1991; Sonea and Panisset, 1976) and from bacteria to plant (Schell et al., 1979).
3. The origin of hundreds of symbiotic complexes, which bring together genomes hitherto highly divergent (see Figure 7-17; Margulis, 1976).
4. The origin of meiotic and dimorphic sexuality, which ensures the regular association of individuals with slightly divergent genomes.
5. The origin of "genes in pieces" organization—in animal cells, ciliate macronuclei, and yeast mitochondria—which permits reassortment of cistrons (usually on the level of gene products or messenger RNA) in the course of ontogenetic development.
6. The origin of new karyotypes by polyploidy and karyotypic fissioning, which brings together new linkage groups in slightly divergent genomes.
7. Colonization of islands and other recolonization, which bring together reproducing individuals to form new communities.

A salient feature of all of these cases is that, as usual in biology, solutions to evolutionary problems are historical. Honed components are recombined in new ways, comparable problems are resolved analogously by different systems, often many times. The principle "reproduce, differentiate, refine, select, and recombine" applies to millions of interactions at many levels—between cells, individuals in societies, communities, and co-evolving species in an ecosystem—in the course of evolution.

The symbiotic concept of the origin of organelles explains the selective advantage of cytoplasmic genes as a legacy. The phrase **cytoplasmic heredity** refers to the inheritance of genetic traits, in sexual organisms, from a single parent only. Considered an anomaly in the history of genetics (Sapp, 1987), cytoplasmic heredity concerns the inheritance of characteristics that are independent of nuclear genes (Jinks, 1964). In many cases, such inheritance is maternal—that is, the trait is transmitted from the female parent. Many phenomena considered first to be cytoplasmic inheritance are now recognized as cases of uniparental or non-Mendelian biparental transmission of mitochondria or plastids. The fact that at least two human diseases (an ophthalmoplegia associated with Kearns-Sayre syndrome, and Leber's hereditary optic neuropathy) are inherited cytoplasmically has greatly helped to enhance the respectability of the study of this kind of genetics! [These illnesses are correlated with mitochondrial DNA deletions or base changes (Moraes et al., 1989; Singh, Lott, and Wallace, 1989).]

An organelle of symbiotic origin can never be reproduced by the action of the host's nuclear genes; therefore, the host and the proto-organelle must have developed ways of retaining at least one copy of the organellar genome at every stage in the host's life cycle (Table 10-6). When fertilization and zygote formation first evolved in protoctists, isogamous and biparental transmission of organelles must have been the rule. Natural selection tended to reduce the inherent redundancy that results from the fusion of gametes each with at least one copy of the symbiont genome. This tendency has produced anisogametes from isogametes in many eukaryotes, reducing the need for both gametes to travel and, ultimately, resulting in the inheritance of the organellar genome with the sedentary gamete — by definition, the female.

Sexual dimorphism has arisen in many different lineages — volvocalean and conjugalean green algae, suctorian ciliates, chytrids, and oomycotes. The female retains the organellar genome, the food supply, and much

TABLE 10-6

Mechanisms for retaining mitochondria through the life cycle of cells.

ORGANISM	MECHANISM
Some ciliates	Numerous mitochondria divide synchronously with nuclear division (Wilson, 1959).
Micromonas (chlorophyte)	One mitochondrion divides synchronously with the cell nucleus (Gibor and Granick, 1964).
Sperm of *Nitella* (charophyte)	Single mitochondrion is surrounded by microtubules produced by one centriole (Turner, 1968).
Trypanosoma brucei	Concatenated DNA of mitochondria is replicated during S phase and segregated by separation of new and old kinetosomes (Robinson and Gull, 1991)
Vicia (bean)	Two groups of mitochondria cluster at opposite poles at time of cell division (Wilson, 1959).
Tetrahymena, Neurospora, yeast, many plant and animal cells, dividing germ cells of vertebrates (cleavage stages)	Many mitochondria are distributed randomly throughout the cell (Wilson, 1959; Clark-Walker and Linnane, 1967; Parsons and Rustad, 1968).
Anaerobic yeast	Neither mature mitochondria nor promitochondria are visible; distribution of genetic potential segregates mitochondria to offspring cells at sub-electron-microscopic levels (Roodyn and Wilkie, 1968).
Hydrometra (insect)	Elongate rods of mitochondria are oriented on spindles; some of the rods are cut by cell equator (Wilson, 1959).
Spermatocytes of *Ascaris* (worm)	Mitochondria are arranged on spindles oriented toward centrioles; they do not divide on spindles, but segregate to offspring cells according to their positions on spindles (Wilson, 1959).
Spermatocytes of *Centrurus* (scorpion)	Primary spermatocyte: mitochondria aggregate into ring-shaped body, oriented on a spindle; cell division cuts this body transversely into two half rings; each half ring forms a rod. Secondary spermatocyte: each rod is carried on a spindle and cut into a half rod by cell division (Wilson, 1959).
Spermatocytes of *Opisthocanthus* (scorpion)	Primary spermatocytes: numerous small mitochondria join to form 24 spheroids.

other cytoplasmic baggage, while the male unloads as much cytoplasmic material as possible without jeopardizing his fertilizing function. The donor and recipient genders are often not directly homologous, but analogous, the result of similar selection pressures—for example, the pollen tube and embryo sac of angiosperms; the motile sperm and the nutrient-containing egg of insects; the antheridium nuclei that pass into the trichogyne and eventually into the ascogonial base of ascomycotous fungi; the antheridium, fertilization tube, and oospore of the peronosporaceous oomycotous water molds; or the egg released from the ovary to be met by the motile sperm in the uterus of mammals.

The strategy of organelles and symbionts has been the same: in the passage from parent to offspring, they tend to be retained by the sedentary partner (the female) and discarded by the motile partner (the male), unless, like the mitochondria of sperm, the organelles themselves are required in the fertilization process. If, like most chloroplasts and crucial insect symbionts, the organelles are not needed in the fertilization process itself, those that best dedifferentiate to the minimal level (one complete genome per individual offspring) are favored by natural selection. Many symbionts are dispensable to the sperm and are transmitted by the egg—either inside or outside the egg membranes—for example, the symbionts of green hydra (Thorington, Berger, and Margulis, 1979). Insects have elaborate mechanisms to ensure transmission of symbionts to the ovary and thence to the egg (Schwemmler, 1991). Thus, maternal inheritance of mitochondria and plastids is not mysterious; it is the consequence of two simultaneous selective pressures—toward anisogamy and against redundancy.

If a mutation of an organelle leads to loss of function, the mutant to be phenotypically distinguishable, must replace all normal organelles in the intracellular population. This replacement rarely occurs, except perhaps immediately after fertilization in a few organisms, such as *Chlamydomonas*, that have one or very few organelles per cell. In *Chlamydomonas* and *Ulva*, plastid traits are inherited uniparentally, and organellar development is apparently under strict control in the zygote (Lewin, 1965a). Such organellar replacement is not expected to occur by chance, especially if the population of organelles is large.

The fact that DNA synthesis in mitochondria is not correlated with the steplike nature of the nuclear growth cycles is also to be expected from their prokaryotic ancestry: DNA synthesis is continuous in bacteria.

Work on organellar development in plastids and mitochondria supports the idea that "cytoplasmic mutagens" and "permanent bleaching agents" act directly on the ability of the organelle to complete its reproductive cycle in the host cytoplasm (Schiff and Lyman, 1982). When exposed to chemical agents, members of uniform populations tend to behave uniformly: the agents are typically benign, lethal, or mutagenic at certain concentrations. From the point of view of the symbiotic theory, the "anomalous 100 percent mutation rate" for cytoplasmic genetic systems (Gibor and Granick, 1964) is a uniform response of the members of a clone of microbes that live within the cells of other microbes. Treatments that produce hosts permanently deprived of normal plastids and mitochondria, with an apparent 100 percent mutation rate, should be interpreted as "effective curative agents," not mutagens. Lederberg (1952) made this point when he claimed that bleaching agents "cured" euglenas of their plastids. Such treatments (streptomycin, heat treatment, ultraviolet treatment, and nitrosoguanidine) are simply regimens more lethal to plastid genomes than to the nucleocytoplasm. This interpretation is consistent with the failure to find back mutations in cytoplasmic genes. The probability of back mutation of ρ^- to ρ^+ in yeast or of recovering the potential for chloroplast development in aplastidic euglenas is as remote as the probability of the spontaneous generation of prokaryotes.

Why has the permanent loss of plastids and plastid DNA been the result of many different treatments of euglenas (Schiff and Lyman, 1982), whereas usually at least some mitochondrial DNA remains in cells in which permanent loss of mitochondrial function has been induced? Most eukaryotes are obligate aerobes; presumably, treatments that induce the complete loss of mitochondria are lethal. Photosynthetic eukaryotes, on the other hand, are facultative autotrophs — loss of photosynthesis can be tolerated. Euglenas lack meiosis and have an idiosyncratic form of mitosis. The relationship between the nucleocytoplasm and the plastids of euglenas is probably less obligate than the organellar – nucleocytoplasmic relationship in meiotic plants and animals — perhaps because euglenas lack Mendelian genetic systems, which tend to centralize the control of organellar functions in the nucleus. Yeasts are meiotic eukaryotes in which the mitochondria are almost entirely integrated with the nucleocytoplasmic genetic system. Antibiotics and other agents reduce but do not eliminate mitochondrial DNA, although phenotypic expression of the mitochondrial genome is permanently impaired in ρ^- cells. No nuclear mutation can restore the ρ^+ "gene"; that is, none can cause the lost or impaired mitochondrial DNA to reappear. The nonfunctional mitochondrial DNA retained by yeasts after treatment was probably originally nuclear DNA.

The symbiotic theory suggests a reinterpretation of the claim that streptomycin causes a large spectrum of cytoplasmic mutations in *Chlamydomonas* — in particular, that it induces acetate-requiring mutants. Streptomycin is not a mutagen like acridines or X rays, but an inhibitor of prokaryotic protein synthesis; it inhibits the release of nascent protein from the ribosomes by binding to small subunits of ribosomes (Kelley and

Schaechter, 1968). Unable to be supported by their chloroplasts, *Chlamydomonas* host cells can no longer nutritionally depend on photosynthesis. After the loss of plastid function, the nucleocytoplasm must be provided with acetate, the same acetate that in the wild-type cell is supplied as photosynthate by the plastid. The inhibition of plastid protein synthesis in euglenas results eventually in the loss of DNA and the permanent loss of the ability to form plastids. In *Chlamydomonas*, although streptomycin inhibition of protein synthesis affects nucleic acid synthesis and produced nonfunctional chloroplasts, chloroplast DNA is not totally lost. That the cytoplasmic mutation is "stable" means that, generations after the drug treatment, a permanently impaired chloroplast remains. Like the unfortunate human victims of some hereditary diseases, these impaired organelles in nature would not survive even a single generation; they owe their continued existence to the nutrient-filled cytoplasmic environment of their hosts — and, of course, to the scientists who care for the "mutant" strains.

Studies of organellar ribosomal RNAs permit us to conclude that mitochondria originated from certain purple eubacteria, the so-called α-subdivision (Palmer, 1985). They also support a polyphyletic origin of, at least, plant mitochondria relative to the others (Gray, 1988).

CHAPTER 11

PHOTOSYNTHESIS IN PLASTIDS

GENETIC CONTINUITY OF PLASTIDS

The origin of an autotrophic plankton-flagellate, as postulated for the commencement of the study of algal phyla, covers a range of time and creative effort as that required for the elaboration of the highest land-types from such a flagellated organism, a period which may be possibly indicated as a thousand million years; and is not likely that such a story may be written in a few pages.

A. H. CHURCH, 1919

All plant cells contain plastids bounded by membranes, in which the photosynthetic pigments, if present, are located. Plant cells such as those of roots, seeds, and pollen, which lack the common green plastids (**chloroplasts**), have these options: to retain the ability to form chloroplasts from proplastids, to mate with cells that have or can form them or to fail to produce offspring. **Proplastids** are inconspicuous, smaller organelles from which plastids develop; **protoplastids** are evolutionary ancestors of plastids. Plastids, generally brightly colored, take many forms; they may even be translucent but they are never absent in cells of members of the plant kingdom. Algae, by definition, are protoctist cells that contain plastids.

Where do plastids come from? Schimper and Meyer, the botanists who first detailed chloroplasts, concluded that plastids are never formed de novo but always by the growth and division of preexisting plastids (Wilson, 1959). Even when first described, chloroplasts were seen to have a persistent individuality; like cells, nuclei, or chromosomes, they followed the general law of genetic continuity. In some algae and bryophytes, (*Zygnema* and *Anthoceros*) mature chloroplasts grow and divide regularly at each cell division. Microscopists have made analogies between plastids and algal endosymbionts since the nineteenth century (Khakhina, 1979).

> In the *Paramecium-Chlorella* symbiosis considerable integration between host and symbionts has occurred. The presence of the algal symbionts profoundly alters the growth pattern of the host. . . . Moreover the division rates of the paramecia and algae have become adjusted so that neither outstrips the other, and thus the hereditary association persists. . . . If metabolites can be exchanged freely between the host and its [algae] as has been suggested by the above experimental evidence natural selection would favor those mutations in either host or symbionts that delete duplicate metabolic pathways, and the two partners would become increasingly dependent upon one another (Karakashian, 1963).

Plastids develop in cells from preexisting chloroplasts or from proplastids (Figure 11-1). They contain DNA, messenger RNA, and protein-synthesizing systems. Plastid proteins are synthesized on small ribosomes sensitive to antibiotics that affect the growth of prokaryotic microbes. These data, consistent with an endosymbiotic origin, suggest that heterotrophic protists symbiotically acquired fully developed, oxygenic-phototrophic prokaryotes that became plastids. The striking similarities between

FIGURE 11-1

Chloroplasts and their development from proplastids. **A–D**, thin-section electron micrographs.
(**A**) Mature chloroplast from mesophyll cell of tobacco leaf, showing chloroplast ribosomes (*r*), numerous small dark grains. The large grain (*g*) shows the incorporation of tritiated uridine into chloroplast RNA. *c* = chloroplast membrane. Bar = 0.1 μm. (**B**) Prolamellar body (*p*) in chloroplasts of seven-day-old barley leaves illuminated for 30 minutes at low light intensities. Bar = 0.5 μm. (**C**) Barley chloroplast 2 hours after being placed in the light. The prolamellar body has dispersed. Bar = 0.5 μm. (**D**) After 12 hours of development in the light, a barley chloroplast is almost mature. Grana (*g*) have formed along primary lamellar layers, and starch grains (*s*) have developed. Bar = 0.5 μm. (**E**) The differentiation of proplastids in etiolated barley leaves incubated in light of low (right) and high (left) intensities. The terms "*albina*," "*xantha*," and "*viridis*" refer to mutants blocked at various stages (arrows) in the development process. Most are nuclear mutations. [Photographs courtesy of D. von Wettstein, K. W. Henningsen, and J. E. Boynton; **E** after von Wettstein (1967)].

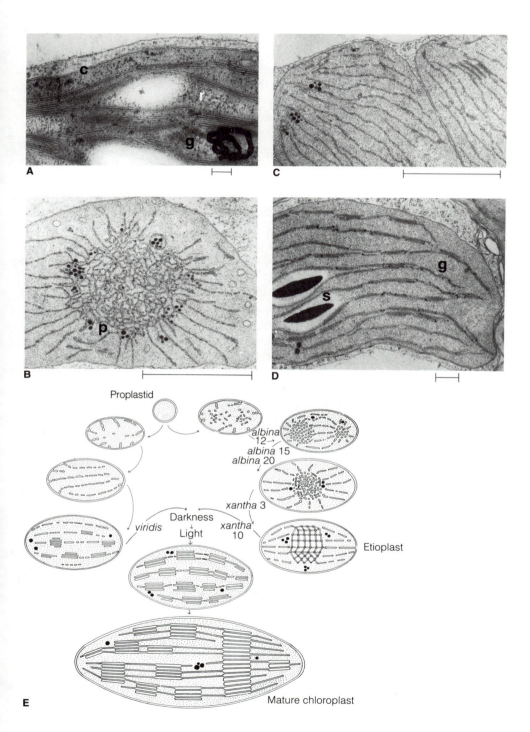

plastids and extant phototrophic bacteria suggest that protoplastids are more recent acquisitions than are protoundulipodia or protomitochondria.

After the presence of extranuclear DNA was correlated with the capacity to form plastids, the search for chloroplast DNA intensified; indeed nucleic acids in chloroplasts are detected easily. Specific nuclease treatments combined with electron microscopy located DNA nucleoids in plastids (Ris and Plaut, 1962). Isolated chloroplast DNA distinguishable from nuclear DNA was detected in density gradient studies (Chun et al., 1963). The presence of nuclear and plastid DNA in the same cell was shown by labeling euglenas with tritiated precursors of nucleic acids and then treating them with DNases and RNases (Sagan et al., 1965). The loss of *Euglena gracilis* chloroplasts and chloroplast DNA—induced by ultraviolet light, elevated temperature, and streptomycin—renders euglenas permanently unable to redifferentiate plastids (Edelman et al., 1964; Edelman, Schiff, and Epstein, 1965). Chloroplast DNA in closed, covalently bound circles was found in all plants and algae in which it was sought. So far, plastids have been observed with 180 to 220 kilobase pairs of DNA, depending on the source. The properties of the DNA are determined by sodium dodecyl sulfate (SDS) equilibrium sedimentation, electron-microscopic measurements, DNA renaturation kinetics and direct isolation and sequencing. At least three plastid genomes have been sequenced entirely (from tobacco, rice, and a liverwort), leaving no doubt that the ancestors of these organelles are cyanobacteria. DNA is synthesized in vitro when precursors are added to the chloroplast fraction. The single, circular, chloroplast DNA molecules contain information for the synthesis of many plastid gene products, including ribosomal RNA, 20 ribosomal proteins, 30 transfer RNAs, peptide subunits of the RNA polymerase, the ATP synthetase, the large subunit of the carbon dioxide–fixing enzyme ribulose-1,5-bisphosphate (RuBP) carboxylase, photosystems I and II proteins, and at least 40 proteins of unknown function. Chloroplast DNA, like that of nearly all prokaryotes, is not complexed to histone or RNA (Tewari, Kolodner, and Dobkin, 1976).

Chloroplasts have many features consistent with an endosymbiotic origin. Isolated chloroplasts perform DNA-dependent RNA synthesis and incorporate amino acids into proteins (Hoober, 1984). The DNA- and RNA-containing plastid, as a reproducing entity, has all the components of an autopoietic system. Plastids are hereditary systems not derived from the nucleus. The differentiation of the mature chloroplast from the proplastid is an adaptive system responsive to visible light (Gillham, 1978). In certain protists, plastids can be differentiated or permanently lost; in many plants they dedifferentiate to the proplastid level. The relationship between the nuclear genome and the plastids of *Euglena* is diagrammed in Figure 11-2;

at the molecular level the nucleocytoplasmic and plastid systems are now inextricably interdependent. Mechanisms that ensure distribution of at least one copy of the plastid to each plant or algal offspring cell are listed in Table 11-1.

The reproduction rates of host and symbiont genomes vary with circumstances, but to maintain any association, they must remain approximately equal. Plastids behave like symbionts: in euglenas that become enormously large when grown in a scarcity of vitamin B_{12}, the number of chloroplasts per euglena increases (Carell, 1969); apparently plastid-controlled synthesis of this vitamin in euglena is sufficient to ensure plastid reproduction but not nucleocytoplasm division. This interpretation is consistent with the prokaryotic distribution of the biosynthesis of vitamin B_{12} (see Table 6-6). In B_{12}-starved euglenas, new pellicular and other microtubules are not polymerized in spite of the large size of the cell; this failure of microtubule polymerization probably contributes to the inhibition of cell division.

Certain protists that lack plastids (or at least have no chlorophyll) are obvious homologues of plastid-containing cells (Figure 11-3). These aplastidic homologues are analogous to formerly symbiotic ciliates, hydroids, platyhelminths, and many other heterotrophs that in nature are found both with and without their photosynthetic symbionts. Despite intensive search for their elusive hypothetical common ancestor (the "uralga"), neither fossil nor extant organisms intermediate between the plastid-lacking cyanobacteria and the plastid-containing algae have ever been found. Rather, more symbioses analogous to the hypothetical one that gave rise to plant cells have come to light—for example, *Cyanophora paradoxa* (Trench, 1991), radiolarian and foraminiferan symbioses (Lee, 1990), and xanthellas and chlorellas of corals (Trench et al., 1978), sponges and ciliates (Taylor, 1979; Esteve et al., 1988), and hydras (Cook, 1980).

Non-Mendelian heredity was first described in plant cells and later recognized to be a consequence of uniparental transmission of chloroplasts (Gillham, 1978). Maternal transmission of plastids to the zygote follows from sexual dimorphism and the genetic continuity of organelles that cannot be made by the nuclear genome. (The maternal transmission of mitochondria is explained similarly; see Chapter 10.) Both the literature and its history have been well reviewed (Gillham, 1978; Sapp, 1987; Sapp, in press, Khakhina, 1979).

Two of the oldest clues to the symbiotic origin of plastids are their morphology and size: plastids and photosynthetic prokaryotes resemble each other greatly. Dinomastigote symbionts of radiolarians are modified by evolution to be distorted, smaller relatives of free-living forms (McLaughlin and Zahl, 1966; Jolley and Smith, 1978). Analogous morpho-

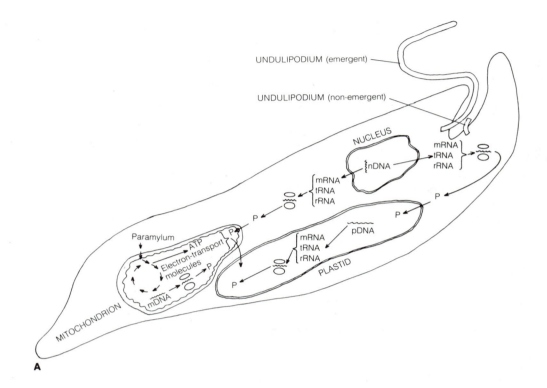

UNDULIPODIUM (emergent)

UNDULIPODIUM (non-emergent)

NUCLEUS

mRNA
tRNA
rRNA

nDNA

mRNA
tRNA
rRNA

P

P

P

Paramylum

ATP

Electron-transport
molecules

mDNA

mRNA
tRNA
rRNA

pDNA

P

PLASTID

MITOCHONDRION

A

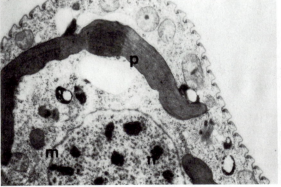

B

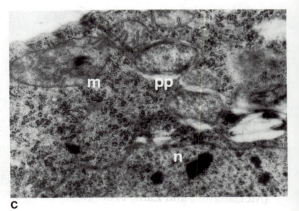

C

FIGURE 11-2

Nucleocytoplasmic – organellar relations in *Euglena*.
(A) Metabolite and gene product transfer between organelles of *Euglena gracilis*. *P* = protein; *ATP* = adenosine triphosphate; *mRNA* = messenger RNA; *tRNA* = transfer RNA; *rRNA* = ribosomal RNA; *pDNA* = plastid DNA; *mDNA* = mitochondrial DNA; *nDNA* = nuclear DNA; paramylum is the carbohydrate photosynthate reserve material characteristic of the euglenids (Bingham and Schiff, 1976). [Diagram courtesy of J. Schiff.] (B) A thin section of *Euglena gracilis* grown in the light, showing the well-developed chloroplast (*p*). *m* = mitochondrion; *n* = nucleus. Transmission electron micrograph, ×24,800. (C) The same strain of *Euglena gracilis* grown for about a week in the absence of light. The chloroplasts dedifferentiate into proplastids (*pp*). This process is reversible: proplastids regenerate and differentiate into mature chloroplasts after about 72 hours of incubation in the light. *m* = mitochondrion; *n* = nucleus. Transmission electron micrograph, ×32,200. [Electron micrographs courtesy of Y. Ben Shaul.]

logical modification of algal symbionts in lichens has been documented (Ahmadjian and Paracer, 1986). Expansion of the thylakoid surface, reduction of the nucleoid or nucleus, and simplification of external morphology accompanies the evolution of autotroph – heterotroph symbioses. Similar changes are expected in phototrophic symbionts that become organelles. Compare, for example, the plant cell and its plastids (Figure 11-4) with heterotrophic protists that harbor symbiotic cyanobacteria (see Figure

TABLE 11-1

Mechanisms for distributing plastids to each offspring in photosynthetic eukaryotes.

PHYLUM	ORGANISM	MECHANISM
Euglenida[a]	*Euglena* (dark-grown)	Distribution of numerous dedifferentiated proplastids 1 μm in diameter (Schiff and Lyman, 1982).
	Euglena (light-grown)	Division of chloroplasts, distribution of 10 to 12 mature chloroplasts.
Chlorophyta[a]	*Micromonas*	One chloroplast dividing synchronously with the nucleus (Gibor and Granick, 1964).
	Chlamydomonas	One large chloroplast cut by cleavage plane (Gillham, 1978).
	Nitella sperm	Enclosure of plastid in bundle of microtubules (product of centriole) packaged into sperm body (Turner, 1968).
Chrysophyta[a]	*Dinobryon*	Enclosure of plastid within a triple membrane, the outer layer of which is an outfolding of the nuclear membrane (Wujek, 1969).
Phaeophyta[a]	*Chorda*	Enclosure of plastid within a triple membrane.
Bacillariophyta[a]	Many diatoms	Small and constant number evenly distributed in mitosis (Darlington, 1958).
Tracheophyte phyla[b]	Most plants	Many chloroplasts randomly distributed at mitosis (Wilson, 1959).

[a]Kingdom Protoctista (Margulis, McKhann, and Olendzenski, 1992).
[b]Kingdom Plantae.

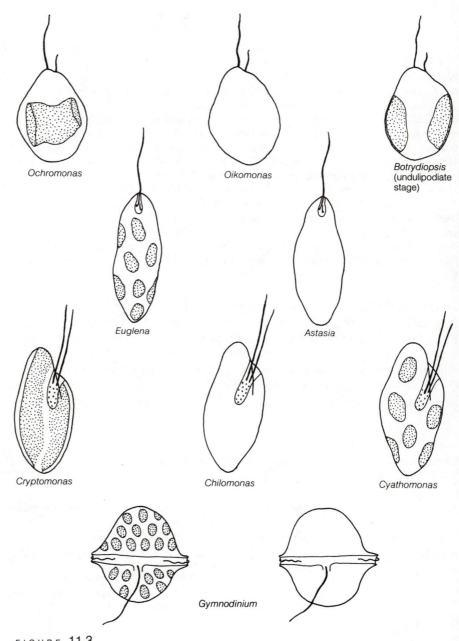

FIGURE 11-3

Protists with and without plastids. *Ochromonas* is a chrysophyte; *Oikomonas* is a heterotrophic mastigote; *Botrydiopsis* is a colonial xanthophyte; *Euglena* and *Astasia* are both euglenids. *Cryptomonas*, *Chilomonas*, and *Cyathomonas* are all cryptomonads whose plastids have brownish pigment, no pigment, and bluish-green pigment; *Gymnodinium* is a dinomastigote that may be plastidic or aplastidic.

7-12). The differences are mainly of degree: plastids are more highly integrated, presumably because they are more ancient.

The conservativeness of microbes already thriving in stable environments supports the inference about the prokaryotic ancestors of algal and plant plastids. The chlorophyll a–phycobilisome-on-thylakoid systems are so similar in both that cyanobacteria may be considered free-living prokaryotic codescendants of red algal plastids. Because of their pigments (chlorophylls a, b and the absence of phycobiliproteins) and their structural configuration, green prokaryotes [chloroxybacteria like *Prochloron*, *Prochlorothrix*, and the tiny marine oxygenic green bacterium of Chisholm et al. (1988), are probably ancestors of the plastids of chlorophytes and plants (Thorne, Newcomb, and Osmonds, 1977). The photosynthetic oxygenic coccoid prokaryote *Prochloron* possesses exactly the features one hypothesizes for protochloroplasts: chlorophylls a and b, small ribosomes, lipids typical of prokaryotes, and a tendency to enter symbiotic associations (Whatley, John, and Whatley, 1979; Lewin and Cheng, 1989).

Continued study of tropical prokaryotes may reveal free-living "brown" or "yellow-green" photosynthetic prokaryotes, ones that contain fucoxanthin or chlorophyll c. Modifications of algal pigment systems may have occurred after symbiotic acquisition of these prokaryotes. Late and independent acquisition of photosynthetic symbionts is almost certainly the explanation for membrane-bounded, anomalous pigment systems in certain motile protists, such as cryptomonads and euglenids (Gibbs, 1981, Whatley, John, and Whatley, 1979). The Phanerozoic adaptive radiation of diatoms, dinomastigotes, and xanthophytes (Lipps, 1970) may also be accountable by plastid polyphyly. Molecular data have not resolved the issue of multiple origins of plastids, although the origins of plastids from cyanobacteria in a plethora of mitochondria-containing heterotrophs by independent symbioses is virtually certain. The chloroxybacteria are part of cyanobacterial diversity; molecular data emphasize the close relations between the green and blue-green oxygenic photoautotrophs without definitive resolution of the "polyphyletic origin" issue (Gray, 1991).

Mitosis, an intracellular motility system of controlled chromosome movements along microtubules that interact with membrane, emerged from symbiont integration. Even if I incorrectly hypothesize symbiogenesis between spirochetes and *Thermoplasma*-like archaebacteria as the predecessor to mitotic–meiotic motility, mitosis (a eukaryotic process) evolved well after photosynthesis, essentially an anaerobic process that, since its inception, supported the entire biosphere. The polyphyly of algae is explained easily if one assumes that different phagotrophs acquired as food different photosynthetic prokaryotes that became protoplastids. These stable symbioses were not established simultaneously; variation both in

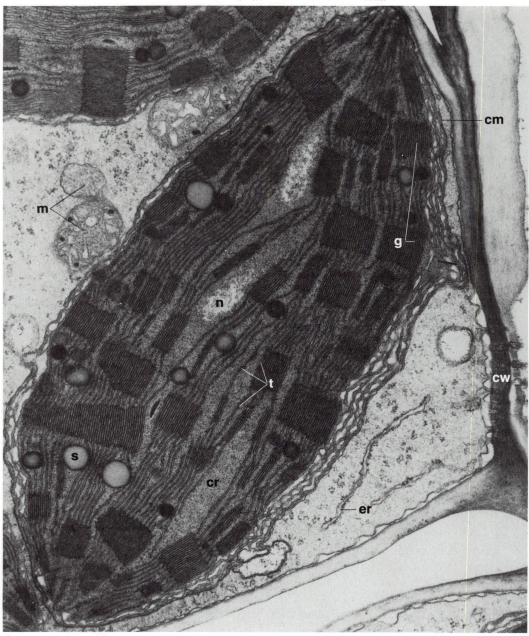

FIGURE 11-4

Chloroplasts.

(A) Ultrastructure of a chloroplast in a leaf mesophyll cell of corn. *cm* = chloroplast membrane; *cr* = chloroplast ribosomes; *cw* = cell wall; *er* = endoplasmic reticulum; *g* = grana lamellae; *m* = mitochondria; *n* = nucleoid; *s* = starch; *t* = thylakoid lamellae. Transmission electron micrograph, ×40,000. [Courtesy of Michael A. Walsh.]

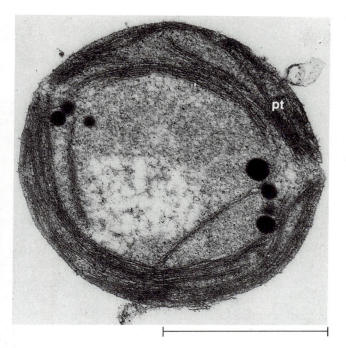

B

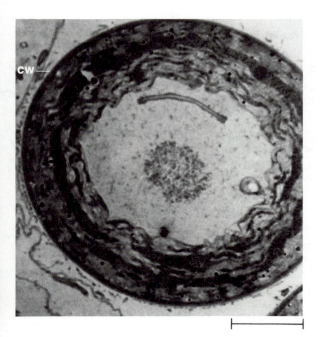

C

(B) *Acetabularia* chloroplast showing parietal thylakoids (*pt*). Electron micrograph, bar = 1 μm. (C) *Prochloron* from Palau didemnid with parietal thylakoids and peptidoglycan cell wall (*cw*). Electron micrograph, bar = 1 μm.

plastids and in the heterotrophic portion of algae suggests that symbiogenesis occurred at various times during the evolution of mitosis. In all algae and plants, the symbionts continued to co-evolve so that, in most, evidence of prior heterotrophy was severely reduced, but not lost. The dependence of the nucleocytoplasm on its plastids is most pronounced in land plants, which are generally unable to survive the loss of photosynthesis. The observation that albino corn plants may be kept alive to maturity by feeding them sterile sugar solutions directly through their leaves illustrates the heterotrophic legacy of even the most photoautotrophic eukaryotes. Acetate — supplied by the nucleocytoplasm in wild-type cells or by a human experimenter, if the mutant is forced to grow photoautotrophically —can compensate for lesions in the photosynthetic apparatus of *Chlamydomonas* ("photosynthetic minus" mutants). Green hydras that harbor chlorellas in their gastrodermal cells maintain the growth of their endosymbiotic algae in the dark indefinitely by supplying them with nutrients and removing their waste (Thorington and Margulis, 1981).

As molecular biological techniques (especially nucleotide and protein sequencing) are applied increasingly to evolutionary problems, the descent of plastids from specific genera of cyanobacteria (for example, *Anacystis* or *Synechococcus*) is recognized more clearly (see Figure 3-9**B**). The nonplastid portion of plant and protoctist cell can be classified with heterotrophic genera on the basis of biochemistry, morphology, life cycle, and mitotic cytology. In *Cryptomonas* the photosynthetic pigments are associated with a structure interpreted to be a vestigial nucleus (the nucleomorph). The demonstration of 18S eukaryotic RNA, like that of red algae, argues unequivocally for the polyphyly of cryptomonads from voracious heterotrophs that ingested but failed to digest red algae (Douglas et al., 1991). The presence or absence of the features of cytokinesis (that is, the phragmoplast or phycoplast), the varying kinetids, and the morphology, especially of cell division, in green algae supports the notion that chlorophyte algae resulted by convergent evolution from several independent symbiotic events. One group, represented by the genus *Klebsormidium*, contains a multilayered body, divides by forming a phragmoplast, and seems related in several other ways to bryophytes and tracheophytes (Pickett-Heaps, 1975). Whether *Klebsormidium*, some charophycean (Graham, 1985), or even some charophycean with an incorporated fungal symbiont genome (Atsatt, 1991) is closest to the ancestors of plants is a question that requires investigation by the molecular biological armamentarium.

The convergent evolution of the green algal habit is analogous to the evolution of lichens, in which the fungi (some 14,000 species) vary enormously. Whereas some 22 chlorophytes, one xanthophyte, one phaeophyte, and 15 cyanobacteria are known as phycobionts of lichens, about 40 per-

cent of all lichens take *Trebouxia* (including *Pseudotrebouxia*) as phycobiont. Evidently, both the protoctists that acquired photosynthetic prokaryotes to become algae and the fungi that acquired algae to become lichens
were limited to only a small number of phycobiont partners.

PLASTID AND HOST GENOMES

Natural Selection, by itself, is not sufficient to determine
the direction of organic evolution. . . . Natural Selection
can only deal with that which has been formed; it has no
creative powers. Any directing influence that Natural
Selection may have in organic evolution, must, in the
nature of the process, be secondary to some other
unknown factor.

I. E. WALLIN, 1927

Since plastids were acquired by symbiosis, plants are genetically and metabolically more complex than many physiologists realize. If eukaryotic heterotrophs are trigenomic by virtue of the nucleocytoplasm, undulipodia
and mitochondria, then plants are at least quadrigenomic by virtue of an
additional organelle, their plastids (Figure 11-5; see also Figure 7-16).
Atstatt (1991) emphasizes the importance of extracellular digestion and
absorptive nutrition in the development of plant tissues and in processes
such as programmed cell death. If he is correct that the understanding of
plant histology requires recognition of the integration of a fungal genome
that accompanied the origin embryophytes from green algae, then all
plants evolved from still another symbiogenesis, and plants have at least
five genomes.

Studies in which plant genomes and their products are homogenized
and their functions reconstructed from isolated chemical reactions are
bound to yield confusing and contradictory results. How successful would
be a biochemist studying the fermentation process in wine-making who
grinds up the "ferment" with its several organisms—all different in age,
physiological condition, degree of interdependence, and many other variables? Secondary plant metabolites, such as flavonoids, alkaloids, aceto-
genins, and lignans, must be sorted out according to whether they are
coded for by one of the genomes alone, or by any two, any three, or all four.
An example of how this sort of analysis may be applied to data in the
physiology literature is outlined in Table 11-2. An analogous analysis of

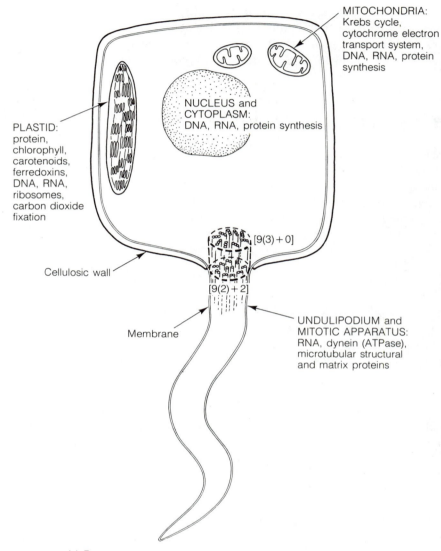

MITOCHONDRIA:
Krebs cycle,
cytochrome electron
transport system,
DNA, RNA, protein
synthesis

NUCLEUS and
CYTOPLASM:
DNA, RNA, protein synthesis

PLASTID:
protein,
chlorophyll,
carotenoids,
ferredoxins,
DNA, RNA,
ribosomes,
carbon dioxide
fixation

[9(3) + 0]

Cellulosic wall

[9(2) + 2]

Membrane

UNDULIPODIUM and
MITOTIC APPARATUS:
RNA, dynein (ATPase),
microtubular structural
and matrix proteins

FIGURE 11-5
Are plant cells quadrigenomic?

plastid-membrane proteins undertaken by Bingham and Schiff (1976) suggests that many of these either directly or indirectly under the genetic
control of plastids. Because the degree of fluidity of plastid and mitochondrial membranes is related to chill resistance and cold sensitivity in agricul-
tural plants (Smillie, 1976), it is of more than academic interest to discern

TABLE 11-2
Organellar genetic control of carotenoids in *Euglena gracilis*.[a]

| | CAROTENOID[b,c] | | | |
PHYSIOLOGY OF CELL	CAROTENE	NEOXANTHIN	ZEAXANTHIN	ANTHERAXANTHIN
Normal (light-grown, green)	+ +	+	+	+ +
Normal (dark-grown, white, chloroplast dedifferentiated)	+	inc.	+ +	+ +
Permanently white	− (?)	−	+	+

[a]After Krinsky et al. (1964).
[b]Predictions: Carotene and neoxanthin are synthesized under control of chloroplast genes. (Neoxanthin production is physiologically connected with photosynthetic activity, as indicated by Krinsky et al., 1964.) Zeaxanthin and antheraxanthin are synthesized under control of nuclear genes.
[c]KEY: + + = at least 10% (by weight) of total carotenoid pigments; + = present; − = absent; inc. = absent, but increases with plastid differentiation (regreening).

the levels of partner integration in these ancient multiple symbioses. The tendency in cell evolution is toward interdependency; organelles become dependent on products of nuclear genomes that require controlled transfer into the maturing organelles (Gray, 1991).

No algae or plants have been discovered that contain plastids but not mitochondria; this fact is interpreted to be a consequence of the order in which these organelles were acquired. Flexible steroid membranes and motility proteins like actin and its associated ATPase are required for the phagocytic process, not only for feeding but for the uptake of protoplastids. Such membrane machination probably requires integrated undulipodial and mitochondrial genomes. (No evidence for phagocytosis exists in any prokaryotes, even in the methane-oxidizing bacteria that can synthesize steroids.) Plastid precursors, however, were ingested by organisms capable of phagocytosis, probably those that contained mitochondria, tubulin, actin, and pathways for steroid synthesis. (If algae evolved directly from cyanobacteria, nucleated organisms having plastids but lacking mitochondria should have survived.) Direct evidence for the acquisition of plastids after that of mitochondria comes from molecular data (Gray, 1991).

FROM SYMBIONT TO ORGANELLE: RARE OR FREQUENT?

Some structures such as mycetocytes, bacteriocytes and mycetomes of insects may be the survival of profound pathological changes.

L. R. CLEVELAND, 1926

Did symbiont integration lead to the formation of organelles and new species only during the evolution of eukaryotic cells in the Proterozoic eon, or is it still occurring? Does the incorporation of symbionts today still lead to speciation? The answer is clear: as such associations form symbiogenesis leads to speciation in some cases, and in others the associations dissolve. During the last 20 years, Jeon has traced the origin of organelles from bacteria (see Figure 10-3). *Peridinium balticum* (a dinomastigote) and *Mesodinium rubrum* (a ciliate; Lindholm, 1985) are each idiosyncratic assemblages of photosynthetic and motile organisms. Based on their ultrastructure and geographical variations, Taylor (1978) interprets the associations as being in the process of formation. Green hydras are symbiotic complexes of aeromonad bacteria, chlorellas, and freshwater hydras; because all three partners are capable of independent growth, this association must be rather young (Margulis et al., 1978; Berger, Thorington, and Margulis, 1979). In some associations, autopoietic systems that are not even whole cells are "borrowed." The exchange of photosynthetic nutrients has arisen between the chloroplasts of siphonaceouos food algae and the tissue cells of the saccoglossan mollusks *Elysia, Tridachia,* and *Plachobranchus oscellatus,* for example. The chloroplasts alone, not the rest of the algal cells, are maintained by the animal, actively photosynthesizing in the diverticulae of the guts of the mollusks (Taylor, 1974). Functional chloroplasts are retained for weeks by these slugs, making the adult mollusks "plant-animals." However, because the gene for the large subunit of the RuBP carboxylase, among others, remains behind in the algal nucleus, the slugs must renew their supply of chloroplasts periodically. That is, the mollusk tissue maintains foreign chloroplasts but is unable to culture them. The chloroplasts of *Euglena* and others with three or four boundary membranes may come not from cyanobacteria but, like those in mollusks, from green algae or other eukaryotic sources (Gibbs, 1981; Gray, 1991).

Some cyclical associations between partners have evolved to such a highly sophisticated state that the exchange of information rather than of physical products is a mainstay of the association. In one association between plants and bacteria, the continued physical proximity of the partners is not required—only a DNA plasmid is exchanged. Several different species of Gram-negative rod bacteria, members of the genus *Agrobacterium,* are found in soil. These agrobacteria attach to the stems or branches of their hosts, most of which are dicotyledonous plants. The bacteria transmit rapidly spreading plasmids to the plant tissue. The plasmids, integrating into the nuclear genome of the plant cells, cause dedifferentiation of the plant tissue and the formation of ugly growths, crown gall tumors, at the site of the wound. The plasmids carry genes not only for their own replication but for the production of carbon-rich and nitrogen-rich nonprotein amino acids such as opines—nopaline or octopine. Opines may be the sole source of carbon and nitrogen for the agrobacteria. When these products of the altered plant metabolism leak into the soil, they support the growth and replication of the agrobacteria that originally donated the plasmids (Schell et al., 1979; Alberts et al., 1989).

A great potential exists for symbiotic associations, gene transfer, and foreign organelle retention because of what Sapp (in press) calls "evolution by association." These phenomena supplement the genetic variation that arises from point mutation and recombination. Whether the new hereditary potential is expressed and retained depends on natural selection. When natural selection has already led to optimized populations of organisms, new sources of variation from symbiosis and gene transfer do not necessarily lead to more rapidly reproducing populations than those already established in the environment in question. When new niches appear, such as the sulfide-rich spreading center at 21° north latitude off the coast of Baja California or the cold seeps found off Florida, strong selection pressures and the potential for symbiogenesis lead to colonization by an assortment of animals capable of growth on sulfide as their source of energy and carbon dioxide as their sole source of carbon (Vetter, 1991). Such **thiotrophic** animals (for example, *Riftia* and other members of Vestimentifera; *Solemya*, etc.) are spectacular offspring of parasexuality. Having marine animals and sulfide-oxidizing bacteria as parents, these holobionts attest that symbiogenesis as a source of evolutionary innovation is not limited to the microcosm. The rate of gain and loss of associations in the biosphere changes according to selection pressures, many of which are exerted by other organisms.

The pre-Phanerozoic world was composed of small organisms but an evolutionarily powerful biota; major trends and innovations of the extant biosphere were established (Bengtson, 1993). Fermentation, photosynthesis, aerobiosis, symbioses, mitosis, meiosis, morphogenesis, and embryogenesis had made possible by then the modulation of the planet's surface by life. The Phanerozoic eon has its own story: the prior appearance of a marine biota in the latest Proterozoic eon (Vendian era), was followed in the Silurian and Devonian periods by the origins of plants and fungi, and the restricted taxa of animals capable of living on dry land. The tale is the familiar story of paleontology. Our final chapter considers a very limited part of this later-day natural history: the consequences of symbiogenesis for the origin of certain metabolic pathways, chromosome behaviors, and aspects of environmental regulation.

CHAPTER 12

PHANEROZOIC
CONSEQUENCES

CALCIUM AND SKELETONS

The base of the Cambrian period (0.6×10^9 years ago) is marked in marine sediments around the world by the appearance of abundant animal life. . . . This sudden appearance of diverse animal stocks has been the most vexing riddle in paleontology.

A. G. FISCHER, 1965

What are made apparent in the earliest Cambrian records are not organisms so much as hard parts. . . . The development of the fossilizable skeleton at the opening the Phanerozoic reflects the initiation of predation, and that before this time the biosphere was largely a peaceable kingdom in which armor was not needed and therefore was not available as an epitaph.

G. E. HUTCHINSON, 1965

Life of the Phanerozoic eon left conspicuous fossils. Evolutionary trends in four great kingdoms, the protoctists, animals, fungi, and plants, can be traced from the record left in the sediments. Multicellularity had evolved independently in many groups of prokaryotes and protists long before the rise of animals and plants. Whether or not most of the soft-bodied fossils are protoctista or animal remains, some metazoa must have evolved by 680 million years ago, the time of deposition of the Vendian biota. The sudden development at the Phanerozoic boundary about 580 million years ago is skeletalization: the appearance of biomineralization on a grand scale.

Is there a relationship between the evolution of cellular mechanisms for handling calcium and the appearance of calcareous skeletons? Explosive evolution of eukaryotes had occurred in the Proterozoic eon, but, because the major patterns of diversification were on the cellular level, and in soft-bodied organisms, their vestiges are subtle. Metazoans may have originated earlier in the Proterozoic eon than usually reckoned (Clemmey, 1976), or they may not have appeared until Ediacaran times. In any case, that eukaryotes having hard parts suddenly appeared in the Tommotian of the early Cambrian still stands. Numerous explanations have been offered, ranging from the obliteration of earlier records by various geologic processes to physiological or environmental factors and ecosystem evolution (Stanley, 1976a,b). The answer to the vexing riddle lies, at least partly, in the nature of the preserved parts. Skeletalization must have been preceded by the internal capacity for mineralization. The key to the mystery may lie in the control of calcium biomineralization (Lowenstam and Margulis, 1980a; Lowenstam and Wiener, 1989; Bengtson, in press).

The inorganic constituents of the hard parts of modern organisms distantly related to the early Phanerozoic eukaryotes are anhydrous calcium carbonates, crystalline carbonates such as calcite and aragonite, calcium phosphates as apatite minerals, and silica in the form of quartz. Calcium phosphates (apatites) were synthesized more widely by late pre-Phanerozoic and earliest Cambrian organisms than by subsequent ones. Calcium carbonate minerals, which were limited initially to a few eukaryotes, became by mid-Cambrian the most common products of biomineralization. They have remained the major components of hard parts through the Phanerozoic eon until today. Biogenic calcium carbonate eventually provided new bases for extensive new communities, microscopic and macroscopic, of which the best-known are coral reefs. Endoliths appeared—cyanobacteria, fungi, and worms that live by penetrating carbonate substrates (Golubic, Perkins, and Lukas, 1975).

Silica was limited initially to the sponges alone (Lowenstam, 1963). Subsequently, it became more widely used—for example, in the tests of chrysophytes, heliozoans (Figure 12-1), radiolarians, and diatoms and in the cell walls of some plants. Although it has always been important in biomineralization, silica has never rivaled the calcium minerals in abundance, diversity, and continuity in the fossil record. Because the cellular mechanisms for using silica are not very well understood, the widespread, polyphyletic appearance of Phanerozoic silicified tests remains unexplained.

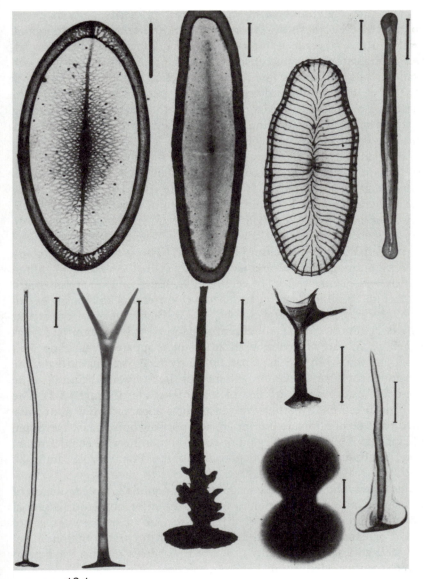

FIGURE 12-1

Silica scales of heliozoans. Top row, left to right: *Raphidiophrys ambigua*, freshwater species, cultures available; *R. marina*, marine species; *R. elegans*, freshwater species; *R. neapolitana*, new species. Bottom row, left to right: *Acanthocystis myriospina* (probably), freshwater species; *A. turfacea*, freshwater species; *A. aculeata*, freshwater species, culture available; *Acanthocystis* spp.; *A. crinoides*, freshwater species, culture available. Electron micrographs, bars = 1 μm. [Courtesy of C. F. Bardele.]

In any case, calcium-bearing compounds, primarily carbonate and phosphate minerals, have always been the most important bioinorganic materials; in recent years, understanding of calcium metabolism has advanced enough to permit new insights into the relationship between cell innovation and fossilization. The rest of this section presents some working hypotheses of the origin and diversification of calcium-bearing hard parts and tissues (Lowenstam and Margulis, 1980b). Information on the earliest fossil record of organisms bearing calcareous hard parts and other mineral constituents is shown in Figure 12-2.

Bioinorganic materials are produced mainly by animals and protoctists. Mineralization is under genetic control; it is often a function of the developmental stage of an organism. Biominerals are most commonly skeletal, but they may also appear in otoliths or statoliths of sense organs, in teeth, and in other hard parts. About fifty different kinds of bioinorganic substances have been identified (Lowenstam and Wiener, 1989). Of these, about two-thirds include calcium (Table 12-1). The greatest quantity and diversity of bioinorganic calcareous materials is in the animal kingdom. Protoctists and plants are less mineralized. In fungi under natural conditions, calcium biomineralization seems to be exceptional and mostly excretory (Graustein, Cromack, and Sollins, 1977). Biosynthesis of calcareous hard parts is almost completely absent in monera (Lowenstam and Margulis, 1980a). Calcareous biomineralized products apparently first appeared (in the Tommotian) in the hard parts of fairly large marine animals, the proterostome coelomates, and only thereafter (in Atdabanian times) in the marine deuterostomes and less complex animal grades (Figure 12-2). The smaller and presumably earlier-evolved protists apparently did not evolve calcified parts before the late Cambrian, and possibly only in late Devonian time (Lipps, 1970). Why have calcium-containing minerals been dominant in biomineralization from the beginning of the Phanerozoic until the present?

Inside eukaryotic cells, and conspicuously in animals, the regulation of calcium ion (Ca^{2+}) concentration is indispensable for numerous essential physiological processes. (Whether this is also true of prokaryotes is not known.) For example, amoeboid cell movement, muscle contraction, secretion, cell adhesion, and many other cell processes require calcium in regulated quantity (Heilbrunn, 1956; Kretsinger, 1976, 1990; Thompson, 1988). Furthermore, calcium plays an informational role. The free intracellular Ca^{2+} concentration in eukaryotes ranges from about $10^{-7} M$ to $10^{-8} M$. Extracellular concentrations of calcium tend to be far higher. The concentration in tissue spaces and plasma, for example, is closer to the concentration in sea water, ranging from about $10^{-2} M$ to $10^{-4} M$. Cells use this difference of some four orders of magnitude informationally: diverse stimuli at cell surfaces induce changes in membrane electropotentials

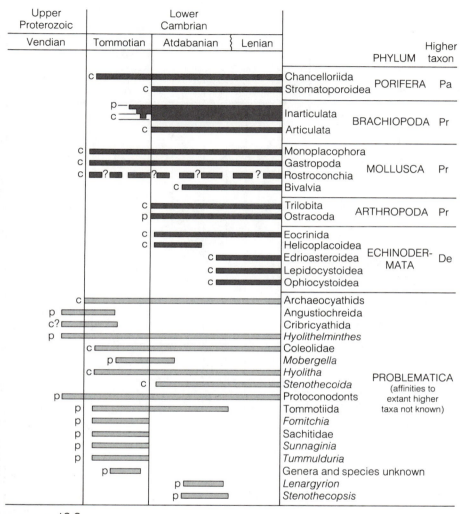

FIGURE 12-2

Appearance of animals with hard parts in the fossil record. c = calcareous material produced; p = phosphatic material produced; Pa = Subkingdom Parazoa; Pr = Series Proterostomia; De = Series Deuterostomia (see Lowenstam and Margulis, 1980b).

followed by rapid influxes of Ca^{2+}. The response to the influx of calcium depends on the nature of the cell; it may lead to movement, to the firing of nerve impulses, to bioluminescence, to certain biosyntheses, and so forth (Alberts et al., 1989). Mitosis itself requires Ca^{2+} regulation. The polymeri-

TABLE 12-1

Calcium biominerals.[a]

	CARBONATES					PHOSPHATES						OXA-LATES		HALIDES		SUL-FATES
	AC	V	A	C	MHC	ACP	AHW	D	FR	CMP	CP	WH	W	AF	F	G
Monera																
Fermenting bacteria								+								
Pseudomonads			+	+												
Proteobacteria				?												
Actinobacteria						?		+								
Protoctista[b]																
Dinomastigota				+												
Prymnesiophyta			?	+												
Phaeophyta			+													
Rhodophyta			+	+												
Conjugaphyta												+				
Chlorophyta			+	+												
Granuloreticulosa (forams)			+	+												
Myxomycota	+			+												
Ciliophora								+								
Plantae																
Bryophyta				+								+				?
Tracheophyta		+	+	+				?				+				+
Fungi																
Basidiomycota												+	+			
Deuteromycota								+				?	?			
Animalia																
Porifera			+	+												
Cnidaria			+	+												+
Platyhelminthes	+			+		+	+									
Bryozoa			+	+												
Brachiopoda				+					+							
Annelida			+	+			+	+		+						
Sipuncula			+	+												
Mollusca	+	+	+	+	+	+	+		+				+	+	+	
Arthropoda	+	+	+	+		+									+	
Echinodermata				+									+			
Chordata	+	+	+	+	+	+		+	+		+		+	+		

[a]Key
AC = amorphous CaCO$_3$ ACP = amorphous carbonate apatite   WH = whewellite
V = vaterite   AHW = amorphous whitlockite   W = weddellite
A = aragonite   D = dahllite   AF = amorphous fluorite
C = calcite   FR = francolite   F = fluorite
MHC = monohydrocalcite   CMP = Ca$_3$Mg$_3$(PO$_4$)$_2$ G = gypsum
    CP = Ca$_8$H$_2$(PO$_4$)$_6 \cdot$ 5H$_2$O

[b]See Margulis, Corliss, Melkonian, and Chapman (1990) for details of about 50 phyla and classes.

zation of microtubule protein, at least in vitro, is very sensitive to Ca^{2+} concentration (Soifer, 1975). Thus, if the Ca^{2+} concentration were to rise internally to levels characteristic of the ambient medium, fundamental processes would be inhibited.

Calcium ion concentration also plays an important role in poriferan cell aggregation. In *Microciona* and *Haliciona*, for example, Ca^{2+} modulates species-specific cell aggregation by means of a uronic acid–rich glycoprotein, weighing about 20,000 kilodaltons, that contains 1150 Ca^{2+}-binding sites per protein molecule. If the Ca^{2+} concentration is reduced to below 10^{-5} M, poriferan cells dissociate; if the Ca^{2+} is restored, they reaggregate into a gel (Kretsinger, 1976).

Thus, intracellular regulation of Ca^{2+} is crucial to functions that are sine qua non of the metazoan condition: mitosis and cell aggregation. But why does calcium play such an important role? Kretsinger assumes that cations rather than anions are favored for rapid transmission of intracellular information because "biological molecules have many more electronegative ligands and hence have a greater range of cation affinity and specificity" (Kretsinger, 1990). Why Ca^{2+} rather than other physiologically significant cations (Na^+, K^+, or Mg^{2+}); this question is explored below.

The late Proterozoic Ediacaran biota, first described from South Australia (Glaessner, 1971), has been found at about two dozen sites around the world: England, Canada (Misra, 1969), Namibia, and elsewhere (McMenamin and McMenamin, 1990). By some 600 million years ago, metazoa from several of the major animal phyla, Cnidaria, Annelida, and Arthropoda, at least, had diversified and were thriving. These animals, most of them marine littoral forms, had attained dimensions as large as several decimeters. At least some were motile. Like their modern counterparts, they probably moved by means of muscle tissue. These metazoa preceded the first biomineralized animals and exceeded them in size, implying that intracellular calcium-regulating mechanisms preceded skeletalization. Muscle tissue is composed of a number of homologous proteins. The light chains of muscle myosin and troponin are both calcium-binding proteins; both are required for the transduction of chemical to mechanical energy. These proteins contain homologous alpha-helical regions that bind directly to calcium and modulate intracellular calcium concentrations (Moncrief, Goodman, and Kretsinger, 1990; Nakayama et al., 1993). The musculature of some Ediacaran biota implies that modulation of intracellular calcium developed at least 150 million years before skeletal hard parts; furthermore, modulation of intracellular calcium must have preceded the evolution of muscle tissue.

Ciliates have an idiosyncratic nuclear genetic organization (Raikov, 1982; Corliss, 1979); they probably evolved before the metazoans. They also use calcium informationally: changes in intracellular Ca^{2+} concentrations determine ciliate behavioral patterns (Schein, Bennet, and Katz, 1976), including mating (Cronkite, 1976). They also affect the behavior of heliozoans strongly (Davidson, 1982) and cause contraction of the rhizoplast of *Platymonas* (Salisbury and Floyd, 1978). These observations are consistent with the hypothesis that calcium modulation evolved in protoctists before metazoan skeletalization. What more was required for the origin of biomineralization in several phyla?

Although no definitive answer is possible, the fossil record is instructive. In strata bearing the oldest mineralized skeletons, one finds the earliest unequivocal evidence for predation—for example, the bored hard tissues of *Colellela billingsi* (Mathews and Missarzhevsky, 1975) and others (McMenamin and McMenamin, 1990). The early skeletalized metazoa are small, which of course implies a larger surface area per unit volume relative to unskeletalized forms. Although elaborately skeletalized early metazoa are well known, they are often present in communities of organisms that are sparingly or weakly skeletalized; examples are the Porifera, Ostracoda, and Trilobita, which appeared slightly later than the Tommotian, in the Atdabanian. In all of these, according to Lowenstam and Wiener (1989), the organic fraction far outweighed the bioinorganic one. Together, these observations imply that the potential for skeletalization, namely the requirement for external extrusion of soluble calcium ions, was widespread in eukaryotes early on, and that the potential was expressed as actual skeletalization in response to other selection pressures, one of which was increased predation, as Hutchinson (1959) and Stanley (1976a) noted. How did these selective pressures operate? Prey, forced to escape from more effective predators, developed highly integrated sensory and motor systems that increased coordination and speed. Muscle contraction responds directly to calcium release. Muscle tissue cells control calcium sequestration and release; they have differentiated calcium storage sites, in the sarcoplasmic reticulum and in mitochondria. Protoctists and small metazoa differentiated such specialized cell complexes near their large surfaces, whereas large metazoa stored calcium inside. Certain groups solved the calcium modulation problems by excreting calcium on cuticles and surface tissues (Lowenstam, 1980).

In algae growing in carbonate-rich water, similar but probably less stringent constraints were put on intracellular calcium concentrations. Because they lack muscular and sensory systems directly responsive to calcium, algae probably have a greater tolerance to variations in intracellular calcium concentration than do metazoa. It would be useful to know the values and ranges of these concentrations in closely related calcified and uncalcified species of red, brown, and green algae. Some groups, such as the coralline algae, either were unable to solubilize the calcium carbonate that formed inside or were selected for because the carbonate deterred animals from eating them. Perhaps the development of waxy cuticular materials in seaweeds was an advanced development, countering the more primitive tendency to become encumbered simply as a consequence of calcium excretion (Lowenstam and Margulis, 1980b).

Kretsinger et al. (1991) distinguishes calcium-modulating and calcium-binding proteins. The former contain the conformation called the "EF hand," a loop that contains six oxygen ligands coordinating Ca^{2+} and the second alpha helix (the "thumb"). Many other proteins, most of them extracellular, bind calcium. Apparently, they bind calcium mainly for their own structural stabilization; they lack EF hands. They do not have a conformation directly sensitive to calcium and therefore do not modulate intracellular calcium concentration. Certain stable calcium-binding proteins are homologous neither to each other nor to the EF hand proteins. Surprisingly, the primary amino acid sequence and three-dimensional structure of independently isolated proteins that were not even suspected to be related shows that they are all descendants of a common ancestor in which the EF hand evolved. These homologous calcium-sensitive proteins include troponin, carp muscle parvalbumin, myosin, vitamin D–induced bovine Ca^{2+}-binding protein, and especially calmodulin, a calcium-sensitive protein apparently ubiquitous in eukaryotes.

The explosive evolution of hard parts in the Phanerozoic was a late development in a long series of innovations, most of which took place within cells and left no trace in the fossil record. These were the appearance of respirable quantities of atmospheric oxygen (Cloud, 1989; Walker, 1977), followed by the symbiotic acquisition of undulipodia, which led to the differentiation of intracellular membrane systems of the nucleus, the endoplasmic reticulum, and, after acquisition of mitochondria, the outer mitochondrial membrane. Such membrane systems imply the capacity for synthesizing steroids and polyunsaturated fatty acids, and, with the motility proteins, were probably preadaptations for endo- and exocytosis. The evolution of mitochondria and microtubule systems was followed by increased selection pressure for stringent regulation of intracellular calcium. Possibly some calcium extrusion had first evolved to protect magnesium enzymes, required for nucleic acid replication and protein synthesis, and to keep apatite from precipitating. However, calcium modulation was mandatory to permit polymerization of microtubules in calcium-rich environments. This modulation was followed by the origin of Ca^{2+}-sequestering vesicles at the bases of cilia and the use of mitochondria in calcium sequestration (Satir, 1974). The origin of mitosis, followed by the origin of haploid–diploid life cycles and the appearance of metazoan multicellularity, histogenesis, and organogenesis, led to the development of specialized intercellular junctions: desmosomes, tight junctions, gap junctions, and so forth. Greater inter- and intracellular regulation and release of calcium were accompanied by a higher degree of specialization of membrane systems, including that of the sarcoplasmic reticulum of muscle tissue. With increasing mobility, including more rapid retraction responses in sessile metazoans, more efficient predators evolved. Predators created selection pressure that stimulated the development of more efficient escape mechanisms. By this time, all of these sensory and motor specializations depended intrinsically on calcium-modulating proteins. The formation and transport of calcite and aragonite crystals in elaborate calcareous skeletons, itself perhaps a product of symbiogenesis as protoctists acquired calcium-precipitating bacterial symbionts, was a great epiphenomenon, evolving only after many other physiological processes were established in Proterozoic eukaryotes.

Symbiogenesis, speciation, and morphogenesis in the Phanerozoic eon

The conception that all living cells are loaded with
bacteria is so startling and antagonistic to our orthodox
notions of the cell, that without further analysis and
reflection the conception appears absurd.

I. E. WALLIN, 1925

Only recently has symbiogenesis been considered important to evolutionary theory by English-speaking scientists [Margulis and Fester (1991), which includes Sapp's (1991) discussion of the historical relationship between the study of cytoplasmic heredity and of symbiogenesis]. The history of symbiogenesis, the early Russian work (reviewed by Khakhina, 1979), and Wallin's notion of **prototaxis** (the response of one living being to another) and of the acquisition of microbial symbionts in speciation (symbionticism) are evaluated by Mehos (in Margulis and McMenamin, 1992). Together, these works make it clear that symbiogenesis is an evolutionary principle of wide application to the origins of species and the diversification of Phanerozoic life, just as its inventor, Mereschkovsky, asserted.

Symbiosis is still underrecognized as a mechanism of evolutionary novelty in animals, plants, and fungi. Zoologists, botanists, and mycologists tend not even to mention symbiogenesis in their discussions of the evolution of the organisms they study. Observations, such as that brittle stars harbor healthy, actively metabolizing bacteria in their cuticles, are over-looked. These nearly "pure-culture" bacteria seem to be ensured genetic continuity by their storage in deep pockets within the genital bursae of these viviparous echinoderms (Walker and Lesser, 1989). Sulfide-oxidizing chemoautotrophic animals at deep-sea vents, families of luminous bacteria, and nematode-ridden lepidopterans (glowworms) owe their existence to symbiogenesis. Entire communities (for example, mycorrhizae of the northern boreal forests and tropical coral reefs) depend unequivocally on microbial symbionts for their existence. The red color of the nidamental gland of sexually active squid is due to crowded pigmented symbiotic bacteria (Bloodgood, 1977). Indeed, incongruities in echinoderm larvae and adults may be due to the symbiotic retention of distantly related embryos by juveniles of different species (Williamson, 1987, 1992)

Symbiotic integration did not cease in the Phanerozoic eon, and sym-biogenesis is not limited to the microbial world. The humpback whale (*Megaptera novianglia*), too, is a product of obligate association between its Darwin-named barnacles (*Coronula diadema*) and its hairy mammalian carcass. The six-sided barnacles, attracted by whale chemistry, settle on the throat and extended pectorals of the large seafaring mammal (Dyer, 1989); growing tissue connects from the barnacles' fluids through the whale's black hair to the arteries and veins of its gigantic circulatory system. Clicking as they eat, the barnacles make noises that help the whale focus on plankton patches; without its six-sided acoustic organs the whale would starve. Humpbacks without vast numbers of barnacles do not exist, nor do barnacles of this sort live without humpbacks. We name the barnacles, we name the whale, and we give the name of the larger complex (whale) to the whole, new autopoietic entity (whale plus barnacle).

The evolution of algivory in surgeonfishes (Acanthuridae) on the hard-reef surfaces in both the Red Sea and the Australian Great Barrier Reef was probably due entirely to the acquisition of huge bacteria capable of multiple fissions. These cigar-shaped, endosporulating prokaryotes may reach sizes up to 417 μm long, the largest known for bacteria. Up to 10^5 bacteria per milliliter of gut fluid was found at a specific location in the gizzard-like stomachs of representatives of 26 different species of these coralline-algae – eating fishes. Although the paper reporting these findings (Clements, Sutton, and Choat, 1989), makes use of obsolete terminology (bacteria are called protists, and the nutritional mode is "herbivory," even though the environment is totally devoid of plants), the photographs make the conclusions inescapable: the coevolution of these huge, unique bacteria with the hard-reef-feeding surgeonfishes. Plankton-eating species lack the microbes —one can imagine that speciation is related directly to symbiont acquisition and dietary change.

The evolution of a striking morphological novelty occurred in certain bivalve mollusks: clams with "windows" (the family Fraginae, to which the heart cockle *Corculum* and the related *Fragum* belong). Symbiosis with dinomastigotes clearly is related to the evolutionary appearance of the windows in the shells of these mollusks, which, in addition to transmitting visible light, perhaps function also as lenses to focus the light. Alteration of the aragonite crystals and decrease of pigmentation are changes in these shells not seen in members of other families. Immediately below the radially arranged, triangular windows at the shell's posterior (in the clam's epithelia and gills) lie densely packed phototrophic symbionts. Although only two recent (and extant) bivalve lineages are known to harbor symbiotic algae (the Fraginae and the tridacnids, which open their shells permitting the entry of light), the origin of the seme is polyphyletic: clam-shell windows seem to have evolved at least twice. Unusual fossil shells (from Permian strata at least 225 million years old) of species belonging to the Alatoconchidae are interpreted to be the earliest evidence for photosynthetic symbionts in bivalve mollusks (Watson and Signor, 1986).

These examples show that symbiogenesis as evolutionary mechanism for generation of morphologically distinct species is not limited to microbial life before the Phanerozoic eon. Of course, the extent to which symbiogenesis may be the major source of evolutionary innovation since the Proterozoic is still unknown. Despite the opinion of some colleagues, it is clear to me that natural selection acting on an accumulation of mutations is inadequate to explain hereditary variation. Fortunately, some modern biochemical and genetic studies about symbiogenesis are under way, at least in marine animals [for example, sulfide-oxidizing tubeworms (Vetter, 1991), luminous fish (McFall-Ngai, 1991); and caterpillars (Nealson, 1991)]. Unfortunately, failure even to review or consider seriously symbiogenesis and symbionticism continues not only to delay the acceptance of the great evolutionary insights of Wallin, Mereschkovsky, and their Russian contemporaries (Khakhina, 1979) but also to retard the integration of microbiology with the rest of evolutionary biology.

The biologist's modern literature about speciation began with Darwin's book *The Origin of Species* (1859), which in fact discusses nearly everything but speciation. I claim that this literature is comparable to that on infectious diseases before the discovery of bacteria: anthropocentric and of limited worth. Speciation, absent in bacteria, is a phenomenon of eukaryotic evolution intimately connected with (if not equivalent to) the acquisition, reproduction, and integration of new microbial genomes. In at least one case, dramatic and complete gender change was correlated with the presence of a virus (the masculinization of a sow bug; Juchault et al., 1992). Speciation occurs by symbionticism, as Wallin observed presciently (Mehos, 1992).

Wallin, like other evolutionists, including Darwin himself, realized that there was a crucial void in evolutionary theory: How, in detail, did new species and new morphologies arise? Defining "symbionticism" as symbiosis on the level of the cell, Wallin wrote that symbionticism "insures the origin of species" and that the acquisition of microbial symbionts offers "the solution to the problem of the origin of species" (see Mehos, 1992). Wallin recognized two principles: prototaxis (a given behavioral, biochemical, metabolic, or even genetic response of one type of living being to another) and microbial symbiogenesis, which he called "symbionticism." The ideas in his most important paper (Wallin, 1923) bear scrutiny today by professional evolutionists, many of whom, to paraphrase the evolutionist–novelist Samuel Butler (Darwin's staunchest critic), have taken the life out of biology. The role of acquisition of hereditary characteristics in the form of microbial and even larger genomes is yet to be evaluated as the major mechanism of evolutionary innovation during the Phanerozoic eon.

PLANTS, ANIMALS, AND CHROMOSOMES

When I view all beings not as special creations, but as the lineal descendants of some few beings which lived long before the first bed of the Silurian system was deposited, they seem to me to become ennobled. Judging from the past, we may safely infer that not one living species will transmit its unaltered likeness to a distant futurity. . . . We can so far take a prophetic glance into futurity as to fortell that it will be the common and widely-spread species, belonging to the larger and dominant groups, which will ultimately prevail and procreate new and dominant species. As all the living forms of life are the lineal descendants of those which lived long before the Silurian epoch, we may feel certain that the ordinary succession by generation has never once been broken, and that no cataclysm has desolated the whole world. Hence we may look with some confidence to a secure future of equally inappreciable length.

C. DARWIN, 1859

The complexities of the Cambrian fauna imply that the Phanerozoic "explosion" was made possible by myriad prior biochemical developments in soft-bodied organisms (Palmer, 1968; McMenamin and McMenamin, 1990). The Phanerozoic eon itself was the time of far more apparent evolutionary trends that can be inferred from studies of live organisms correlated with paleontological observations. Tissue aggregation, organ and organ-system differentiation, complex sexual dimorphisms, polymorphic life cycles, attainment of immense size, transitions from aquatic to terrestrial and aerial environments, and complex intraspecific behaviors and social organization were among the conspicuous evolutionary phenomena that appeared during the last half billion years (Mayr, 1976). During the Phanerozoic eon, these phenomena overwhelmed most innovations on the cellular level. The rate of evolution may be accelerating as new levels of biological, including social, organization emerge (Pettersson, 1977). New tissue symbionts did not appear until their hosts evolved, and complex phenomena depended on cellular and other developments that preceded them at lower levels of organization. Extensive plant biosynthetic pathways required the prior evolution of polygenomic plant cells; circulatory systems in animals and vascular plants evolved upon substrate laid by the prior appearance of structural materials—cartilage, bone, and lignin. Phanerozoic animals and plants populated the Earth only because of the precursor world created by the evolution of terrestrial, subterranean, aerial, and aquatic microbial communities that had dominated the earlier eons.

Although it became less conspicuous, cell evolution itself did not cease. Meiotic cell division evolved in mitotic organisms probably to rid incompletely cannibalized protists of their inadvertent diploidy, just as Cleveland (1947) claimed and we described (Margulis and Sagan, 1986). Symbioses were still important in the origins of new groups, such as the cycads, cone-bearing plants that maintain symbiotic nostocalean cyanobacteria in their roots (Figure 12-3). The replicative nature of the genetic determinants of undulipodia (microtubular organizing centers; MCs) provides an intellectual framework for Phanerozoic evolutionary phenomena—for example, the cellular basis for certain episodes of speciation in animals and plants (Stebbins, 1966). Adaptive radiations underlain by **polyploidy** in plants and by **karyotypic fissioning** in animals (Todd, 1970) (Figure 12-4) can be analyzed as an emergent property of symbiogenesis.

Alteration of kinetochore/chromosome ratios seems to underlie some evolutionary phenomena. Several groups of mammals have excellent fossil records—they underwent episodes of prodigious speciation. Todd (1970, 1992) noted a striking correlation between these adaptive radiations and an increase in number and decrease in size of chromosomes (both by about a factor of two). Todd suggests that karyotypic fissioning events underlie the episodes of mammalian speciation and that such chromosomal changes preadapted the mammalian lineages for adaptive radiations resulting from geographical and other isolating mechanisms. Karyotypic fissioning is a

FIGURE 12-3

Cycad symbionts.
(A) Seedling of *Macrozamia communis* showing corraloid roots (*c*), which harbor cyanobacterial symbionts. [Courtesy of C. P. Nathaniels and I. A. Staff.]
(B) Transverse section through the corraloid roots showing the inner (*o*), outer (*i*), and cyanobacterial (*cb*) layers of the cortex. Bar = 10 μm.

duplication of the kinetochores, producing two half-chromosomes from each chromosome. Like mitosis, fissioning can be understood as a consequence of a change in the nucleocytoplasm/MC ratio; that is, an alteration of the chromatin/kinetochore ratio from 1/1 to 1/2. If a chromosome is mediocentric, the extra replication of kinetochores produces two new chro-

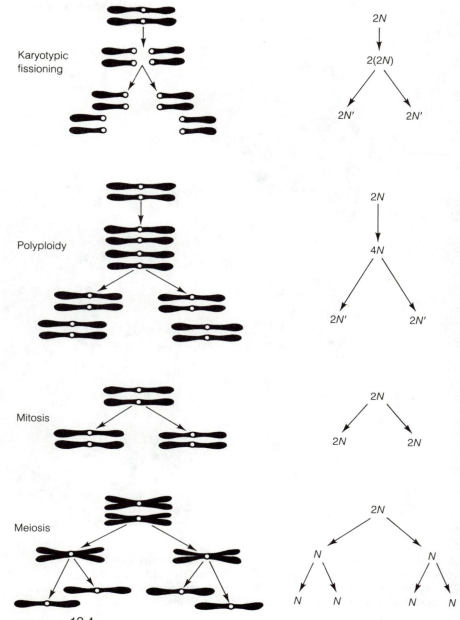

FIGURE 12-4

Chromosomal mechanisms that preadapt for speciation. Karyotypic fissioning doubles the number of chromosomes while maintaining the same quantity of genetic material in a diploid cell; each chromosome becomes smaller. A round of polyploidy will double both the number of chromosomes and the quantity of genetic material in a diploid cell; the size of each chromosome does not change. Chromosome behavior in mitosis and meiosis are shown for comparison.

mosomes (see Figure 12-4); acrocentric chromosomes remain unchanged. Now that many mammalian karyotypes and fossil histories have been studied, evolutionary relationships between the animals can be inferred more properly. To quote Todd (1967), the concept has been

> prevalent that karyotype evolution proceeds from high diploid numbers of many acrocentric chromosomes to low diploid numbers with many elements (chromosomes) having subterminal centrometers. The Robertsonian centric fusion hypothesis is generally invoked to explain the necessary transformation. However, as karyological data accumulates, this theory becomes increasingly irreconcilable with classical mammalian phylogenies based on comparative anatomy and paleontology. Hence, an alternative theory, that through fissioning of the centromeric regions of the chromosome deriving two small telocentrics from a metacentric [see Figure 12-4], high diploid numbers evolve from low numbers, must be considered, despite a paucity of cytological evidence for any underlying mechanism.

From the point of view of cell history as set forth in Chapter 8 and 9, the underlying cytological mechanisms do exist, although they are seldom discussed in the mammalian-revolution literature.

Fissioning is envisioned as a rare event that occurs in spermatogenesis in all the chromosomes of one cell that will become a germ cell of one individual. The theory requires that, after each new fissioning, not all the new chromosomes become incorporated into all individuals in the general population. For example, the many coadapted linkages carried by the large X chromosomes are essential for too many mammalian functions to survive fissioning, so that in the new populations they remain unchanged. The evidence for karyotypic fissioning in carnivore evolution—for example, in canine speciation—supports Todd's ideas (Todd, 1992). Old-world monkeys and apes probably arose from ancestors in which the germ-plasm karyotype had fissioned by centrometric duplication (Giusto and Margulis, 1980; Figure 12-5). The closely related platyrrhine monkeys—for example, the spider monkey ($2N = 34$) and the woolly monkey ($2N = 62$)—are related by fissioning. Karyotypic fissioning, consistent with the fossil record of these animals, is important in mammalian sympatric speciation and should come under closer scrutiny.

The formation of new plant species by polyploidy (Stebbins, 1966) can be seen as suppressed reproduction of MCs relative to cytoplasmic division. In the formation of polyploids, chromatin and kinetochores reproduce, but cytokinesis is delayed, probably because spindles, and thus centrioles or equivalent centriolar MCs, fail to form. Usually, two new complete diploid karyotypes within one nuclear membrane are produced; that is, tetraploids

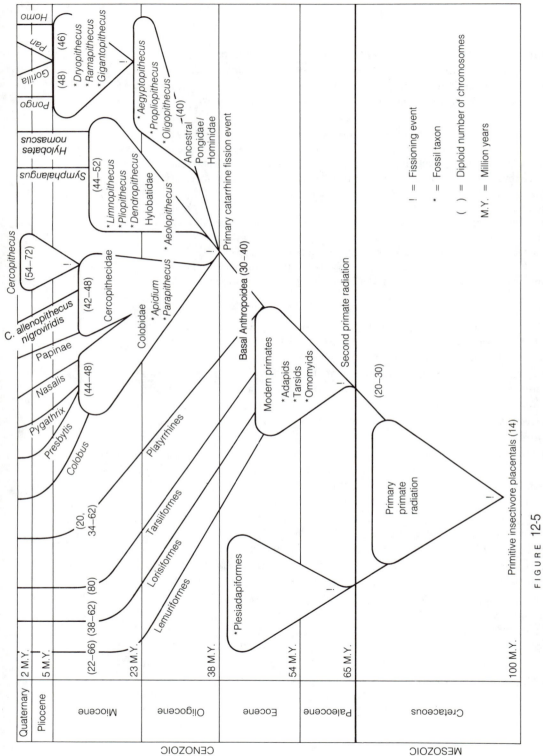

FIGURE 12-5

The karyotypic evolution of humans and related primates (Giusto and Margulis, 1980).

(4*N*) from diploids. With subsequent chromatin differentiation, a new diploid karyotype (2*N'*) results (see Figure 12-4). Like karyotypic fissioning, polyploidization is preadaptive for episodes of speciation.

Karyotypic fissioning prevails in animals over polyploidization because it is far gentler to these obligate diploids. The fissioned individual differs from its immediate predecessor only in the number of linkage groups and kinetochores. Precisely the same quantity and quality of genetic material is present, but it is organized into about twice as many linkage groups. Polyploidy approximately doubles the total amount of genetic material, with severe and immediate consequences for many gene products and metabolites. Such consequences are apparently lethal for mammals but tolerated by plants. The extent to which karyotypic fissioning may have contributed speciation of plants and other animals, especially reptiles and birds, is yet to be discovered (Todd, 1992).

The Phanerozoic eon could usher in the "age of animals and plants" because the monerans and protoctists had long since supplied the component parts: the genetic systems, metabolism, and ecological transformations required to make latter-day evolutionary innovations possible.

GAIA AND THE EXTANT BIOSPHERE

The atmosphere, therefore, is the mysterious link that connects the animal with the vegetable, the vegetable with the animal kingdom.

> J. DUMAS and J. BOUSSINGAULT, 1844
> (in Aulie, 1970)

Forests, coral reefs, and great grasslands that punctuate the Phanerozoic fossil record are epiphenomenal in that they followed from the radical innovations in cell metabolism and behavior that had occurred earlier in the Archean and Proterozoic eons. We look briefly here at the evolution of Phanerozoic ecosystems from the vantage point of those that preceded them. We may question that life has progressed in the last 3000 million years, but certainly it has expanded. The biosphere has enlarged, the biota has diversified, and the inhabited territory has expanded. Microbes and fish are found in the ocean abyss over 10 kilometers below the surface. The biosphere extends to an altitude of some 77 kilometers; at that extreme height are highly pigmented spores of fungi (*Circinella, Aspergillus, Penicillium, Papulaspora*) and aerobic bacteria (*Mycobacterium, Micrococcus*) (Imshenetsky, Lysenko, and Kazakov, 1978). The Antarctic dry valleys yield endolithic algae and fungi, their water requirements met by condensation of dew (Friedman and Ocampo, 1976). Boiling hot springs maintain low species diversity, but thermophilic microbes survive and grow. Sulfuric acid waters at pH 1 select an idiosyncratic community; ammonia springs at pH 9 contain quantities of peculiar tolerant forms (Deal, Souza, and Mack, 1975). Some physical limits to the growth and survival of life are listed in Table 12-2.

The biota at the fringes of the biosphere depends totally on the core: the thriving tropical and temperate regions. Monerans—better adapted than eukaryotes to physical extremes of heat, desiccation, and starvation—tend to be the pioneers. As in the Archean and Proterozoic eons, monerans today supply the rest of life in the hydrosphere, soil, and atmosphere with critical elements in usable form and quantity. The cycling of carbon, nitrogen, phosphorus, sulfur, oxygen, and hydrogen depends on prokaryotes, as it has since the Lower Archean; eukaryotes could not survive and grow at all in a world without bacteria.

Scientists assume that environmental conditions are determined by geological and climatological factors and that life is forced to adapt or perish. However, with a deeper understanding of the troposphere, and with insights due largely to the English scientist and inventor James E. Love-

TABLE 12-2
Physical limits to growth and survival of organisms.[a]

FACTOR	LIMITS FOR GROWTH	LIMITS FOR SURVIVAL (1-HOUR OF EXPOSURE)	
		Growing cells	Spores
Temperature	−15°C–85°C	−269°C–105°C	−269°C–60°C
Water activity (a_w)[b]	0.9–1.0[c]	0–1	0–1
Pressure	600 bars	3000 bars	20,000 bars
pH	1.5–11.5	1–12	Maximum 13–14 in 15.3 M NH_4OH for about 3 weeks (Bacillus subtilis)
Ultraviolet radiation (260 nm)	Not known	0.1 joules/cm²	0.1 joules/cm²
Ionizing radiation	Not known	2–4 mrad	2–4 mrad
Nutrients		Varies with species; more than 25 elements required in optimal quantities	None

[a]After Mazur et al. (1978).
[b]Water activity (a_w) = $\rho_{solution}/\rho_{pure\ liquid\ water}$. Depends only on quantity of solute, temperature, and pressure, not on the nature of the solute. Liquid water is always required for growth.
[c]0.6–1.0 for halophilic bacteria.

lock, it becomes more obvious that the surface and lower atmosphere of the Earth is modified greatly by life in ways that support more life, though this life may differ in detail. Life makes its own environment. The concept that the temperature, acidity, oxidation state, and other aspects of the reactive chemistry of the Earth's surface are kept relatively constant by interacting life forms is the "Gaia hypothesis," the subject of two recent books by Lovelock (1988, 1991).

Simply stated, the Gaia hypothesis is that selected aspects of the atmosphere, sediments, and hydrosphere are controlled by growth, metabolic activities, and other interactions in the biosphere. The composition of the Earth's atmosphere, especially the nonnoble gases, is clearly anomalous when compared with the carbon dioxide atmospheres of Mars and Venus (Table 12-3). The concentration of oxygen in the Earth's atmosphere remains constant in the presence of nitrogen, methane, hydrogen, and

TABLE 12-3
Atmosphere of Venus, Earth, and Mars: major and trace constituents.[a]

	VENUS Observed	EARTH		MARS	
		Observed	Reconstructed[b]	Observed	Reconstructed[b]
Major constituents					
Carbon dioxide	98%	0.03%	98%	95%	98%
Nitrogen	1.9% Ve[c]	79%	1.8%	2.7% Vi[c]	<1.8%
Oxygen	Trace	21%	Trace	0.13%	Trace
Water (precipitable)	5 μm above clouds 10 μm below clouds	5 × 10⁴ μm in atmosphere, 3 × 10⁹ μm in oceans	Same as observed	10–20μm	500 × 10⁶ μm
Temperature	750 ± 10 K	290 ± 50 K	—	220 ± 40 K	—
Ratio: mass of outgrassed volatiles/mass of planet (×10⁶)	100	—	60	—	0.7[d]
Total atmospheric pressure	90 bars	1 bar	65 bars	0.0064 bar	≥1 bar
Observed trace constituents					
Methane (CH₄)	Undetected	1–2 ppm		Undetected (<0.02 ppm)	
Formaldehyde (CH₂O)	Undetected	0.0001 ppm		Undetected (<0.6 ppm)	
Hydrogen sulfide (H₂S)	?	0.0002 ppm		Undetected (<0.1 ppm)	

[a]The aid of T. C. Owen in the design of this table is acknowledged gratefully. (Data from Levine, 1989; Goldsmith and Owen, 1992 and personal communication.)
[b]Calculated prediction based on the return of lithospheric volatiles (for example, ice and carbonates) to the atmosphere.
[c]Ve = Venera spacecraft, USSR, 1975; Vi = Viking spacecraft, USA, 1976.
[d]Mars may have been deficient in volatiles; alternatively, water-derived surface features suggest that before 2.5 × 10⁹ years ago, Mars had more than 0.5 kilometers of precipitable water.

many other strong potential reactants. Stable alkalinity and lack of extreme temperature excursions are porperties of the Earth's surface that are probably determined largely by life. Although there is no evidence that life has altered the atmosphere or sediments of Mars (Mazur et al., 1978) or Venus, extinct early life is sought on Mars (McKay and Stoker, 1992). Gaia apparently now lives only on Earth.

The anomalies of Earth's atmosphere are far from random. At least at the core, the tropical and temperate regions, surface and atmosphere are skewed from the values deduced by interpolating between values for Mars and Venus, and the deviations are in directions favored by extant organisms. Oxygen is maintained at about 20 percent, the mean temperature of the lower atmosphere is about 18° C, and the pH is just over 8. These planet-wide anomalies have persisted for a very long time; the chemically bizarre composition of the Earth's atmosphere has prevailed for millions of years, even though the residence times of the reactive gases can be measured in months and years (Lovelock and Margulis, 1974; Lovelock, 1988).

The Gaia hypothesis has heuristic value; it has already united disparate approaches to the study of life on Earth (Barlow, 1991). Although first developed in the context of planetary control of surface temperature, composition of reactive gases, and alkalinity, the Gaia concept can be expanded to explain other Earth anomalies. The possibility exists that lateral plate-tectonic movement on the third planet is a unique Gaian phenomenon (Mackenzie and Agegian, 1989; Mackenzie, 1990); that retention of planetary water and therefore granite (Taylor and McLennan, 1985; Taylor, 1987) is the consequence of the tenacious thirst of the biota; and that the distribution of many economically important mineral deposits — iron (Lovley, 1991; Nealson and Myers, 1992), gold (Mossman and Dyer, 1985), and phosphorites (Holland and Schidlowski, 1982) — is owed to the interaction of organisms with sediment.

Chance alone cannot account for the fact that temperature, pH, and the concentration of nutrient elements, at least during the Phanerozoic eon, have been just those that are optimal for life. The major perturbers of atmospheric gases are organisms themselves — primarily microbes. More likely solar energy is co-opted, it seems to us, by the biota in ways that keep the Earth's surface conditions relatively constant. The biological maintenance of environmental homeorrhesis* has been reviewed (Schneider and Boston, 1992; Lovelock, 1991). We are as yet unable to detail the multiple interacting feedback systems that maintain the gases of the Earth's atmo-

---

*Homeorrhesis is physiological regulation (of temperature, ion concentration, etc.) around a moving set-point. Homeostasis is the special case of homeorrhesis where the set-point is constant, or nearly constant.

sphere in a condition so utterly different from the atmospheres of Venus and Mars. The United States Viking mission to Mars (1975 to 1982; Horowitz, 1986) and the USSR Venera missions to Venus (1975 to the present) are our proprioceptors: the dry, desolate, carbon dioxide–swept neighbors contrast bleakly with the home planet. Very early here, organisms developed the capacity to produce and remove reactive gases; atmospheric equilibria of the present sort have prevailed on our planet as far back as the fossil record can be read.

In the Archean eon, environmental regulatory mechanisms were probably different: regulation was maintained around anoxic or low oxygen values and, perhaps, higher temperatures. The Sun's luminosity has increased during the past three billion years (Dilke and Gough, 1972). However, always capable of exponential growth and big chemical diversification, life during this time responded: episodes of adaptive radiation led to the appearance of new species that gave rise to more-inclusive taxa. Conditions have been geophysiologically regulated so that an evolving biota has inhabited an expanding biosphere through time. The total number of species has probably increased, although in many groups there were great demises (Gould, 1989). Diversity, in the Gaian context of regulation of atmospheric gases, is not a luxury. Millions of differently functioning types of organisms are necessary for the maintenance of flexibility, responsiveness, and redundancy in the face of inevitable astronomical and geological perturbations on the planet (Harborne, 1988).

Gaia is only symbiosis as seen from space, Gregory Hinkle emphasizes: the physical continuity connecting all living symbionts is provided by the fluid atmosphere and hydrosphere—a 25-kilometer ring at the planetary surface (Hinkle and Margulis, 1990). Life, distributed patchily as a veritable scum reaching mainly to within a few meters of the Earth's watery or solid surface, extends over an area of about 500 million square kilometers. The next advance in the expansion of life may be above the troposphere to the stark cold reaches of the rest of the solar system and, perhaps, beyond. But we should be warned: to survive as it produces waste gases and liquids, life must moisten and modify its environment. Without diversity—the capacity for rapid reproduction, interactions, flexibility, and redundancy—life does not maintain itself. That life can survive in space has been demonstrated amply, but survival is not reproduction. Whether Gaia will reproduce (Sagan, 1991), that is, whether the incessantly growing, metabolizing and environment-regulating biota will ever expand off the Earth's surface is surely an open question.

APPENDIX

The Phyla and Classes of the Modified Whittaker Five-kingdom System.*

In a sense, therefore, the whole organic world constitutes a single, great individual, vague and badly coordinated it is true, but none the less a continuing whole with interdependent parts.

J. HUXLEY, 1912

All life — that is, all "organisms" or living entities on Earth — can be classified phylogenetically by the number of once-independent genomes (genes and associated protein-synthesizing systems) that comprise them. Members of the Prokaryotae, whether archaebacteria or eubacteria, have one such system, whereas eukaryotes are all heterogenomic and hence have more than one. The names and defining characteristics of the higher taxa (the two superkingdoms or domains and the five kingdoms they comprise) reflect this knowledge. The higher taxa for all life are summarized in Table A-1. The details of this classification system, from superkingdom to sample genera, are listed in the appendix that follows (Margulis, 1992).

---

*Extinct forms omitted. Classes of only major phyla are listed. Viruses are not cellular and thus are excluded from this list; see Table 2-8.

Superkingdom Prokaryota (chromonemal organization)

KINGDOM PROKARYOTAE (MONERA)

Prokaryotic cells, bacteria. Nutrition absorptive, chemosynthetic, photohetero-trophic, or photoautotrophic. Metabolism anaerobic, facultative, or aerobic. Reproduction asexual and chromonemal; recombination unidirectional or mediated by viruses. Nonmotile or motile either by gliding or by bacterial flagella composed of flagellin protein. Solitary unicellular or multicellular: filamentous, colonial, or mycelial. Some possess sheaths; some produce endospores or sorocarps (reproductive stalks with differentiated propagules).

SUBKINGDOM ARCHAEBACTERIA

Division Mendosicutes (walls distinctive, lack peptidoglycan)

PHYLUM 1. Methanocreatrices: methane-synthesizing bacteria; anaerobic chemotrophs; use CO_2 as electron acceptor for anaerobic respiration, reducing it to CH_4 (*Methanococcus, Methanosarcina, Methanobacterium*)

PHYLUM 2. Halophilic (salt-requiring) and thermoacidophilic (high-temperature– and low-pH–tolerant) bacteria: distinctive ether-linked lipids and 16S ribosomal RNA nucleotide sequences (*Desulfurococcus, Halobacterium, Halococcus, Sulfolobus, Thermoplasma, Thermoproteus*)

SUBKINGDOM EUBACTERIA

Division Tenericutes (lack walls)

PHYLUM 3. Aphragmabacteria: unable to form cell walls; mycoplasmas (*Mycoplasma, Anaplasma, Bartonella*)

Division Gracilicutes (Gram-negative bacteria)

PHYLUM 4. Spirochaetae (spirochetes)

Class 1. Spirochaetaceae: facultative or obligate anaerobes (*Spirochaeta, Treponema, Borrelia*)

Class 2. Pillotinaceae: complex symbionts, many flagella (*Cristispira, Pillotina, Hollandina, Diplocalyx*)

Class 3. Leptospiraceae: obligate aerobes (*Leptospira, Leptonema*)

PHYLUM 5. Thiopneutes: anaerobic reducers of sulfate or sulfur to H_2S; limited synthesis of heme proteins

TABLE A - 1

Five-kingdom classification of life.

KINGDOM	CHARACTERISTICS AND ESTIMATED NUMBER[a] OF SPECIES
Superkingdom Prokaryota (Domain PROKARYOTAE) KINGDOM MONERA	Bacterial cell organization (chromonema, small ribosomes, continuous DNA synthesis; rotary motor flagella[b]). Single homologous genomes. All modes of metabolism represented.[c]
Archaebacteria	Archaebacterial ribosomes and 16S rRNA; ether-linked lipids. (Examples: methanogenic, thermoacidophilic, and halophilic bacteria) 500
Eubacteria	Eubacterial 16S rRNA and lipids; peptidoglycan cell walls. (Examples: enterobacteria, cyanobacteria and other phototrophs, mitochondria and plastids, spirochetes and other Gram-negative bacteria, Gram-positive bacteria) 10,000
Superkingdom Eukaryota (Domain EUKARAYOTA)	Nucleated cell organization with chromatin (histones, nucleosomes, nuclear pore complexes); mitotic karyokinesis; actin-based cytokinesis; microtubule-based, intracellular motility systems; intermittent DNA synthesis. Proteinaceous, cellulosic, and lignaceous walls, or lack walls. Two modes of metabolism.[c]
KINGDOM PROTOCTISTA	Microorganisms and their larger descendants composed of multiple heterologous genomes. Variations on mitosis and meiosis; many have undulipodiated[b] cells. (Examples: algae, chytrids, slime molds, ciliates) 250,000
KINGDOM FUNGI MYCHOTA[d]	Haploid or dikaryotic, osmotrophic heterotrophs that develop from spores and display zygotic meiosis; chitinous walls; no undulipodiated cells. (Examples: yeasts, mushrooms, lichens) 70,000
KINGDOM ANIMALIA	Diploids develop from products of anisogamous (sperm/egg) fertilization into blastula embryos. Ingestive nutrition; complex cell connections (e.g., desmosomes, septate junctions); muscle tissue; no cell walls. (Examples: porifera, mollusks, chordates) 30,000,000
KINGDOM PLANTAE	Embryophytes. Gametophytes develop from haploid spores; fertilization of eggs by undulipodiated sperm produces sporophyte embryos that are retained in maternal tissue. Plasmodesmata between cells; cellulosic walls. (Examples: bryophytes, tracheophytes) 400,000

[a]Upper estimates (May, 1993; and Margulis, 1992.)

[b]Flagella are extracellular, composed of flagellin proteins, and powered by rotary motors. Undulipodia are intracellular, underlain by kinetosomes, and powered by microtubule-associated protein attached to 24-nm tubules that are organized into [9(2) + 0] arrays.

[c]Chemoorganoheterotrophy to photolithoautotrophy, all modes represented in prokaryotes. Chemo/photo refers to source of energy; organo/litho refers to source of electrons; hetero/auto refers to source of carbon. Chemoorganoheterotrophy and photolithoautotrophy are the two modes of metabolism associated with eukaryotes. For the bacterial alternatives, see Table I-2 in Margulis, Corliss, Melkonian, and Chapman, 1990, or Margulis et al., 1992.

[d]Preferred formal name of kingdom (Melvin Fuller, personal communication).

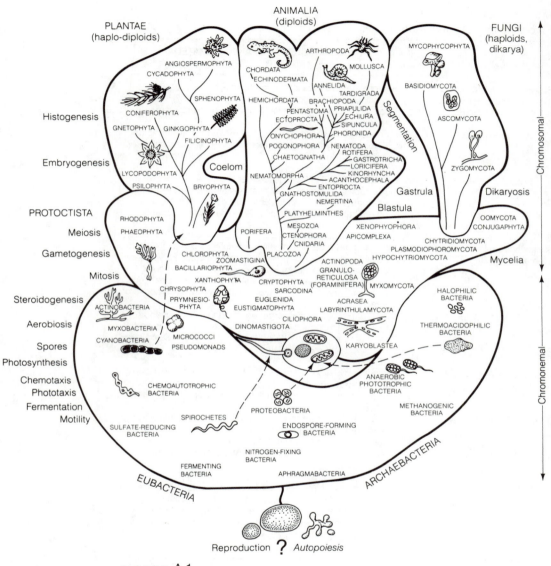

FIGURE A-1

Phyla of the five kingdoms.

Class 1. Non–spore-forming sulfate reducers (*Desulfovibrio*)

Class 2. Endospore-forming sulfate reducers (*Desulfotomaculum*)

Class 3. Non–spore-forming elemental-sulfur reducers (*Desulfuromonas*)

PHYLUM 6. Anaerobic phototrophic bacteria: synthesis of iron- and magnesium-chelated tetrapyrroles, chlorophylls, and carotenoids; photosystem I. Phototrophic green and purple bacteria.

Class 1. Green sulfur bacteria: use H_2S as hydrogen donor (*Chlorobium, Chloroflexus*)

Class 2. Purple sulfur bacteria: use H_2S as hydrogen donor (*Chromatium*)

Class 3. Purple nonsulfur bacteria: photoheterotrophs; use organic hydrogen donors (*Rhodopseudomonas, Rhodomicrobium, Rhodospirillum*)

PHYLUM 7. Cyanobacteria (blue-green bacteria) aerobic or facultatively aerobic photosynthesizers; oxygen-eliminating photosynthesis; chlorophyll *a* and phycobiliproteins on phycobilisomes; photosystem I and II; H_2O or H_2S as hydrogen donor

Class 1. Coccogoneae: coccoid cyanobacteria (*Chroococcus, Chamaesiphon*)

Class 2. Hormogoneae: filamentous cyanobacteria (*Nostoc, Oscillatoria*)

Class 3. Chloroxybacteria (Prochlorophyta*): prokaryotic green bacteria that use H_2O as hydrogen donor; oxygen-eliminating photosynthesis; chlorophylls *a* and *b*; no phycobiliproteins (*Prochloron, Prochlorothrix*)

PHYLUM 8. Nitrogen-fixing aerobic bacteria (*Azotobacter, Beijerinckia, Rhizobium*)

PHYLUM 9. Pseudomonads: aerobic heterotrophs; Gram-negative; respiratory metabolism (*Pseudomonas, Hydrogenomonas*)

PHYLUM 10. Omnibacteria: aerobic heterotrophic proteobacteria

Class 1. Enterobacteria: coliforms, facultative aerobes (*Escherichia, Salmonella, Serratia*)

Class 2. Prosthecate bacteria (*Caulobacter*)

Class 3. Budding bacteria (hyphomicrobia) (*Hyphomicrobium*)

Class 4. Aggregated bacteria: form distinctive cell aggregates (*Sphaerotilus, Leptothrix*)

---

*Because it confuses these cyanobacterial prokaryotes with algae and plants, this term should be discouraged (Urbach, 1992).

PHYLUM 11. Chemoautotrophic bacteria

Class 1. Nitrogen-oxidizing bacteria (*Nitrobacter, Nitrosomonas, Nitrocystis*)

Class 2. Sulfur-oxidizing bacteria (*Thiobacillus*)

Class 3. Methylomonads: methane- and methanol-oxidizing bacteria (*Methylomonas*)

PHYLUM 12. Myxobacteria: heterotrophic aerobic gliding bacteria

Class 1. Myxobacteriales: sorocarp-forming myxobacteria (*Myxococcus, Chondromyces, Polyangium*)

Class 2. Cytophagales: filamentous and unicellular gliding bacteria: cytophages and flexibacters (*Beggiatoa, Saprospira*)

Division Firmicutes (Gram-positive bacteria)

PHYLUM 13. Fermenting bacteria: unable to synthesize porphyrins

Class 1. Lactic acid bacteria (*Lactobacillus, Streptococcus*)

Class 2. Clostridia (*Clostridium*)

Class 3. Peptococcaceae (*Peptococcus, Peptostreptococcus, Ruminococcus*)

PHYLUM 14. Aeroendospora: aerobic endospore-forming bacteria (*Bacillus*)

PHYLUM 15. Micrococci: Gram-positive aerobes; Krebs cycle present (*Paracoccus, Sarcina, Gaffkya*)

PHYLUM 16. Actinobacteria: Gram-positive coryneform and mycelial bacteria

Class 1. Coryneforms (*Arthrobacter, Propionibacterium*)

Class 2. Actinobacteria *sensu stricto* (once called actinomycetes) (*Mycobacterium, Actinomyces, Nocardia, Streptomyces*)

For there is no middle ground between symbiosis and
nonsymbiosis. Either symbiosis with Cyanophyceae
exists, and then one has plants, or it does not exist, and
then we have animals.

K. S. MERESCHKOVSKY, 1909
(in Khakhina, 1979)

Superkingdom Eukaryota (chromosomal organization)

KINGDOM PROTOCTISTA

Eukaryotic cells. Nutrition ingestive, absorptive, or, if photoautotrophic, by photosynthetic plastids. Reproduction asexual, premitotic, or eumitotic sexual.* In eumitotic forms, meiosis and fertilization are present, but details of cytology, life cycle, and ploidy level vary from group to group. Solitary unicellular, colonial unicellular, or multicellular. All lack embryos and complex cell junctions (for example, desmosomes). Most species aquatic. Pore-studded nuclei; undulipodia (including cilia) and mitochondria present in most species. Primarily unicellular forms are sometimes classified as Protista, a more restricted kingdom.

I. Phyla in which members lack undulipodia at all stages and lack sexual life cycles

PHYLUM 1. Rhizopoda: rhizopod naked and shelled amoebae (*Entamoeba, Arcella*) A*

PHYLUM 2. Haplosporidia: haplosporidian symbiotrophs of vertebrates (*Haplosporidium, Minchinia*) A

PHYLUM 3. Paramyxea: paramyxean symbiotrophs of marine organisms (*Paramyxa, Marteilia*) A

PHYLUM 4. Myxozoa: myxozoan symbiotrophs of fish (*Myxobolus, Henneguya*) A

PHYLUM 5. Microspora: microsporidian symbiotrophs of vertebrates (*Nosema*) A

II. Phyla in which members lack undulipodia at all stages and display complex sexual life cycles

PHYLUM 6. Acrasea: acrasid cellular slime molds (*Guttulinopsis, Acrasis*) F

PHYLUM 7. Dictyostelida: dictyostelid cellular slime molds (*Dictyostelium, Polysphondylium*) F

---

*Premitotic = variation on standard mitosis, no meiotic sexuality; eumitotic = meiotic reduction of chromosome number followed by syngamy.
*Classification according to the obsolete taxonomy: A = Kingdom Animalia; F = Kingdom Fungi; Pl = Kingdom Plantae. For a complete list of classes, see Margulis, McKhann, and Olendzenski, 1992. For a comprehensive list of genera, see Margulis, Corliss, Melkonian, and Chapman, 1990.

PHYLUM 8. Rhodophyta: red seaweeds (*Bangia, Porphyra, Nemalion*)   Pl

PHYLUM 9. Conjugaphyta: conjugating green algae, gamophyta (*Spirogyra, Zygnema,* desmids)   Pl

III. Phyla in which members display reversible formation of undulipodia and lack complex sexual life cycles

PHYLUM 10. Xenophyophora: deep-sea   xenophyophorans   (*Psammetta, Stannoma*)   A

PHYLUM 11. Cryptophyta: cryptomonads (*Cryptomonas, Cyanomonas, Chilomonas, Cyathomonas*)   A

PHYLUM 12. Glaucocystophyta: glaucocystids   (*Cyanophora,   Cyanidium, Gloeochaete*)   Pl

PHYLUM 13. Karyoblastea: amitotic amoebae (*Pelomyxa*)   A

PHYLUM 14. Zoomastigina: "protozoa," "animal flagellates"; mastigotes   A

　Class 1. Amoebomastigota: amoebomastigotes (*Naegleria, Willaertia*)

　Class 2. Bicosoecids: shelled bimastigotes; one undulipodium attached, one unattached (*Bicosoeca*)

　Class 3. Choanomastigota: collared mastigotes; some bear plastids (*Monosiga, Desmarella*)

　Class 4. Diplomonadida: diplomonads; doubled nuclei and organelles (*Hexamita, Giardia, Enteromonas*)

　Class 5. Pseudociliata (*Stephanopogon*)

　Class 6. Kinetoplastida: bodos, trypanosomes (*Bodo, Crithidia, Leishmania, Trypanosoma*)

　Class 7. Opalinata: binucleate; numerous undulipodia; opalinids (*Opalina, Cepedea, Zelleriella*)

　Class 8. Proteromonadida: proteromonads (*Proteromonas, Karotomorpha*)

　Class 9. Parabasalia: distinctive parabasal bodies (Golgi apparatus) (*Calonympha, Trichomonas*)

　Class 10. Retortamonadida: retortamonads (*Retortamonas, Chilomastix*)

　Class 11. Pyrsonymphida: no parabasal bodies; long axostyles (*Oxymonas, Pyrsonympha, Notila, Saccinobaculus*)

PHYLUM 15. Euglenida: euglenids (*Euglena, Peranema, Astasia*)   Pl, A

PHYLUM 16. Chlorarachnida: green amoebae (*Chlorarachnion*)   Pl

PHYLUM 17. Prymnesiophyta: prymnesiophytes, haptomonads (*Chrysochromulina, Emiliania*)   Pl

PHYLUM 18. Raphidophyta: raphidophytes (*Olisthodiscus, Vacuolaria*)   Pl

PHYLUM 19. Eustigmatophyta: eustigmatophytes (*Pleurochloris, Vischeria*) Pl

PHYLUM 20. Actinopoda: actinopods Pl

    Class 1. Polycystina: polycystine radiolarians (*Thalassicolla, Halosphaera, Diplosphaera*)

    Class 2. Phaeodaria: phaeodarian radiolarians (*Aulacantha, Auloceros, Protocystis*)

    Class 3. Heliozoa: heliozoans or sun-animalcules (*Echinosphaerium, Actinophrys*)

    Class 4. Acantharia: acantharians (*Acanthometra, Acanthocystis*)

PHYLUM 21. Hyphochytriomycota: hyphochytrids; anterior mastigonemate single undulipodium (*Rhizidiomyces, Anisolpidium*) F

PHYLUM 22. Labyrinthulomycota: slime nets (*Labyrinthula, Labyrinthorhiza*) and thraustochytrids (*Thraustochytrium*) A, F

PHYLUM 23. Plasmodiophoromycota: plasmodiophorids (*Plasmodiophora, Spongospora, Woronina, Polymyxa*) F

IV. Phyla in which members display reversible formation of undulipodia and complex sexual life cycles

    PHYLUM 24. Dinomastigota (Dinoflagellata): dinomastigotes (dinoflagellates) (*Gymnodinium, Peridinium*) A, Pl

    PHYLUM 25. Chrysophyta: chrysophytes, yellow-brown or golden-yellow algae, silicoflagellates (*Ochromonas, Dinobryon, Mallomonas, Dictyocha*) Pl

    PHYLUM 26. Chytridiomycota: chytridiomycotes, chytrids; posterior single undulipodium, aquatic molds (*Neocallimastix, Olpidium, Allomyces, Blastocladiella*) F

    PHYLUM 27. Plasmodial slime molds A, F, P

        Class 1. Myxomycota: myxomycotes (*Physarum, Didymium, Stemonitis*)

        Class 2. Protostelida: protostelids (*Protostelium, Ceratiomyxella*)

    PHYLUM 28. Ciliophora: ciliates (*Tetrahymena, Sorogena, Paramecium*) A

    PHYLUM 29. Granuloreticulosa: naked and testate reticulose amoebae (*Reticulomyxa*) and foraminiferans (*Globigerina, Nodosaria*) A

    PHYLUM 30. Apicomplexa: parasites (*Eimeria, Plasmodium, Toxoplasma*) A

    PHYLUM 31. Bacillariophyta: diatoms (*Surirella, Navicula, Planktoniella*) Pl

    PHYLUM 32. Chlorophyta: green algae (*Chlamydomonas, Chara*) Pl

        Class 1. Prasinophyceae

Class 2. Chlorophyceae

Class 3. Charophyceae

Class 4. Ulvophyceae

PHYLUM 33. Oomycota: oomycotous water molds (*Saprolegnia, Albugo, Achylya, Pythium, Phytophthora*)   F

PHYLUM 34. Xanthophyta: yellow-green algae (*Tribonema, Botrydium*)   Pl

PHYLUM 35. Phaeophyta: brown seaweeds (*Fucus, Dictyota*)   Pl

Incertae sedis

PHYLUM 36. Ellobiopsida: ellobiopsids (*Ellobiopsis, Thalassomyces*)   F

PHYLUM 37. Ebridians: ebriids (*Ebria, Hermesinum*)   A

KINGDOM FUNGI (MYCHOTA)*

Haploid or dikaryotic. Organisms are mycelial or secondarily unicellular, with chitinous walls, and always use absorptive nutrition. Cells always contain mitochondria but lack [9(2) + 2] undulipodia. Body plan is often branched, composed of hyphae, coenocytic filaments that may be divided by perforate septa. No pinocytosis or phagocytosis; extensive cytoplasmic streaming. Zygotic meiosis; reproduction by usually haploid, uni- or multicellular resistant propagules (spores).

PHYLUM 1. Zygomycota: zygomycotous molds (*Phycomyces, Rhizopus, Mucor*)

PHYLUM 2. Ascomycota: sac fungi or ascomycotes, molds and morels

Class 1. Hemiascomycetae: yeasts (*Saccharomyces*); leaf-curl fungi

Class 2. Euascomycetae: molds (*Neurospora*); cup fungi, morels, truffles

Class 3. Loculoascomycetae: ascostromatic fungi (*Elsinoe*)

Class 4. Laboulbeniomycetae: insect parasites (*Amorphomyces, Rhizomyces*)

PHYLUM 3. Basidiomycota: club fungi

Class 1. Heterobasidiomycetae: jelly fungi, rusts, smuts (*Puccinia, Ustilago*)

Class 2. Homobasidiomycetae: mushrooms, shelf fungi, coral fungi, earth stars, stink horns, bird's-nest fungi (*Agaricus, Schizophyllum*)

---

*Preferred name of kingdom (M. Fuller, University of Georgia, personal communication.)

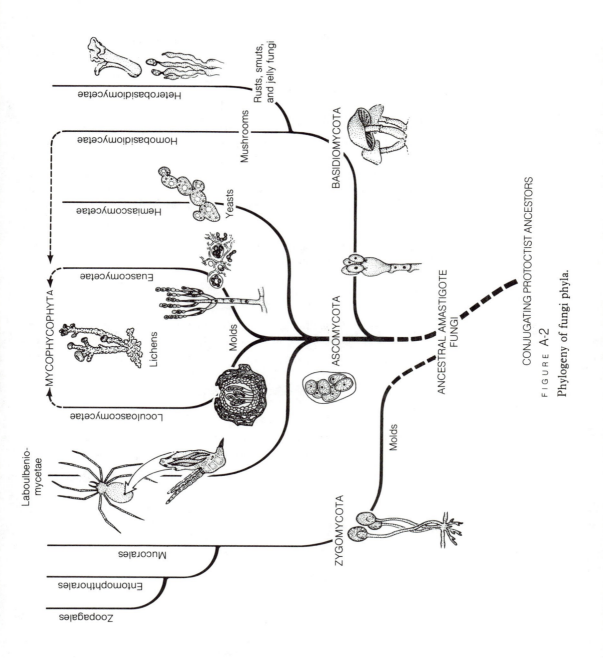

PHYLUM 4. Deuteromycota: fungi imperfecti (*Cryptococcus, Candida, Histo-plasma, Rhizoctonia*)

PHYLUM 5. Mycophycophyta: lichens; fungal component + cyanobacterial component (often, *Nostoc*) or fungal component + chlorophyte component (often, *Trebouxia* or *Pseudotrebouxia*)*

KINGDOM ANIMALIA

Gametic meiosis; anisogametous fertilization; sperm and egg form a zygote, which cleaves to form a diploid blastula; gastrulation generally follows. Multicellular heterotrophs: nutrition heterotrophic; sometimes ingestive by phagocytosis and pinocytosis, sometimes absorptive. Extensive cellular and tissue differentiation; desmosomes, gap junctions, and other differentiated connections between cells.

PHYLUM 1. Placozoa: diploblastic, dorsoventral organization; no polarity or bilaterality (*Trichoplax*)

PHYLUM 2. Porifera: sponges

Class 1. Calcarea: calcitic spicules (*Scycon, Leucosolenia*)

Class 2. Desmospongiae: spongin network with or without siliceous spicules (*Microciona, Haliclona*)

Class 3. Sclerospongiae: spongin network with or without siliceous spicules; basal skeleton of aragonite

Class 4. Hexactinellida: siliceous spicules with three axes (*Euplectella*)

PHYLUM 3. Cnidaria: coelenterates

Class 1. Hydrozoa: hydras (*Craspedacusta, Hydra, Obelia*)

Class 2. Scyphozoa: true jellyfish (*Cyanea, Aurelia, Pelagia*)

Class 3. Anthozoa: corals and sea anemones (*Antipathes, Heliopora, Porites, Renilla*)

Class 4. Cubozoa: sea wasps and several other tropical and subtropical genera (*Carybdea*)

PHYLUM 4. Ctenophora: comb jellies (*Beröe, Mnemiopsis, Pleurobrachia*)

PHYLUM 5. Mesozoa: mesozoans (*Dicyema, Salinella, Pseudicyema*)

---

*Many lichenologists prefer to classify symbionts separately; that is, they simply place each mycobiont (fungus) with its relatives, ignoring the algal component, which is simply classified with its own relatives.

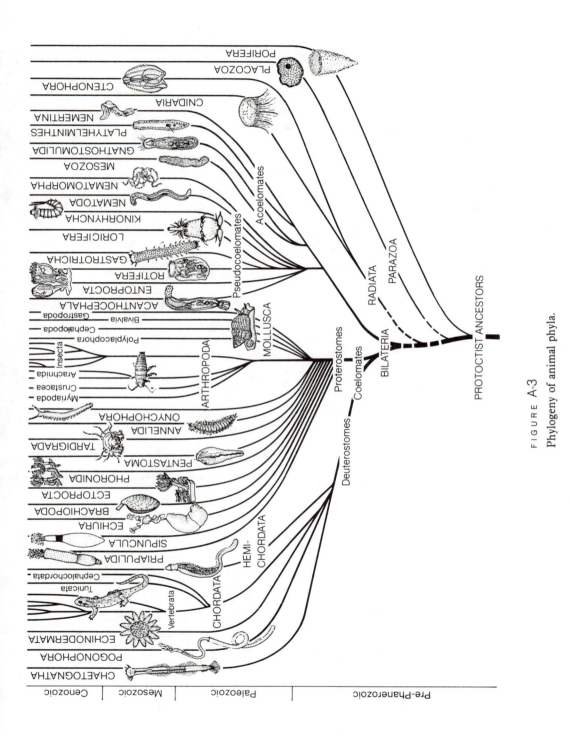

PHYLUM 6. Platyhelminthes: flatworms

Class 1. Turbellaria: free-living flatworms; planarians (*Dugesia, Procotyla, Afronta*)

Class 2. Trematoda: flukes (*Schistosoma, Polystomoidella, Bucephalopsis*)

Class 3. Cestoda: tapeworms (*Taenia, Rhopalura, Gyrocotyle*)

PHYLUM 7. Nemertina: ribbon worms (*Cerebratulus, Lineus, Tubulanus*)

PHYLUM 8. Gnathostomulida: microscopic acoelomate marine worms (*Haplognathia*)

PHYLUM 9. Gastrotricha: gastrotrichs (*Chaetonotus, Macrodasys, Tetranchyroderma*)

PHYLUM 10. Rotifera: rotifers (*Asplanchna, Brachionus, Euclanis, Philodena, Notommata*)

PHYLUM 11. Kinorhyncha: kinorhynchs   (*Centroderes,   Echinoderes, Pycnophyes*)

PHYLUM 12. Loricifera: spiny-headed   marine   animals   (*Nanaloricus, Pliciloricus*)

PHYLUM 13. Acanthocephala: spiny-headed worms; gut parasites of vertebrates (*Leptorhynchoides, Moniliformis, Acanthogyrus*)

PHYLUM 14. Entoprocta: entoprocts or kamptozoa (*Barentsia, Loxosoma, Urantella*)

PHYLUM 15. Nematoda: nematodes; roundworms (*Ascaris, Caenorhabditis, Wilsonema, Trichinella*)

PHYLUM 16. Nematomorpha: horsehair worms (*Gordius, Nectonema*)

PHYLUM 17. Ectoprocta: ectoprocts (*Bugula, Pectinatella, Cristatella, Plumatella*)

PHYLUM 18. Phoronida: phoronid worms (*Phoronopsis, Phoronis*)

PHYLUM 19. Brachiopoda: brachiopods, or lamp shells

Class 1. Inarticulata: lack hinges between valves (*Lingula, Glottidia*)

Class 2. Articulata: hinged valves; lack anus (*Hemithyris*)

PHYLUM 20. Mollusca: mollusks

Class 1. Monoplacophora: monoplacophorans (*Neopilina*)

Class 2. Aplacophora: solenogasters (*Neomenia, Proneomenia*)

Class 3. Caudofoveata: wormlike mollusks (*Chaetoderma, Scutopus*)

Class 4. Polyplacophora: chitons (*Lepidochitonia, Nuttallina, Ischnochiton*)

Class 5. Pelecypoda: bivales (*Anodonta, Tagelus, Mytilus, Ostea*)

Class 6. Gastropoda: snails (*Busycon, Crepidula, Vermicularia, Helicodiscus*)

Class 7. Scaphopoda: tooth shells (*Cadulus, Dentalium*)

Class 8. Cephalopoda: squids, octopuses (*Nautilus Sepia, Loligo*)

PHYLUM 21. Priapulida: priapulids (*Halicryptus, Priapulus*)

PHYLUM 22. Sipuncula: sipunculans, or peanut worms (*Phascolopsis, Aspidosiphon, Dendrostomum, Golfingia*)

PHYLUM 23. Echiura: echiurans, or spoon worms (*Bonellia, Ikeda, Thallassema*)

PHYLUM 24. Annelida: segmented worms

Class 1. Polychaeta: marine bristle worms (*Aphrodite, Nereis, Nephtys, Salmacina*)

Class 2. Oligochaeta: terrestrial bristle worms (*Lumbricus, Ripistes, Stylaria*)

Class 3. Hirudinea: leeches (*Glossiphonia, Hirudo, Piscicola*)

PHYLUM 25. Tardigrada: water bears (*Hypsibius, Echiniscus, Macrobiotus*)

PHYLUM 26. Pentastoma: tongue worms (*Armillifera, Cephalobaena, Waddycephalus, Porocephalus, Linguatula*)

PHYLUM 27. Onychophora: velvet worms (*Peripatus, Speleoperipatus*)

PHYLUM 28. Arthropoda: animals with segmented bodies and appendages (*Callinectes, Eupagurus, Homerus*)

SUBPHYLUM 1. Crustacea: crustaceans

SUBPHYLUM 2. Uniramia

Class 1. Diplopoda: millipedes (*Julus, Polyxenus, Glomeris*)

Class 2. Chilopoda: centipedes (*Scutigera, Lithobius*)

Class 3. Pauropoda: centipede-like animals with 9 or 10 pairs of legs (*Pauropus*)

Class 4. Symphyla: centipede-like animals with 10 to 12 pairs of legs (*Scutigerella, Hanseniella*)

Class 5. Insecta: insects (*Apis, Kalotermes, Drosophila*)

SUBPHYLUM 3. Chelicerata

Class 1. Pycnogonida: sea spiders

Class 2. Merostomata: horseshoe crabs (*Limulus*)

Class 3. Arachnida: scorpions, daddy longlegs or harvestmen, spiders, some mites and ticks (*Thelyphonus, Androctonus, Liphistius*)

PHYLUM 29. Pogonophora: beard worms (*Riftia, Ridgeia, Lamellisabella, Siboglinum, Spirobrachia*)

PHYLUM 30. Echinodermata: echinoderms

SUBPHYLUM 1. Pelmatozoa

Class 1. Crinoidea: sea lilies (*Neometra, Ptilocrinus*)

SUBPHYLUM 2. Eleutherozoa

Class 1. Holothuroidea: sea cucumbers (*Thyone, Holothuria, Euapta*)

Class 2. Echinoidea: sea urchins, sand dollars (*Echinus, Arbacia, Strongylocentrotus*)

Class 3. Stelleroidea: starfish or sea stars, brittle stars, basket stars (*Astronyx, Ophiocoma, Ophiothrix*)

PHYLUM 31. Chaetognatha: arrow worms (*Bathybelos, Spadella, Sagitta*)

PHYLUM 32. Hemichordata: acorn or tongue worms (*Saccoglossus, Balanoglossus*); pterobranchs (*Rhabdopleura, Cephalodiscus*)

PHYLUM 33. Chordata: notochord-bearing animals

SUBPHYLUM 1. Tunicata: tunicates

Class 1. Larvacea: tadpole-like adults (*Oikopleura, Appendicularia*)

Class 2. Ascidiacea: adults grow a typical tunic (*Halocynthia, Cyathocormus, Perophora*)

Class 3. Thaliacea: chain tunicates or salps (*Pyrosoma, Doliolum*)

SUBPHYLUM 2. Cephalochordata: cephalochordates, or lancelets (*Asymmetron, Amphioxus, Branchiostoma*)

SUBPHYLUM 3. Agnatha: agnathids

Class 1. Cyclostomata: lampreys, hagfish, slime eels (*Tubuliporu, Crisia, Lichenopora*)

SUBPHYLUM 4. Gnathostomata: jawed chordates

SUPERCLASS 1. Pisces: fish

Class 1. Chondrichthyes: sharks, skates, rays

Class 2. Osteichthyes: bony fish (*Amia, Gadus, Fundulus*)

SUPERCLASS 2. Tetrapoda: animals with four limbs

Class 1. Amphibia: frogs, toads, salamanders (*Ambystoma, Bufo*)

Class 2. Reptilia: turtles, lizards, snakes, crocodiles (*Chrysemys, Crotalus*)

Class 3. Aves: birds (*Cygnus, Anas, Aptenodytes*)

Class 4. Mammalia: mammals (*Gorilla, Balaenoptera*)

KINGDOM PLANTAE

Gametic meiosis, anisogamous fertilization; diploid phase develops from embryo; haploid phase develops from spores. Multicellular autotrophs; chloroplasts with chlorophylls *a* and *b*; exhibit extensive tissue differentiation. Primarily terrestrial. Production of complex secondary compounds (for example, anthocyanins and terpenoids) is common.

PHYLUM 1. Bryophyta: bryophytes

Class 1. Hepaticae: liverworts (*Marchantia*)

Class 2. Anthocerotae: hornworts (*Anthoceros*)

Class 3. Musci: mosses (*Polytrichum, Bryum, Andreaea*)

PHYLUM 2. Psilophyta: psilophytes (*Psilotum, Tmesipteris*)

PHYLUM 3. Lycopodophyta: club mosses (*Lycopodium, Selaginella, Isoetes*)

PHYLUM 4. Sphenophyta (Equisetophyta): horsetails (*Equisetum*)

PHYLUM 5. Filicinophyta (Pteridophyta, Filicinae): ferns (*Polypodium, Osmunda*)

PHYLUM 6. Cycadophyta: cycads (*Zamia, Cycas, Microcycas*)

PHYLUM 7. Ginkgophyta (*Ginkgo*)

PHYLUM 8. Coniferophyta: conifers (*Taxus, Pinus, Tsuga*)

PHYLUM 9. Gnetophyta: cone-bearing desert plants (*Ephedra, Gnetum, Welwitschia*)

PHYLUM 10. Angiospermophyta: Anthophyta, Magnoliophyta, angiosperms, flowering plants (*Aster, Brassica, Zea, Acer, Allium, Lilium*)

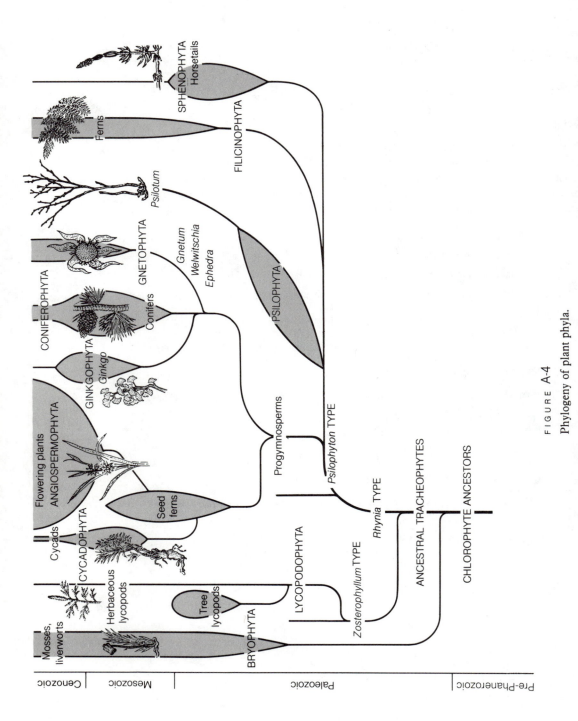

FIGURE A-4

Phylogeny of plant phyla.

REFERENCES

Abelson, P. H. 1963. Organic geochemistry. *J. Wash. Acad. Sci.* 53:105–115.

Ahmadjian, V. 1967. *The Lichen Symbiosis*. Waltham, Mass.: Blaisdell.

Ahmadjian, V. 1980. Separation and artificial synthesis of lichens. In *Cellular Interactions in Symbiosis and Parasitism*. C. B. Cook, P. W. Pappas, and E. D. Rudolph, eds. Columbus: Ohio State University Press, pp. 3–29.

Ahmadjian, V., and S. Paracer. 1986. *Symbiosis: An Introduction to Biological Associations*. Hanover, N. H.: Published for Clark University by University Press of New England.

Alberts, B., D. Bray, J. Lewis, M. Raff, K. Roberts, and J. D. Watson. 1989. *Molecular Biology of the Cell*. 2nd ed. New York: Garland.

Alexopoulos, C. J. 1962. *Introductory Mycology*. 2nd ed. New York: Wiley.

Allsopp, A. 1969. Phylogenetic relationships of the procaryota and the origin of the eucaryotic cell. *New Phytol.* 68:591–612.

Almassy, R. J., and R. E. Dickerson. 1978. *Pseudomonas* cytochrome *c* 551 at 2.0 Å resolution: enlargement of the cytochrome *c* family. *Proc. Natl. Acad. Sci. U. S. A.* 75:2674–2678.

Altman, P. L., and D. S. Dittmer, eds. 1972–1974. *Biology Data Book*. 3 vols. Bethesda, Md.: Federation of American Societies for Experimental Biology.

Ashen, J. B. 1992. Ultrastructure of new microbial mat and termite spirochetes and the symbiotic origin of undulipodia. M. S. thesis. University of Massachusetts, Amherst.

Atema, J. 1975. Stimulus transmission along microtubules in sensory cells: an hypothesis. In *Microtubules and Microtubule Inhibitors*. M. Borgers and M. De Brabander, eds. Amsterdam: North-Holland, pp. 247–257.

Atsatt, P. R. 1991. Fungi and the origin of land plants. In *Symbiosis as a Source of Evolutionary Innovation: Speciation and Morphogenesis*. L. Margulis and R. Fester, eds. Cambridge, Mass.: MIT Press, pp. 301–315.

Aulie, R. P. 1970. Boussingault and the nitrogen cycle. *Proc. Am. Philos. Soc.* 114:435–478.

Awramik, S., L. Margulis, and E. S. Barghoorn. 1976. Evolutionary processes in the formation of stromatolites. *Dev. Sedimentol.* 20:149–162.

Awramik, S. M. 1980. The pre-Phanerozoic biosphere: three billion years of crises and opportunities. In *Biotic Crises in Ecologic and Evolutionary Time*. M. H. Nitecki, ed. New York: Academic Press, pp. 83–102.

Awramik, S. M., and E. S. Barghoorn. 1977. The Gunflint microbiota. *Precambrian Res.* 5:121–143.

Awramik, S. M., and E. S. Barghoorn. 1978. *Bibliography of Precambrian paleontology and paleobiology.* Harvard University, Botanical Museum Leaflet No. 26.

Awramik, S. M., A. Haupt, H. J. Hofmann, and M. R. Walter. 1979. Stromatolite bibliography 2. *Precambrian Res.* 9:105–166.

Awramik, S. M., J. W. Schopf, and M. R. Walter. 1983. Filamentous fossil bacteria 3.5×10^9 years old from the Archean of Western Australia. *Precambrian Res.* 20:357–374.

Baker, K. P., and G. Schatz. 1991. Mitochondrial proteins essential for viability mediate protein import into yeast mitochondria. *Nature* 349:205–214.

Balows, A., H. G. Trüper, M. Dworkin, W. Harder, and K.-H. Schleifer, eds. 1991. *The Prokaryotes. A Handbook on the Biology of Bacteria: Ecophysiology, Isolation, Identification, Applications.* 4 vols. 2nd ed. New York: Springer-Verlag.

Bamje, M. S., and N. I. Krinsky. 1965. Carotenoid de-epoxidation in algae. II. Enzymatic conversion of antheraxanthin to zeoxanthin. *J. Biol. Chem.* 240:467–470.

Banks, H. P. 1972a. *Evolutionary History of Plants.* Belmont, Calif.: Wadsworth.

Banks, H. P. 1972b. The stratigraphic occurrence of early land plants. *Palaeontology* 15:365–377.

Barath, Z., and H. Kuntzel. 1972. Cooperation of mitochondrial and nuclear genes specifying the mitochondrial genetic apparatus in *Neurospora crassa. Proc. Natl. Acad. Sci. U. S. A.* 69:1371–1374.

Barghoorn, E. S. 1970. [Oral remarks]. In *Origins of Life: Proceedings of the First Conference.* L. Margulis, ed. New York: Published for Interdisciplinary Communication Associates by Gordon and Breach, pp. 105–106.

Barghoorn, E. S. 1971. The oldest fossils. *Sci Am.* 224(5):30–42.

Barghoorn, E. S., A. H. Knoll, H. Dembicki, Jr., and W. G. Meinschein. 1977. Variation in stable carbon isotopes in organic matter from the Gunflint Iron Formation. *Geochim. Cosmochim. Acta* 41:425–430.

Barghoorn, E. S., and S. A. Tyler. 1965. Microorganisms from the Gunflint chert. *Science* 147:563–577.

Barlow, C., ed. 1991. *From Gaia to Selfish Genes: Selected Writings in the Life Sciences.* Cambridge, Mass.: MIT Press.

Barth, A., J. Stricker, and L. Margulis. 1991. Search for eukaryotic motility proteins in spirochetes: immunological detection of a tektin-like protein in *Spirochaeta halophila. BioSystems* 24:313–319.

Bassham, J. A. 1980. Photosynthesis and biosynthetic pathways to chemicals. In *Future Sources of Organic Raw Materials (IUPAC).* L. E. St. Pierre and G. R. Brown, eds. New York: Pergamon, pp. 601–612.

Beale, G. 1954. *The Genetics of Paramecium.* Cambridge. Cambridge University Press.

Beale, G. H., A. Jurand, and J. R. Preer. 1969. The classes of endosymbiont of *Paramecium aurelia. J. Cell Sci.* 5:65–91.

Beisson, J., and T. M. Sonneborn. 1965. Cytoplasmic inheritance of the organization of the cell cortex in *Paramecium aurelia. Proc. Natl. Acad. Sci. U. S. A.* 53:275–282.

Belar, K. 1915a. Protozoenstudien 1. *Arch. Protistenkd.* 36:13–51.

Belar, K. 1915b. Protozoenstudien 2. *Arch. Protistenkd.* 36:241–302.

Belar, K. 1923. Untersuchungen an *Actinophrys sol* Ehrenberg. 1. Die Morphologie des Formwechsels. *Arch. Protistenkd.* 46:1–96.

Bengtson, S., ed. In press. *Early Life on Earth: Nobel Symposium No. 84*. New York: Columbia University Press.

Bereiter-Hahn, J. 1990. Behavior of mitochondria in the living cell. *Int. Rev. Cytol. 122*:1–63.

Berg, H. C. 1975. Bacterial behaviour. *Nature 254*:389–392.

Berger, B., G. Thorington, and L. Margulis. 1979. Two aeromonads: growth of symbionts from *Hydra viridis. Curr. Microbiol. 3*:5–10.

Berkner, L. V., and L. Marshall. 1965. On the origin and rise of oxygen concentration in the Earth's atmosphere. *J. Atm. Sci. 22*:225–261.

Bermudes, D., D. Chase, and L. Margulis. 1988. Morphology as a basis for taxonomy of large spirochetes symbiotic in wood-eating cockroaches and termites. *Int. J. Syst. Bacteriol. 38*:291–302.

Bermudes, D., L. Margulis, and G. Tzertzinis. 1987. Prokaryotic origins of undulipodia: application of the Panda Principle to the centriole enigma. *Ann. N. Y. Acad. Sci. 503*:187–197.

Bermudes, D. G. 1987. Distribution and immunocytochemical localization of tubulin-like proteins in spirochetes. Ph.D. thesis. Boston University, Boston.

Bernal, J. D. 1967. *The Origin of Life*. Cleveland: World; London: Weidenfeld and Nicolson.

Bessey, E. A. 1950. *Morphology and Taxonomy of Fungi*. Philadelphia: Blakiston.

Bingham, S., and J. A. Schiff. 1976. Cellular origins of plastid membrane polypeptides in *Euglena*. In *Genetics and Biogenesis of Chloroplasts and Mitochondria*. T. Bücher, W. Neupert, W. Sebald, and S. Werner, eds. Amsterdam: North-Holland, pp. 79–86.

Bisalputra, T., B. R. Oakley, D. C. Walker, and C. M. Shields. 1975. Microtubular complexes in blue-green algae. *Protoplasma 86*:19–28.

Bloodgood, R. A. 1977. The squid accessory nidamental gland: ultrastructure and association with bacteria. *Tissue & Cell 9*:197–208.

Bloodgood, R. A., and T. P. Fitzharris. 1977. Specific association of prokaryotes with symbiotic flagellate protozoa from the hindgut of the termite *Reticulitermes* and the wood-eating roach *Cryptocercus. Cytobios 17*:103–122.

Bloodgood, R. A., and K. R. Miller. 1974. Freeze-fracture of microtubules and bridges in motile axostyles. *J. Cell Biol. 62*:660–671.

Bloodgood, R. A., K. R. Miller, T. P. Fitzharris, and J. R. McIntosh. 1974. The ultrastructure of *Pyrsonympha* and its associated microorganisms. *J. Morph. 143*:77–106.

Borgers, M., and M. DeBrabander, eds. 1975. *Microtubules and Microtubule Inhibitors*: Proceedings of the International Symposium on Microtubules and Microtubule Inhibitors, Beerse, Belgium, 2–5 September 1975. Amsterdam: North-Holland.

Brennen, C., and H. Winet. 1977. Fluid mechanics of propulsion by cilia and flagella. *Annu. Rev. Fluid Mech. 9*:339–398.

Breznak, J. A. 1973. Biology of nonpathogenic, host-associated spirochetes. *Crit. Rev. Microbiol. 2*:457–489.

Breznak, J. A., W. J. Brill, J. W. Mertins, and H. C. Coppel. 1973. Nitrogen fixation in termites. *Nature 244*:577–580.

Brock, T. D., and M. T. Madigan. 1991. *Biology of Microorganisms*. 6th ed. Englewood Cliffs, N. J.: Prentice-Hall.

Broda, E. 1975. *The Evolution of the Bioenergetic Process*. Oxford: Pergamon.

Brokaw, C. J., and I. R. Gibbons. 1974. Mechanisms of movement in flagella and cilia. In *Swimming and Flying in Nature*, vol. 1. T. Y.-T. Wu, C. J. Brokaw, and C. Brennen, eds. New York: Plenum, pp. 89–126.

Bronchart, R., and V. Demoulin. 1977. Unusual mitosis in the red alga *Porphyridium purpureum*. *Nature* 268:80–81.

Bryant, M. P., E. A. Wolin, M. J. Wolin, and R. S. Wolfe. 1967. *Methanobacillus omelianskii*, a symbiotic association of two species of bacteria. *Arch. Mikrobiol.* 59:20–31.

Buat-Menard, P., and R. Chesselet. 1979. Variable influence of the atmospheric flux on the trace metal chemistry of oceanic suspended matter. *Earth and Planetary Sci. Lett.* 42:399–411.

Bücher, T., W. Neupert, W. Sebald, and S. Werner, eds. 1976. *Genetics and Biogenesis of Chloroplasts and Mitochondria*: Interdisciplinary Conference on the Genetics and Biogenesis of Chloroplasts and Mitochondria, Munich, Germany, August 2–7, 1976. Amsterdam: North-Holland.

Buchner, P. 1965. *Endosymbiosis of Animals with Plant-like Microorganisms*. New York: Wiley Interscience.

Butterfield, N. J., A. H. Knoll, and K. Swett. 1990. A bangiophyte red alga from the Proterozoic of Arctic Canada. *Science* 250:104–107.

Cachon, J., and M. Cachon. 1974. Les systèmes axopodiaux. *Anneé Biol.* 13:523–560.

Cachon, J., M. Cachon, L. G. Tilney, and M. S. Tilney. 1977. Movement generated by interactions between the dense material at the ends of microtubules and non-actin-containing microfilaments in *Sticholonche zanclea*. *J. Cell Biol.* 72:314–338.

Calvin, M. 1969a. *Chemical Evolution: Molecular Evolution Towards the Origin of Living Systems on the Earth and Elsewhere*. Oxford: Clarendon.

Calvin, M. 1969b. Molecular paleontology. *Perspect. Biol. Med.* 13:45–62.

Campbell, I. H., and S. R. Taylor. 1983. No water, no granites—no oceans, no continents. *Geophys. Res. Lett.* 10:1061–1064.

Campbell, S. E. 1979. Soil stabilization by a prokaryotic desert crust: implications for Precambrian land biota. *Origins Life* 9:335–348.

Canale-Parola, E. 1992. Free-living saccharolytic spirochetes: The genus *Spirochaeta*. In *The Prokaryotes. A Handbook on the Biology of Bacteria: Ecophysiology, Isolation, Identification, Applications*. A. Balows, H. G. Trüper, M. Dworkin, W. Harder, and K.-H. Schleifer, eds. 4 vols. 2nd ed. New York: Springer-Verlag, pp. 3524–3536.

Cantino, E. C., and R. B. Myers. 1973. The gamma particle and intracellular interactions in *Blastocladiella emersoni*. *Brookhaven Symp. Biol.* 25:51–74.

Carell, E. F. 1969. Studies on chloroplast development and replication in *Euglena*. I. Vitamin B_{12} and chloroplast replication. *J. Cell Biol.* 41:431–440.

Cavalier-Smith, T. 1975. The origin of nuclei and of eukaryotic cells. *Nature* 256:463–468.

Cheung, A. T. W., and T. L. Jahn. 1975. Helical nature of the continuous ciliary beat of *Opalina*. *Acta Protozool.* 14:219–232.

Chisholm, S. W., R. J. Olson, E. R. Zettler, R. Goericke, J. B. Waterbury, and N. A. Welschmeyer. 1988. A novel free-living prochlorophyte abundant in the oceanic euphotic zone. *Nature* 334:340–343.

Chun, E. H. L., M. H. Vaughn, Jr., and A. Rich. 1963. The isolation and characterization of DNA associated with chloroplast preparations. *J. Mol. Biol.* 7:130–141.

Church, A. H. 1919. The building of an autotrophic flagellate. *Oxford Bot. Mem.* 1:4–27.

Clark-Walker, G. D., and A. W. Linnane. 1967. The biogenesis of mitochondria in *Saccharomyces cerevisiae*. A comparison between cytoplasmic respiratory-deficient mutant yeast and chloramphenicol-inhibited wild type cells. *J. Cell Biol. 34*:1–14.

Clements, K. D., and S. Bullivant. 1991. An unusual symbiont from the gut of surgeonfishes may be the largest known prokaryote. *J. Bacteriol. 173*:5359–5362.

Clements, K. D., D. C. Sutton, and J. H. Choat. 1989. Occurrence and characteristics of unusual protistan symbionts from surgeonfishes (*Acanthuridae*) of the Great Barrier Reef, Australia. *Mar. Biol. 102*:403–412.

Clemmey, H. 1976. World's oldest animal traces. *Nature 261*:576–578.

Cleveland, L. R. 1926. Some problems which may be studied by oxygenation. *Science 63*:168–170.

Cleveland, L. R. 1935. The centriole and its role in mitosis as seen in living cells. *Science 81*:598–600.

Cleveland, L. R. 1938. Longitudinal and transverse division in two closely related flagellates. *Biol. Bull. 74*:1–24.

Cleveland, L. R. 1947. The origin and evolution of meiosis. *Science 105*:287–289.

Cleveland, L. R. 1951. Hormone-induced sexual cycles of flagellates. *J. Morphol. 88*:385–440.

Cleveland, L. R. 1956. *The Flagellates of Termites.* (Parts 1 and 2, North American termites; Part 3, Australian termites.) 16 mm films. Amherst: Dept. of Zoology, University of Massachusetts.

Cleveland, L. R. 1957. Types and life cycles of centrioles of flagellates. *J. Protozool. 4*:230–241.

Cleveland, L. R. 1963. Functions of flagellate and other centrioles in cell replication. In *The Cell in Mitosis:* Proceedings of the First Annual Symposium held under the Provisions of the Wayne State Fund Research Recognition Award. L. Levin, ed. New York: Academic Press, pp. 3–31.

Cleveland, L. R., and A. V. Grimstone. 1964. The fine structure of the flagellate *Mixotricha paradoxa* and its associated micro-organisms. *Proc. R. Soc. Lond., Ser. B. Biol. Sci. 159*:668–686.

Cloud, P. E., Jr. 1968a. Atmospheric and hydrospheric evolution on the primitive Earth. *Science 160*:729–736.

Cloud, P. E., Jr. 1968b. Premetazoa evolution and the origin of Metazoa. In *Evolution and Environment.* E. T. Drake, ed. New Haven: Yale University Press, pp. 1–72.

Cloud, P. E., Jr. 1974. Evolution of ecosystems. *Am. Sci. 62*:54–66.

Cloud, P. E., Jr. 1978. *Cosmos, Earth and Man: A Short History of the Universe.* New Haven: Yale University Press.

Cloud, P. E., Jr. 1988. *Oasis in Space: Earth History From the Beginning.* New York: W. W. Norton.

Cloud, P. E., Jr., H. D. Holland, B. Commoner, C. F. Davidson, A. G. Fischer, L. V. Berkner, and L. C. Marshall. 1965. Symposium on the evolution of the Earth's atmosphere. *Proc. Natl. Acad. Sci. U. S. A. 53*:1169–1226.

Cohen, Y., E. Padan, and M. Shilo. 1975. Facultative anoxygenic photosynthesis in the cyanobacterium *Oscillatoria limnetica*. *J. Bacteriol. 123*:855–861.

Cohen, Y., and E. Rosenberg. 1989. *Microbial Mats: Physiological Ecology of Benthic Microbial Communities.* Washington, D. C.: American Society for Microbiology.

Coleman, W. 1971. *Biology in the Nineteenth Century: Problems of Form, Function, and Transformation.* New York: Wiley.

Cook, C. B. 1980. Infection of invertebrates with algae. In *Cellular Interactions in*

Symbiosis and Parasitism. C. B. Cook, P. W. Pappas, and E. D. Rudolph, eds. Columbus: Ohio State University Press, pp. 47–74.

Cook, C. B., P. W. Pappas, and E. D. Rudolph, eds. 1980. *Cellular Interactions in Symbiosis and Parasitism.* Columbus: Ohio State University Press.

Coonen, L. P. 1977. Aristotle's biology. *BioScience* 27:733–738.

Cooper, G., and L. Margulis. 1977. Delay in migration of symbiotic algae in *Hydra viridis* by inhibitors of microtubule protein polymerization. *Cytobios* 19:7–19.

Copeland, H. F. 1956. *The Classification of Lower Organisms.* Palo Alto, Calif.: Pacific Books.

Corliss, J. O. 1979. *The Ciliated Protozoa: Characterization, Classification, and Guide to the Literature.* 2nd ed. Oxford: Pergamon.

Correns, C. 1909. Vererbungsversuche mit blass (gelb) grünen und buntblätrigen Sippen bei *Mirabilis jalapa, Urtica pilulifera* und *Lunaria annua. Z. Indukt. Abstammungs-Vererbungsl. 1:*291–329.

Crick, F. H. C. 1968. The origin of the genetic code. *J. Mol. Biol. 38:*367–379.

Cronkite, D. 1976. A role of calcium ions in chemical induction of mating in *Paramecium tetraurelia. J. Protozool. 23:*431–433.

Curtis, H. 1968. *The Science of Biology.* New York: Worth.

Daniels, E. W., and E. P. Breyer. 1967. Ultrastructure of the giant amoeba *Pelomyxa palustris. J. Protozool. 14:*167–179.

Daniels, E. W., E. P. Breyer, and R. R. Kudo. 1966. *Pelomyxa palustris* Greef. II. Its ultrastructure. *Z. Zellforsch. Mikrosk. Anat. 73:*367–383.

Darland, G., T. D. Brock, W. Samsenoff, and S. F. Conti. 1970. A thermophilic acidophilic mycoplasma isolated from a coal refuse pile. *Science 170:*1416–1418.

Darlington, C. D. 1958. *The Evolution of Genetic Systems.* 2nd ed. New York: Basic Books.

Darwin, C. 1859. *The Origin of Species by Means of Natural Selection, or the Preservation of Favoured Races in the Struggle for Life.* London: John Murray.

Davidson, C. F. 1965. Geochemical aspects of atmospheric evolution. *Proc. Natl. Acad. Sci. U. S. A. 53:*1194–1204.

Davidson, L. A. 1982. Ultrastructure, behavior and algal flagellate affinities of the helioflagellate *Ciliophrys marina* and the classification of the helioflagellates (Protista-Actinopoda-Heliozoa). *J. Protozool. 29:*19–29.

Day, W. 1984. *Genesis on Planet Earth: The Search for Life's Beginning.* 2nd ed. New Haven: Yale University Press.

Deal, P. H., K. A. Souza, and H. M. Mack. 1975. High pH, ammonia toxicity, and the search for life on the Jovian planets. *Origins Life 6:*561–573.

Deamer, D. 1992. Origins of membranes: structures and functions. In *Environmental Evolution: The Effects of the Origin and Evolution of Life on Planet Earth.* L. Margulis and L. Olendzenski, eds. Cambridge, Mass.: MIT Press, pp. 41–54.

DeBrugerolle, G., and J.-P. Mignot. 1984. The cell characters of two helioflagellates related to the centrohelidian lineage: *Dimorpha* and *Tetradimorpha. Origins Life 13:*305–314.

DeDuve, C. 1991. *Blueprint for a Cell: The Nature and Origin of Life.* Burlington, N. C.: Neil Patterson Publishers, Carolina Biological Supply Company.

DeLey, J. 1968. Molecular biology and bacterial phylogeny. In *Evolutionary Biology.* T. Dobzhansky, M. K. Hecht, and W. C. Steere, eds. New York: Appleton-Century-Crofts, pp. 104–156.

DeRosa, F., D. Faber, C. Williams, and L. Margulis. 1978. Inhibitory effects of the herbicide trifluralin on the establishment of the clover root nodule symbiosis. *Cytobios 21*:37–43.

Desportes, I. 1970. Ultrastructure des gregarines du genre *Stylocephalus*; la phase enkystée. *Ann. Sci. Nat. Zool. Biol. Anim.*, 12th Ser. *12*:73–170.

Dévai, I., L. Felföldy, I. Wittner, and S. Plósz. 1988. Detection of phosphine: new aspects of the phosphorus cycle in the hydrosphere. *Nature 333*:343–345.

Deysson, G. 1968. Antimitotic substances. *Int. Rev. Cytol. 24*:99–148.

Dilke, F. W. W., and D. O. Gough. 1972. The solar spoon. *Nature 240*:262–264, 293–294.

Dillon, L. S. 1962. Comparative cytology and the evolution of life. *Evolution 16*:102–117.

Dimroth, E., and M. M. Kimberley. 1976. Precambrian atmospheric oxygen: evidence in the sedimentary distributions of carbon, sulfate, uranium, and iron. *Can. J. Earth Sci. 13*:1161–1185.

Dippell, R. V. 1976. Effects of nuclease and protease digestion on the ultrastructure of *Paramecium* basal bodies. *J. Cell Biol. 69*:622–637.

Dirksen, E. R. 1991. Centriole and basal body formation during ciliogenesis revisited. *Biol. Cell 72*:31–38.

Dirksen, E. R., and T. T. Crocker. 1965. Centriole replication in differentiating ciliated cells of mammalian respiratory epithelium. An electron microscopic study. *J. Microsc. (Paris) 5*:629–644.

Dobell, C. 1912. Researches on the spirochetes and related organisms. *Arch. Protistenkd. 26*:117–240.

Dobell, C. 1958. *Antony van Leeuwenhoek and His "Little Animals"*. New York: Russell & Russell.

Dodge, J. D. 1966. The Dinophyceae. In *The Chromosomes of Algae*. M. B. E. Godward, ed. London: Edward Arnold, pp. 96–115.

Dodson, E. O. 1971. The kingdoms of organisms. *Syst. Zool. 20*:265–281.

Doflein, F. 1916. *Lehrbuch der Protozoenkunde: eine Darstellung der Naturgeschichte der Protozoen mit besonderer Berücksichtigung der parasitischen und pathogenen Formen*. 4th ed. Jena: Gustav Fischer.

Doflein, F., and E. Reichenow. 1929. *Lehrbuch der Protozoenkunde, eine Darstellung der Naturgeschichte der Protozoen mit besonderen Berücksichtigung der parasitischen und pathogenen Formen*. Jena: Gustav Fischer.

Dougherty, E. C., and M. B. Allen. 1958. The words protist and protista. *Experientia 14*:78.

Dougherty, E. C., and M. B. Allen. 1960. Is pigmentation a clue to protistan phylogeny? In *Comparative Biochemistry of Photoreactive Systems*. M. B. Allen, ed. New York: Academic Press, pp. 129–144.

Douglas, S. E., C. A. Murphy, D. F. Spencer, and M. W. Gray. 1991. Cryptomonad algae are evolutionary chimaeras of two phylogenetically distinct unicellular eukaryotes. *Nature 350*:148–151.

Dubosq, O., and P. Grassé. 1927. Flagelles et schizophytes de *Calotermes (Glyptotermes) iridipennis* Frogg. *Arch. Zool. Exp. Gen. 66*:451–496.

DuBuy, H. G., C. F. T. Mattern, and F. L. Riley. 1965. Isolation and characterization of DNA from kinetoplasts of *Leishmania enriettii*. *Science 147*:754–756.

Dunlop, J. S. R., M. D. Muir, V. A. Milne, and D. I. Groves. 1978. A new microfossil assemblage from the Archaean of Western Australia. *Nature 274*:676–678.

Dustin, P. 1978. *Microtubules*. Berlin: Springer-Verlag.

Dustin, P., J.-P. Hurbert, and J. Flament-Durand. 1975. Centriole and cilia formation in rat pituicytes after treatment with colchicine and vincristine. Studies in vivo and in vitro. In *Microtubules and Microtubule Inhibitors*. M. Borgers and M. DeBrabander, eds. Amsterdam: North-Holland, pp. 289–296.

Dworkin, M., 1992. Prokaryotic life cycles. In *The Prokaryotes. A Handbook on the Biology of Bacteria: Ecophysiology, Isolation, Identification, Applications*. A. Balows, H. G. Trüper, M. Dworkin, W. Harder, and K.-H. Schleifer, eds. 4 vols. 2nd ed. New York: Springer-Verlag, pp. 209–240.

Dyer, B. D. 1989. Symbiosis and organismal boundaries. *Am. Zool. 29*:1085–1093.

Echlin, P., and I. Morris. 1965. The relationship between blue-green algae and bacteria. *Biol. Rev. Camb. Philos. Soc. 40*:143–187.

Edelman, M., C. A. Cowan, H. T. Epstein, and J. A. Schiff. 1964. Studies of chloroplast development in *Euglena*. VIII. Chloroplast associated DNA. *Proc. Natl. Acad. Sci. U. S. A. 52*:1214–1219.

Edelman, M., J. A. Schiff, and H. T. Epstein. 1965. Studies of chloroplast development in *Euglena*. XII. Two types of satellite DNA. *J. Mol. Biol. 11*:769–774.

Edelman, M., D. Swinton, J. A. Schiff, H. T. Epstein, and B. Zeldin. 1967. Deoxyribonucleic acid of the blue-green algae (*Cyanophyta*). *Bacteriol. Rev. 31*: 315–331.

Edmundson, W. T., ed. 1966. *Marine Biology 3*: Proceedings of the Third International Interdisciplinary Conference (Ecology of Invertebrates). N. Y. Acad. Sci. Interdisciplinary Communications Program.

Edwards, P. 1976. A classification of plants into higher taxa based on cytological and biochemical criteria. *Taxon 25*:529–542.

Eglinton, G., and M. Calvin. 1967. Chemical fossils. *Sci. Am. 216* (1):32–43.

Eglinton, G., and M. T. J. Murphy. 1969. *Organic Geochemistry: Methods and Results*. Berlin: Springer-Verlag.

Ehrensvard, G. 1962. *Life: Origin and Development*. Chicago: University of Chicago Press.

Enzien, M. 1990. Cyanobacteria or Rhodophyta? Interpretation of a Precambrian microfossil. *BioSystems 24*:245–251.

Ephrussi, B. 1953. *Nucleo-Cytoplasmic Relations in Microorganisms, Their Bearing on Cell Heredity and Differentiation*. Oxford: Clarendon.

Epstein, H. T., and E. Alloway. 1967. Properties of selectively starved euglena. *Biochim. Biophys. Acta 142*:195–207.

Esteve, I., J. Mir, N. Gaju, H. McKhann, and L. Margulis. 1988. Green endosymbiont of *Coleps* from Lake Cisó identified as *Chlorella vulgaris*. *Symbiosis 6*:197–210.

Famintzyn, A. S. 1891. Concerning the symbiosis of algae with animals. *Trans. Bot. Lab. Acad. Sci. 1*:1–22. [In Russian; cited in Khakhina (1979).]

Fedonkin, M. A. 1993. Vendian body fossils and trace fossils. In *Early Life on Earth: Nobel Symposium No. 84*. S. Bengtson, ed. New York: Columbia University Press (in press).

Fenchel, T., and B. J. Finlay. 1991. The biology of free-living anaerobic ciliates. *Eur. J. Protistol. 26*:201–215.

Ferguson-Wood, E. J. 1967. *Microbiology of Oceans and Estuaries*. (Elsevier Oceanography Series, vol. 3) Amsterdam: Elsevier.

Fischer, A. G. 1965. Fossils, early life, and atmospheric history. *Proc. Natl. Acad. Sci. U. S. A. 53*:1205–1215.

Fleischaker, G. R. 1990. Origins of life: an operational definition. *Origins Life Evol. Biosphere 20*:127–137.

Florkin, M., and H. S. Mason, eds. 1960–1964. *Comparative Biochemistry. A Comprehensive Treatise.* 7 vols. New York: Academic Press.

Fox, D. L., and R. A. Lewin. 1963. A preliminary study of the carotenoids of some flexibacteria. *Can. J. Microbiol. 9:*753–768.

Fox, S. W., ed. 1965. *Origins of Prebiological Systems and Their Molecular Matrices.* New York: Academic Press.

Fox, S. W. 1977. The nature of life. In *The Origin and Nature of Protolife.* W. H. Heidcamp, ed. Baltimore: University Park Press, pp. 23–92.

Fox, S. W., and R. J. McCauley. 1968. Could life originate now? *Nat. Hist.* (Sept.):26–33.

Fracek, S., and L. Margulis. 1979. Colchicine, nocodazole and trifluralin: different effects of microtubule polymerization inhibitors on the uptake and migration of endosymbiotic algae in *Hydra viridis. Cytobios 25:*7–15.

Fracek, S. P., Jr. and J. F. Stolz. 1985. *Spirochaeta bajacaliforniensis* sp. n. from a microbial mat community at Laguna Figueroa, Baja California Norte, Mexico. *Arch. Microbiol. 142:*317–325.

Francis, S., E. S. Barghoorn, and L. Margulis. 1978. On the experimental silicification of microorganisms. III. Implications of the preservation of the green prokaryotic alga *Prochloron* and other coccoids for interpretation of the microbial fossil record. *Precambrian Res. 7:*377–383.

Francis, S., L. Margulis, and E. S. Barghoorn. 1978. On the experimental silicification of microorganisms. II. On the time of appearance of eukaryotic organisms in the fossil record. *Precambrian Res. 6:*65–100.

Fridovitch, I. 1977. Oxygen is toxic. *BioScience 27:*462–466.

Friedman, E. I., and R. Ocampo. 1976. Endolithic blue-green algae in the dry valleys: primary producers in the Antarctic desert ecosystem. *Science 193:*1247–1249.

Fritsch, F. E. 1935. *Structure and Reproduction of the Algae.* 2 vols. Cambridge, Engl.: Cambridge University Press.

Fuerst, J. A., and R. I. Webb. 1991. Membrane-bounded nucleoid in the eubacterium *Gemmata obscuriglobus. Proc. Natl. Acad. Sci. U. S. A. 88:*8184–8188.

Gabriel, M. 1960. Primitive genetic mechanisms and the origin of chromosomes. *Am. Nat. 94:*257–269.

Gharagozlou, I. D. 1968. Aspect infrastructural de *Diplocalyx calotermitidis* nov. gen. nov. sp., spirochaetale de l'intestin de *Calotermes flavicollis. C. R. Seances Acad. Sci., Ser. D. 266:*494–496.

Ghosh, A., J. Maniloff, and D. A. Gerling. 1978. Inhibition of mycoplasm cell division by cytochalasin B. *Cell. 13:*57–64.

Gibbs, S. P. 1981. The chloroplasts of some algal groups may have evolved from endosymbiotic eukaryotic algae. *Ann. N. Y. Acad. Sci. 361:*193–208.

Gibor, A., and S. Granick. 1964. Plastids and mitochondria: heritable systems. *Science 145:*890–897.

Gillham, N. W. 1978. *Organelle Heredity.* New York: Raven.

Girbardt, M., and H. Hadrich. 1975. Ultrastruktur des Pilzkernes. III. Genese des kern-assorziierten Organells (NAO-"KCE"). *Z. Allg. Mikrobiol. 15:*157–173.

Giusto, J. P., and L. Margulis. 1980. Karyotypic fission theory and the evolution of old world monkeys and apes. *BioSystems 13:*267–302.

Glaessner, M. F. 1968. Biological events and the Precambrian time scale. *Can. J. Earth Sci. 5:*585–590.

Glaessner, M. F. 1971. Geographic distribution and time range of the Ediacara Precambrian fauna. *Geol. Soc. Am. Bull. 82:*509–514.

Glaessner, M. F. 1976. Early Phanerozoic annelid worms and their geological and biological significance. *J. Geol. Soc. Lond. 132*:259–275.

Goff, L. J. 1983. *Algal Symbiosis: A Continuum of Interaction Strategies*. Cambridge, Engl.: Cambridge University Press.

Goff, L. J. 1991. Red algal parasites. In *Symbiosis as a Source of Evolutionary Innovation: Speciation and Morphogenesis*. L. Margulis and R. Fester, eds. Cambridge, Mass.: MIT Press, pp. 550–590.

Goksøyr, J. 1967. Evolution of eucaryotic cells. *Nature 214*:1161.

Goldschmidt, R., and M. Popoff. 1907. Die Karyokinese der Protozöen und der Chromidialapparat der Protozöen und Metazöenzelle. *Arch. Protistenkd. 8*:321–343.

Goldsmith, D., and T. C. Owen, 1992. *The Search for Intelligent Life in the Universe*. 2nd ed. New York: W. A. Benjamin.

Golubic, S., and E. S. Barghoorn. 1977. Interpretation of microbial fossils with special reference to the Precambrian. In *Fossil Algae: Recent Results and Developments*. E. Flugel, ed. Berlin: Springer-Verlag, pp. 1–14.

Golubic, S., and S. E. Campbell. 1979. Analogous microbial forms in recent subaerial habitats and in Precambrian cherts: *Gloeothece coerulea* (Geitler) and *Eosynechococcus moorei* (Hofmann). *Precambrian Res. 8*:201–217.

Golubic, S., R. D. Perkins, and K. J. Lukas. 1975. Boring microorganisms and microborings in carbonate substrates. In *The Study of Trace Fossils: A Synthesis of Principles, Problems, and Procedures in Ichnology*. R. W. Frey, ed. New York: Springer-Verlag, pp. 229–259.

Goodenough, U. W. 1989. Basal body chromosomes? *Cell 59*:1–3.

Goodwin, A. M., J. Monster, and H. G. Thode. 1976. Carbon and sulfur isotope abundances in Archean iron formations and early Precambrian life. *Econ. Geol. 71*:870–891.

Goreau, T. J. 1976. Coral skeletal chemistry: physiological and environmental regulation of stable isotopes and trace metals in *Monoastrea annularis*. *Proc. R. Soc. Lond., Ser. B. Biol. Sci. 196*:291–315.

Gorovsky, M. A., and J. Woodard. 1968. Radioautographic studies of RNA synthesis in macro- and micronuclei of *Tetrahymena* (abs. 126). *J. Cell Biol. 39*:54a.

Gould, G. W., and G. J. Dring. 1979. On a possible relationship between bacterial endospore formation and the origin of eukaryotic cells. *J. Theor. Biol. 81*:47–53.

Gould, S. J. 1989. *Wonderful Life: The Burgess Shale and the Nature of History*. New York: W. W. Norton.

Goulian, M., A. Kornberg, and R. L. Sinsheimer. 1967. Enzymatic synthesis of DNA. XXIV. Synthesis of infectious phage ΦX174 DNA. *Proc. Natl. Acad. Sci. U. S. A. 58*:2321–2328.

Graham, L. E. 1985. The origin of the life cycle of land plants. *Am. Sci. 73*:178–186.

Grant, V. 1971. *Plant Speciation*. New York: Columbia University Press.

Graustein, W. C., K. Cromack, Jr., and P. Sollins. 1977. Calcium oxalate: occurrence in soils and effect on nutrient and geochemical cycles. *Science 198*:1252–1254.

Gray, M. W. 1983. The bacterial ancestry of plastids and mitochondria. *BioScience 33*:693–699.

Gray, M. W. 1988. Organellar origins and ribosomal RNA. *Biochem. Cell Biol. 66*:325–348.

Gray, M. W. 1989. Origin and evolution of mitochondrial DNA. *Annu. Rev. Cell Biol. 5*:25–50.

Gray, M. W. 1991. Origin and evolution of plastid genomes and genes. In *Cell Culture and Somatic Cell Genetics of Plants*, vol. 7A. San Diego: Academic Press, pp. 303–330.

Grell, K. G. 1967. Sexual reproduction in protozoa. In *Research in Protozoology*, vol. 2. T. T. Chen, ed. Oxford: Pergamon, pp. 147–213.

Grimstone, A. V., and I. R. Gibbons. 1966. The fine structure of the centriolar apparatus and associated structures in the complex flagellates *Trichonympha* and *Pseudotrichonympha*. *Philos. Trans. R. Soc. Lond. B. Biol. Sci. 250*:215–242.

Grompe, M., J. Versalovic, T. Koeuth, and J. R. Lupski. 1991. Mutations in the *Escherichia coli dna-G* gene suggest coupling between DNA replication and chromosome partitioning. *J. Bacteriol. 173*:1268–1278.

Grosovsky, B. D. D., and L. Margulis. 1982. Termite microbial communities. In *Experimental Microbial Ecology*. R. G. Burns and J. H. Slater, eds. Oxford: Blackwell, pp. 519–532.

Guerrero, R. 1991. Predation as prerequisite to organelle origin: *Daptobacter* as example. In *Symbiosis As a Source of Evolutionary Innovation: Speciation and Morphogenesis*. L. Margulis and R. Fester, eds. Cambridge, Mass.: MIT Press, pp. 106–117.

Guerrero, R., J. Ashen, and L. Margulis. 1992. *Spirosymplokos deltaeiberi* g. nov. sp. nov., multiple-fission large spirochete from microbial mats. *Arch. Microbiol.* (in press).

Guerrier-Takada, C., K. Gardiner, T. Marsh, N. Pace, and S. Altman. 1983. The RNA moiety of ribonuclease P is the catalytic subunit of the enzyme. *Cell 35*:849–857.

Haeckel, E. 1866. *Generalla Morphologie der Organismen*. 2 vols. Berlin: Reimer.

Haeckel, E. 1878. *Das Protistenreich*. Leipzig: Gunther.

Haldane, J. B. S. 1929. *The Origin of Life*. Reprinted in Bernal, J. D. 1967. *The Origin of Life*. Cleveland: World; London: Weidenfeld and Nicolson, pp. 242–249.

Hale, M. E., Jr. 1967. *The Biology of Lichens*. London: Edward Arnold.

Hall, J. L., Z. Ramanis, and D. J. L. Luck. 1989. Basal body/centriolar DNA: molecular genetic studies in *Chlamydomonas*. *Cell 59*:121–132.

Hall, W. T., and G. Claus. 1967. Ultrastructural studies on the cyanelles of *Glaucocystis nostochinearum*. *J. Phycol. 3*:37–51.

Hallbauer, D. K., and K. T. Van Warmelo. 1974. Fossilized plants in thucholite from Precambrian rocks of the Witwatersrand, So. Africa. *Precambrian Res. 1*:199–212.

Hanson, E. D. 1977. *The Origin and Early Evolution of Animals*. Middletown, Conn.: Wesleyan University Press.

Harborne, J. B. 1988. *Introduction to Ecological Biochemistry*. 3rd ed. London: Academic Press.

Harold, F. M. 1986. *The Vital Force: A Study of Bioenergetics*. New York: W. H. Freeman.

Hartman, H. 1975. The centriole and the cell. *J. Theor. Biol. 51*:501–509.

Hartman, H., J. G. Lawless, and P. Morrison, eds. 1987. *Search for the Universal Ancestors: The Origins of Life*. Palo Alto, Calif.: Blackwell.

Harvey, E. N. 1952. *Bioluminescence*. New York: Academic Press.

Haugaard, N. 1968. Cellular mechanisms of oxygen toxicity. *Physiol. Rev. 48*:311–373.

Hayes, J. M. 1967. Organic constituents of meteorites: a review. *Geochim. Cosmochim. Acta 31*:1395–1440.

Heath, I. B. 1980. Variant mitosis in lower eukaryotes: indicators of the evolution of mitosis? *Int. Rev. Cytol. 64*:1–80.

Heath, I. B. 1981. Mechanisms of nuclear division in fungi. In *The Fungal Nucleus:*

Symposium of the British Mycological Society. K. Gull and S. Oliver, eds. Cambridge, Engl.: Cambridge University Press, pp. 85–112.

Heidemann, S. R., G. Sander, and M. W. Kirschner. 1977. Evidence for a functional role of RNA in centrioles. *Cell 10*:337–350.

Heilbrunn, L. V. 1956. *The Dynamics of Living Protoplasm.* New York: Academic Press.

Hemmingsen, S. M., C. Woolford, S. M. Van Der Vies, K. Tilly, D. T. Dennis, C. P. Georgopoulos, R. W. Hendrix, and R. J. Ellis. 1988. Homologous plant and bacterial proteins chaperone oligomeric protein assembly. *Nature 333*:330–334.

Henneguy, L. F. 1923. *La Vie Cellulaire.* Paris: Collection Payot.

Hepler, P. K., and W. T. Jackson. 1968. Microtubules and early stages of cell-plate formation in the endosperm of *Haemanthus katherinae* Baker. *J. Cell Biol. 38*:437–446.

Herdman, M., and R. Y. Stanier. 1977. The cyanelle: chloroplast or endosymbiotic prokaryote? *Fed. Eur. Microbiol. Soc. 1*:7–12.

Herring, P. J., ed. 1978. *Bioluminescence in Action.* New York: Academic Press.

Hess, R. T., and D. B. Menzel. 1968. Rat kidney centrioles: vitamin E intake and oxygen exposure. *Science 159*:985–987.

Hinkle, G. 1991. Status of the theory of the symbiotic origin of undulipodia (cilia). In *Symbiosis As a Source of Evolutionary Innovation: Speciation and Morphogenesis.* L. Margulis and R. Fester, eds. Cambridge, Mass.: MIT Press, pp. 135–142.

Hinkle, G. 1992. Symbiosis and organelles: undulipodia and the origins of eukaryotes. Ph.D. thesis. Boston University, Boston.

Hinkle, G., and L. Margulis. 1990. Global ecology and the Gaia hypothesis. *Physiol. Ecol. Japan 27*:53–62.

Hodgeson, G. W., and C. Ponnamperuma. 1968. Prebiotic porphyrin genesis: porphyrins from electric discharge in methane, ammonia, and water vapor. *Proc. Natl. Acad. Sci. U. S. A. 59*:22–28.

Hoering, T. 1967. The organic geochemistry of Precambrian rocks. In *Researches in Geochemistry*, vol. 2. P. Abelson, ed. New York: Wiley, pp. 87–111.

Hofmann, H. J. 1993. Proterozoic carbonaceous compressions ("metaphytes" and "worms"). In *Early Life on Earth: Nobel Symposium No. 84.* S. Bengtson, ed. New York: Columbia University Press, pp. 271–286 (in press).

Hogg, J. 1861. On the distribution of a plant and an animal and on a fourth kingdom of nature. *Edinb. New Philos. J. (New Series) 12*:216–225.

Holland, H. D., and M. Schidlowski, eds. 1982. *Mineral Deposits and the Evolution of the Biosphere*: Report of the Dahlem Workshop on Biospheric Evolution and Precambrian Metallogeny, Berlin 1980, September 1–5. Berlin: Springer-Verlag.

Hollande, A., and J. Carruette-Valentin. 1970. Interpretation générale des structures rostrales des Hypermastigines et modalités de la pleuromitose chez les flagelles du genre *Trichonympha. C. R. Seances Acad. Sci., Ser. D 270*:1476–1490.

Hollande, A., and J. Carruette-Valentin. 1971. Les atractophores, l'induction du fuseau et la division cellulaire chez les Hypermastigines. Étude infrastructurale et révision systématique des Trichonymphines et des Spirotrichonymphines. *Protistologica 7*:5–100.

Hollande, A., and I. Gharagozlou. 1967. Morphologie infrastructurale de *Pillotina calotermitidis* nov. gen., nov. sp., spirochaetale de l'intestin de *Calotermes praecox. C. R. Seances Acad. Sci., Ser. D 265*:1309–1312.

Holt, J. G., ed. 1984–1989. *Bergey's Manual of Systematic Bacteriology.* 4 vols. Baltimore: Williams & Wilkins.

Holt, J. G., and R. A. Lewin. 1968. *Herpetosiphon aurantiacus* gen. et sp. n. A new filamentous gliding organism. *J. Bacteriol.* 95:2407–2408.

Holton, R. W., H. H. Blecker, and T. S. Stevens. 1968. Fatty acids in blue-green algae: possible relation to phylogenic position. *Science 160*:545–547.

Honegger, R. 1991. Fungal evolution: symbiosis and morphogenesis. In *Symbiosis As a Source of Evolutionary Innovation: Speciation and Morphogenesis.* L. Margulis and R. Fester, eds. Cambridge, Mass.: MIT Press, pp. 319–340.

Hoober, J. K. 1984. *Chloroplasts.* New York: Plenum.

Hooker, R. 1597. *The Laws of Ecclesiastical Polity.* Cited in Mencken, H. L. 1942. *A New Dictionary of Quotations on Historical Principles from Ancient and Modern Sources.* New York: Knopf, p. 830.

Hoops, H. J., and G. B. Witman. 1983. Outer doublet heterogeneity reveals structural polarity related to beat direction in *Chlamydomonas* flagella. *J. Cell Biol.* 97:902–908.

Horodyski, R. J. 1977. *Lyngbya* mats at Laguna Mormona, Baja California, Mexico: comparison with Proterozoic stromatolites. *J. Sediment. Petrol.* 47:1305–1320.

Horodyski, R. J., and S. P. Vonder Haar. 1975. Recent calcareous stromatolites from Laguna Mormona (Baja California, Mexico). *J. Sediment. Petrol.* 45:894–906.

Horowitz, N. H. 1986. *To Utopia and Back: The Search for Life in the Solar System.* New York: W. H. Freeman.

Hovind-Hougen, K. 1976. Determination by means of electron microscopy of morphological criteria of value for classification of some spirochetes, in particular treponemes. *Acta Pathol. Microbiol. Scand., Sect. B,* Suppl. 255.

Hsu, K. J. 1992. Is Gaia endothermic? *Geol. Mag.* 129:129–141.

Hutchinson, G. E. 1957. *Treatise on Limnology,* vol. 2. New York: Wiley.

Hutchinson, G. E. 1959. Homage to Santa Rosalia or why are there so many kinds of animals? *Am. Nat.* 93:145–159.

Hutchinson, G. E. 1965. *The Ecological Theatre and the Evolutionary Play.* New Haven: Yale University Press.

Huxley, J. 1912. *The Individual in the Animal Kingdom.* Cambridge, Engl.: Cambridge University Press.

Imshenetsky, A. A., S. V. Lysenko, and G. A. Kazakov. 1978. Upper boundary of the biosphere. *Appl. Environ. Microbiol.* 35:1–5.

Inoué, S., and R. E. Stephens, eds. 1975. *Molecules and Cell Movement.* New York: Raven.

Irvine, W. M., M. Ohishi, and N. Kaifu. 1991. Chemical abundances in cold, dark interstellar clouds. *Icarus 91*:2–6.

Jacob, F., and E. L. Wollman. 1961. *Sexuality and Genetics of the Bacteria.* New York: Academic Press.

James, S. W., L. P. W. Ranom, C. D. Silflow, and P. A. LeFebvre. 1988. Mutants resistant to antimicrotubule herbicides map to a locus on the unilinkage group in *Chlamydomonas reinhardtii. Genetics 118*:141–147.

Jensen, T. E. 1989. Thylakoids in aged cyanobacterial cells suggest origin of eukaryotic nuclear membrane. *Cytobios 60*:47–61.

Jeon, K. W. 1991. *Amoeba* and x-bacteria: symbiont acquisition and possible species change. In *Symbiosis As a Source of Evolutionary Innovation: Speciation and*

Morphogenesis. L. Margulis and R. Fester, eds. Cambridge, Mass.: MIT Press, pp. 118–131.

Jeon, K. W., and T. I. Ahn. 1978. Temperature sensitivity: a cell character determined by obligate endosymbionts in amoebas. *Science 202*:635–637.

Jinks, J. L. 1964. *Extrachromosomal Inheritance.* Englewood Cliffs, N. J.: Prentice-Hall.

John, P., and F. R. Whatley. 1977a. The bioenergetics of *Paracoccus denitrificans. Biochim. Biophys. Acta 463*:129–153.

John, P., and F. R. Whatley. 1977b. *Paracoccus denitrificans* Davis (*Micrococcus denitrificans* Beijerinck) as a mitochondrion. *Adv. Bot. Res. 4*:51–115.

Johnson, D. E., and S. K. Dutcher. 1991. Molecular studies of linkage group XIX of *Chlamydomonas reinhardtii*: evidence against a basal body location. *J. Cell Biol. 113*:339–346.

Johnson, K. A., and J. L. Rosenbaum. 1990. The basal bodies of *Chlamydomonas reinhardtii* do not contain immunologically detectable DNA. *Cell 62*:615–619.

Johnson, K. A., and J. L. Rosenbaum. 1991. Basal bodies and DNA. *Trends in Cell Biol. 1*:145–149.

Joklik, W. K. 1974. Evolution in viruses. In *Evolution in the Microbial World:* Twenty-Fourth Symposium of the Society for General Microbiology. M. J. Carlile and J. J. Skehel, eds. Cambridge, Engl.: Cambridge University Press, 293–320.

Jolley, E., and D. C. Smith. 1978. The green hydra symbiosis. I. Isolation, culture and characteristics of the *Chlorella* symbiont of 'European' *Hydra viridis. New Phytol. 81*:637–645.

Jordan, P. M. 1990. Biosynthesis of δ-aminolevulinic acid and its transformation into coproporphyrinogen in animals and bacteria. In *Biosynthesis of Heme and Chlorophylls.* H. A. Dailey, ed. New York: McGraw-Hill, pp. 55–121.

Juchault, P., C. Louis, G. Martin, and G. Noulin. 1992. Masculinization of female isopods (Crustacea) correlated with non-Mendelian inheritance of cytoplasmic viruses. *Proc. Natl. Acad. Sci. U. S. A.* (in press).

Karakashian, M. W. 1975. Symbiosis in *Paramecium bursaria. Symp. Soc. Exp. Biol. 29*: 145–173.

Karakashian, S. J. 1963. Growth of *Paramecium bursaria* as influenced by the presence of algal symbionts. *Physiol. Zool. 36*:52–68.

Karakashian, S. J., M. W. Karakashian, and M. A. Rudzinska. 1968. Electron microscopic observations on the symbiosis of *Paramecium bursaria* and its intracellular algae. *J. Protozool. 15*:113–128.

Karakashian, S. J., and R. W. Siegel. 1965. A genetic approach to endocellular symbiosis. *Exp. Parasitol. 17*:103–122.

Kates, M., B. Palameta, C. N. Joo, D. J. Kushner, and N. E. Gibbons. 1966. Aliphatic diether analogs of glyceride-derived lipids. IV. The occurrence of D_1-O-dihydrophytylglycerol ether–containing lipids in extremely halophilic bacteria. *Biochemistry 5*:4092–4099.

Kelley, W. S., and M. Schaechter. 1968. The "life cycle" of bacterial ribosomes. In *Advances in Microbial Physiology*, vol. 2. A. H. Rose and J. F. Wilkinson, eds. New York: Academic Press, pp. 89–142.

Kendrick, B. 1991. Fungal symbioses and evolutionary innovations. In *Symbiosis as a Source of Evolutionary Innovation: Speciation and Morphogenesis.* L. Margulis and R. Fester, eds. Cambridge, Mass.: MIT Press, pp. 249–261.

Keosian, J. 1968. *The Origin of Life.* 2nd ed. New York: Reinhold.

Khakhina, L. N. 1979. *Concepts of Symbiogenesis*. Leningrad: Akademii Nauk. (In Russian; English translation of L. N. Khakhina in L. Margulis and M. McMenamin, eds. 1992. *Concepts of Symbiogenesis: History of Symbiosis as an Evolutionary Mechanism*. New Haven: Yale University Press.)

Kirby, H., Jr. 1936. Two polymastigote flagellates of the general *Pseudodevescovina* and *Caduceia. Q. J. Microsc. Sci. 19*:309–335.

Kirby, H., Jr. 1941. Organisms living on and in protozoa. In *Protozoa in Biological Research*. G. N. Calkins and F. M. Summers, eds. New York: Columbia University Press, pp. 1009–1113.

Klein, R., and A. Cronquist. 1967. A consideration of the evolutionary and taxonomic significance of some biochemical, micromorphological and physiological characters in the thallophytes. *Q. Rev. Biol. 42*:105–296.

Knoll, A. H. 1979. Archean photoautotrophy: some alternatives and limits. *Origins Life 9*:313–327.

Knoll, A. H. 1990. Precambrian evolution of prokaryotes and protists. In *Palaeobiology: A Synthesis*. D. E. G. Briggs and P. R. Crowther, eds. Oxford: Blackwell, pp. 9–16.

Knoll, A. H., and E. S. Barghoorn. 1977. Archean microfossils showing cell division from the Swaziland System of South Africa. *Science 198*:396–398.

Kohlmeyer, J. 1975. New clues to the possible origin of ascomycetes. *BioScience 25*:86–93.

Kornberg, A., and T. Baker. 1992. *DNA Replication*. 2nd ed. New York: W. H. Freeman.

Kozo-Polyansky, B. M. 1924. *New Principles of Biology*. Moscow: Puchina Publishing House. (In Russian.)

Kretsinger, R. H. 1976. Calcium-binding proteins. *Annu. Rev. Biochem. 45*:239–266.

Kretsinger, R. H. 1990. Why cells must export calcium. In *Intracellular Calcium Regulation*. F. Bronner, ed. New York: Liss, pp. 439–457.

Kretsinger, R. H., D. Tolbert, S. Nakayama, and W. Pearson. 1991. The EF-hand, homologs and analogs. In *Novel Calcium Binding Proteins*. C. Heizmann, ed. New York: Springer-Verlag, pp. 17–38.

Krinsky, N. I. 1966. The role of carotenoid pigments as protective agents against photosensitized oxidations in chloroplasts. In *Biochemistry of Chloroplasts*, vol. 1. T. W. Goodwin, ed. London: Academic Press, pp. 423–430.

Krinsky, N. I., A. Gordon, and A. Stern. 1964. The appearance of neoxanthin during the regreening of dark grown *Euglena. Plant Physiol. 39*:441–445.

Kroon, A. M., P. Terpstra, M. Holtrop, H. de Vries, C. van den Bogert, J. je Jonge, and E. Agsteribbe. 1976. The mitochondrial RNAs of *Neurospora crassa:* Their function in translation and their relation to the mitochondrial genome. In *Genetics and Biogenesis of Chloroplasts and Mitochondria*. T. Bücher, W. Neupert, W. Sebald, and S. Werner, eds. Amsterdam: North-Holland, pp. 685–699.

Krumbein, W. E., Y. Cohen, and M. Shilo. 1977. Solar Lake (Sinai). 4. Stromatolitic cyanobacterial mats. *Limnol. Oceanogr. 22*:635–656.

Kubai, D. F. 1975. The evolution of the mitotic spindle. *Int. Rev. Cytol. 43*:167–227.

Kuroiwa, T., T. Yorihuzi, N. Yabe, T. Ohta, and H. Uchida. 1990. Absence of DNA in the basal body of *Chlamydomonas reinhardtii* by fluorimetry using a video-intensified photon-counting system. *Protoplasma 158*:155–164.

Kusel, J. P., K. E. Moore, and M. M. Weber. 1967. The ultrastructure of *Crithidia fasciculata* and morphological changes induced by growth in acriflavin. *J. Protozool.* *14*:283–296.

Kuznicki, L., T. L. Jahn, and J. R. Fonseca. 1970. Helical nature of the ciliary beat of *Paramecium multimicronucleatum. J. Protozool. 17*:16–24.

La Berge, G. L. 1967. Microfossils and Precambrian iron formations. *Geol. Soc. Am. Bull. 78*:331–342.

Lange, R. T. 1966. Bacterial symbiosis with plants. In *Symbiosis. Volume I: Associations of Microorganisms, Plants, and Marine Organisms.* S. M. Henry, ed. New York: Academic Press, pp. 99–170.

Lanham, U. N. 1968. The Blochmann bodies: hereditary intracellular symbionts of insects. *Biol. Rev. Camb. Philos. Soc. 43*:269–286.

Lapo, A. 1987. *Traces of Bygone Biospheres.* Oracle, Arizona: Synergetic Press.

Lanyi, J. K. 1980. Physical chemistry and evolution of salt tolerance in halobacteria. In *Limits of Life:* Proceedings of the Fourth College Park Colloquium on Chemical Evolution. C. Ponnamperuma and L. Margulis, eds. Dordrecht: Reidel, pp. 61–67.

Lapworth, C. 1879. On the tripartite classification of the Lower Paleozoic rocks. *Geol. Mag. 6*:1–15.

Lauterborn, R. 1896. *Untersuchungen über Bau, Kernteilung, und Bewegung der Diatomeen.* Leipzig: W. Engleman.

Lavette, A. 1967. Recherches sur les constituants cytoplasmiques des flagellés termiticoles et sur la digestion du bois. *Ann. Sci. Nat. Zool. Biol. Anim., 12th Ser. 9*:457–528.

Lazcano, A. 1992. Origins of life: historical development of recent theories. In *Environmental Evolution: The Effects of the Origin and Evolution of Life on Planet Earth.* L. Margulis and L. Olendzenski, eds. Cambridge, Mass.: MIT Press, pp. 57–69.

Lazcano, A. 1993. The transition from non-living to living. In *Early Life on Earth: Nobel Symposium* 84. S. Bengtson, ed. New York: Columbia University Press (in press).

Lazcano, A., R. Guerrero, L. Margulis, and J. Oró. 1988. The evolutionary transition from RNA to DNA in early cells. *J. Mol. Evol. 27*:283–290.

Lederberg, J. 1952. Cell genetics and hereditary symbiosis. *Physiol. Rev. 32*:403–430.

Lee, J. J. 1990. Phylum Granuloreticulosa (Foraminifera). In *Handbook of Protoctista.* L. Margulis, J. O. Corliss, M. Melkonian, D. J. Chapman, eds. Boston: Jones and Bartlett, pp. 524–548.

Leedale, G. F. 1974. How many are the kingdoms of organisms? *Taxon 23*:261–270.

Leidy, J. 1881. The parasites of the termites. *J. Acad. Nat. Sci. (New Ser.) 8*:425–447.

Levine, J. S. 1989. Photochemistry of biogenic gases. In *Global Ecology: Towards a Science of the Biosphere.* M. B. Rambler, L. Margulis, and R. Fester, eds. Boston: Academic Press, pp. 51–74.

Lewin, R. 1982. RNA can be a catalyst. *Science. 218*:872–874.

Lewin, R. A. 1965a. Freshwater species of *Saprospira. Can. J. Microbiol. 11*:135–139.

Lewin, R. A. 1965b. Isolation and some physiological features of *Saprospira thermalis. Can. J. Microbiol. 11*:77–86.

Lewin, R. A. 1976. Prochlorophyta as a proposed new division of algae. *Nature 261*:697–698.

Lewin, R. A. 1977. *Prochloron*, type genus of the Prochlorophyta. *Phycologia 16*:217.

Lewin, R. A. and L. Cheng. 1989. *Prochloron: A microbial enigma.* New York: Chapman and Hall.

Lewis, D. H. 1973a. Concepts in fungal nutrition and the origin of biotrophy. *Biol. Rev. Camb. Philos. Soc. 48*:261–278.

Lewis, D. H. 1973b. The relevance of symbiosis to taxonomy and ecology with particular reference to mutualistic symbiosis and the exploitation of marginal habitats. In *Taxonomy and Ecology*. V. H. Heywood, ed. London: Academic Press, pp. 151–172.

Licari, G. R., and P. E. Cloud, Jr. 1968. Reproductive structures and taxonomic affinities of some nannofossils from the Gunflint Iron Formation. *Proc. Natl. Acad. Sci. U. S. A. 59*:1053–1060.

Lighthill, M. J. 1976. Flagellar hydrodynamics. *SIAM Rev. 18*:161–230.

Little, M., and T. Seehaus. 1988. Comparative analysis of tubulin sequences. *Comp. Biochem. Physiol. 90B*:655–670.

Linck, R. W., and R. E. Stephens. 1987. Biochemical characterization of tektins from sperm flagellar doublet microtubules. *J. Cell Biol. 104*:1069–1075.

Lindholm, T. 1985. *Mesodinium rubrum*—a unique photosynthetic ciliate. *Adv. Aquat. Microbiol. 3*:1–48.

Lipps, J. H. 1970. Plankton evolution. *Evolution 24*:1–22.

Loomis, W. F. 1988. *Four Billion Years: An Essay on the Evolution of Genes and Organisms*. Sunderland, Mass.: Sinauer Associates.

Lovelock, J. E. 1988. *The Ages of Gaia: A Biography of Our Living Earth*. New York: W. W. Norton.

Lovelock, J. E. 1991. *Healing Gaia: Practical Medicine for the Planet*. New York: Harmony Books.

Lovelock, J. E., and L. Margulis. 1974. Homeostatic tendencies of the Earth's atmosphere. *Origins Life 5*:93–103.

Lovley, D. R. 1991. Dissimilatory Fe (III) and Mn (IV) reduction. *Microbiol. Rev. 55*:259–287.

Lowenstam, H. A. 1963. Biologic problems relating to the composition and diagenesis of sediments. In *The Earth Sciences: Problems and Progress in Current Research*. T. W. Donnelly, ed. Chicago: University of Chicago Press, pp. 137–195.

Lowenstam, H. A. 1980. Bioinorganic constituents of hard parts. In *Biogeochemistry of Amino Acids*. P. E. Hare, T. C. Hoering, and K. King, Jr., eds. New York: Wiley, pp. 1–16.

Lowenstam, H. A., and L. Margulis. 1980a. Calcium regulation and the appearance of calcareous skeletons in the fossil record. In *The Mechanisms of Biomineralization in Animals and Plants*: Proceedings of the Third International Biomineralization Symposium. M. Omori and N. Watabe, eds. Tokyo: Tokai University Press, pp. 289–300.

Lowenstam, H. A., and L. Margulis. 1980b. Evolutionary prerequisites for early Phanerozoic calcareous skeletons. *BioSystems 12*:27–41.

Lowenstam, H. A., and S. Wiener. 1989. *On Biomineralization*. New York: Wiley.

Lucretius, T. 1st century B.C. In *De Rerum Natura*. W. E. Leonard and S. B. Smith, eds. Madison: University of Wisconsin Press, 1965.

Lumière, A. 1919. *Le Myth des Symbiotes*. Paris: Masson.

Lund, J. W. G. 1964. Classical and modern criteria used in algal taxonomy with special reference to genera of microbial size. In *Microbial Classification*: Twelfth Symposium of the Society for General Microbiology. G. C. Ainsworth and P. H. A. Sneath, eds. Cambridge, Engl.: Cambridge University Press, pp. 68–110.

Luykx, P. 1970. *Cellular Mechanisms of Chromosome Distribution* (*Int. Rev. Cytol.* Suppl. 2) G. H. Bourne and J. F. Danielli, eds. New York: Academic Press.

Lwoff, A. 1950. *Problems of Morphogenesis in Ciliates: The Kinetosomes in Development, Reproduction, and Evolution.* New York: Wiley.

Lynn, D. H., and E. B. Small. 1990. Phylum Ciliophora. In *Handbook of Protoctista.* L. Margulis, J. O. Corliss, M. Melkonian, and D. J. Chapman, eds. Boston: Jones and Bartlett, pp. 498–523.

MacDonald, K. 1972. The ultrastructure of mitosis in the marine red alga *Membranoptera platyphylla. J. Phycol. 8*:156–166.

Machin, K. E. 1963. The control and synchronization of flagellar movement. *Proc. R. Soc. Lond., Ser. B. Biol. Sci. 158*:88–104.

Mackenzie, F. T. 1990. Sea level change, sediment mass and flux and chemostratigraphy. In *Cretaceous Resources, Events and Rhythms.* R. N. Ginsburg and B. Beaudoin, eds. Dordrecht: Kluwer, pp. 289–304.

Mackenzie, F. T., and C. R. Agegian. 1989. Biomineralization and tentative links to plate tectonics. In *Origin, Evolution, and Modern Aspects of Biomineralization in Plants and Animals.* R. E. Crick, ed. New York: Plenum, pp. 11–27.

Malawista, S. E., H. Sato, and K. G. Bensch. 1968. Vinblastine and griseofulvin reversibly disrupt the living mitotic spindle. *Science 160*:770–771.

Margulis, L. 1970. *Origin of Eukaryotic Cells.* New Haven: Yale University Press.

Margulis, L. 1976. A review: the genetic and evolutionary consequences of symbiosis. *Exp. Parasitol. 39*:277–349.

Margulis, L. 1980. Undulipodia, flagella and cilia. *BioSystems 12*:105–108.

Margulis, L. 1988. Systematics: the view from the origin and early evolution of life. Secession of the protoctista from the animal and plant kingdoms. In *Prospects in Systematics.* D. Hawksworth and R. G. Davies, eds. Oxford: Clarendon, pp. 430–443.

Margulis, L. 1990. Words as battlecries—Symbiogenesis and the new field of endocytobiology. *Bioscience 40*:673–677.

Margulis, L. 1991. Symbiogenesis and symbionticism. In *Symbiosis as a Source of Evolutionary Innovation: Speciation and Morphogenesis.* L. Margulis and R. Fester, eds. Cambridge, Mass.: MIT Press.

Margulis, L. 1992. Biodiversity: Molecular biological domains, symbiosis, and kingdom origins. *BioSystems* (in press).

Margulis, L., J. Ashen, M. Solé, and R. Guerrero. 1993. Large multicellular organization and desiccation resistance in microbial mat spirochetes. *Proc. Natl. Acad. Sci.* (submitted).

Margulis, L., D. Chase, and L. To. 1979. Possible evolutionary significance of spirochaetes. *Proc. R. Soc. Lond., Ser. B. Biol. Sci. 204*:189–198.

Margulis, L., J. O. Corliss, M. Melkonian, and D. J. Chapman, eds. 1990. *Handbook of Protoctista: The Structure, Cultivation, Habitats and Life Histories of the Eukaryotic Microorganisms and Their Descendants Exclusive of Animals, Plants and Fungi.* Boston: Jones and Bartlett.

Margulis, L., M. Enzien, and H. I. McKhann. 1990. Revival of Dobell's "chromidia" hypothesis: chromatin bodies in the amoebomastigote *Paratetramitus jugosus. Biol. Bull. 178*:300–304.

Margulis, L., and R. Fester, eds. 1991. *Symbiosis as a Source of Evolutionary Innovation: Speciation and Morphogenesis.* Cambridge, Mass.: MIT Press.

Margulis, L., and G. Hinkle. 1992. Large symbiotic spirochetes: *Clevelandina, Cristispira, Diplocalyx, Hollandina,* and *Pillotina.* In *The Prokaryotes. A Handbook on the Biology of Bacteria: Ecophysiology, Isolation, Identification, Applications,* vol. 4. 2nd ed. A. Balows, H. G. Trüper, M. Dworkin, W. Harder, and K.-H. Schleifer, eds. New York: Springer-Verlag, pp. 3965–3978.

Margulis, L., G. Hinkle, J. Stolz, F. Craft, I. Esteve, and R. Guerrero. 1990. *Mobilifilum chasei*: Morphology and ecology of a spirochete from an intertidal, stratified microbial mat community. *Arch. Microbiol. 153*:422–427.

Margulis, L., H. I. McKhann, and L. Olendzenski. 1992. *Illustrated Glossary of Protoctista*. Boston: Jones and Bartlett.

Margulis, L., and M. McMenamin. 1990. Kinetosome-centriolar DNA significance for endosymbiosis theory. *Treb. Soc. Cat. Biol. 41*:5–16.

Margulis, L., and M. McMenamin, eds. 1992. *Concepts of Symbiogenesis: History of Symbiosis as an Evolutionary Mechanism*. New Haven: Yale University Press.

Margulis, L., L. Nault, and J. M. Sieburth. 1991. *Cristispira* from oyster styles: complex morphology of large symbiotic spirochetes. *Symbiosis 11*:1–17.

Margulis, L., J. A. Neviackas, and S. Banerjee. 1969. Cilia regeneration in *Stentor*: inhibition, delay and abnormalities induced by griseofulvin. *J. Protozool. 16*:660–667.

Margulis, L., L. Olendzenski, and B. Afzelius. 1990. Endospore-forming filamentous bacteria symbiotic in termites: ultrastructure and growth in culture of *Arthromitus*. *Symbiosis 8*:95–116.

Margulis, L., and D. Sagan. 1986. *Origins of Sex: Three Billion Years of Genetic Recombination*. New Haven: Yale University Press.

Margulis, L., and D. Sagan. 1991. *Microcosmos: Four Billion Years of Evolution from Our Microbial Ancestors*. New York: Simon & Schuster.

Margulis, L., and K. V. Schwartz. 1988. *Five Kingdoms—An Illustrated Guide to the Phyla of Life on Earth*. 2nd ed. New York: W. H. Freeman.

Margulis, L., G. Thorington, B. Berger, and J. Stolz. 1978. Endosymbiotic bacteria associated with the intracellular green algae of *H. viridis*. *Curr. Microbiol. 1*:227–232.

Margulis, L., L. To, and D. Chase. 1978. Microtubules in prokaryotes. *Science 200*:1118–1124.

Margulis, L., L. P. To, and D. Chase. 1981. Microtubules, undulipodia and *Pillotina* spirochetes. *Ann. N. Y. Acad. Sci. 361*:356–368.

Margulis, L., J. C. G. Walker, and M. Rambler. 1976. Reassessment of roles of oxygen and ultraviolet light in Precambrian evolution. *Nature 264*:620–624.

Marler, J. E., and C. van Baalen. 1965. Role of H_2O_2 in single cell growth of the blue-green algae, *Anacystis nidulans*. *J. Phycol. 1*:180–185.

Mathews, S. C., and V. V. Missarzhevsky. 1975. Small shelly fossils of late Precambrian and early Cambrian age: a review of recent work. *J. Geol. Soc. Lond. 131*: 289–304.

Matthews, C. 1992. Origins of life: polymers before monomers? In *Environmental Evolution: The Effects of the Origin and Evolution of Life on Planet Earth*. L. Margulis and L. Olendzenski, eds. Cambridge, Mass.: MIT Press, pp. 29–38.

Maul, G. G. 1977. The nuclear and the cytoplasmic pore complex: structure, dynamics, distribution, and evolution. In *Studies in Ultrastructure*. (*Int. Rev. Cytol.* Suppl. 6) G. H. Bourne, J. F. Danielli, and K. W. Jeon, eds. New York: Academic Press, pp. 75–186.

May, R. M. 1993. Past efforts and future prospects towards understanding how many species there are. In *Biological Diversity and Global Change*: IUBS Twenty-fourth General Assembly, Sept. 1991 (in press).

Mayr, E. 1976. *Evolution and the Diversity of Life: Selected Essays*. Cambridge, Engl.: Belknap.

Mazia, D. 1967. Fibrillar structure in the mitotic apparatus. In *Formation and Fate of Cell Organelles*. K. B. Warren, ed. New York: Academic Press, pp. 39–54.

Mazia, D. 1975. Microtubule research in perspective. *Ann. N. Y. Acad. Sci. 253*:7–13.

Mazur, P., E. S. Barghoorn, H. O. Halvorson, T. H. Jukes, I. R. Kaplan, and L. Margulis. 1978. Biological implications of the Viking mission to Mars. *Space Sci. Rev. 22*:3–34.

McElroy, W. D., and H. H. Seliger. 1962. Origin and evolution of bioluminescence. In *Horizons in Biochemistry*. M. Kasha and B. Pullman, eds. New York: Academic Press, pp. 91–101.

McFall-Ngai, M. J. 1991. Luminous bacterial symbiosis in fish evolution: adaptive radiation among the leiognathid fishes. In *Symbiosis As a Source of Evolutionary Innovation: Speciation and Morphogenesis*. L. Margulis and R. Fester, eds. Cambridge, Mass.: MIT Press, pp. 381–409.

McFall-Ngai, M. J., and E. G. Ruby. 1991. Symbiont recognition and subsequent morphogenesis as early events in an animal-bacterial mutualism. *Science 254*: 1491–1494.

McIntosh, R. J., P. K. Hepler, and D. G. Van Wie. 1969. Model for mitosis. *Nature 224*:659–663.

McKay, C. P., and C. R. Stoker. 1992. Gaia and life on Mars. In *Scientists on Gaia*. S. Schneider and P. Boston, eds. Cambridge, Mass: MIT Press, pp. 375–381.

McLaughlin, J. J. A., and P. A. Zahl. 1966. Endozoic algae. In *Symbiosis. Volume I: Associations of Microorganisms, Plants, and Marine Organisms*. S. M. Henry, ed. New York: Academic Press, pp. 257–297.

McMenamin, M. A. S., and D. L. S. McMenamin. 1990. *The Emergence of Animals: The Cambrian Breakthrough*. New York: Columbia University Press.

Mehos, D. 1992. Ivan E. Wallin's theory of symbionticism. In *Concepts of Symbiogenesis: History of Symbiosis as an Evolutionary Mechanism*. L. N. Khakhina, L. Margulis, and M. McMenamin, eds. New Haven: Yale University Press, Appendix pp. 149–163.

Mereschkovsky, K. S. 1909. *The Theory of Two Plasms as the Foundation of Symbiogenesis: New Knowledge Concerning the Origins of Organisms*. Kazan: Publishing Office of the Imperial University. [In Russian; cited in Khakhina (1979).]

Mereschkovsky, K. S. 1910. *Outline Course on General Botany*. Memoirs of the University of Kazan, vol. 77. [In Russian; cited in Khakhina (1979).]

Mereschkovsky, K. S. 1920. La plante considerée comme un complexe symbiotique. *Societé des Sciences Naturelles de l'Ouest de la France, Nantes, Bulletin 6*:17–98. [Cited in Khakhina (1979).]

Milhaud, M., and G. D. Pappas. 1968. Cilia formation in the adult cat brain after pargyline treatment. *J. Cell Biol. 37*:599–602.

Miller, M. W. 1961. *The Pfizer Handbook of Microbial Metabolites*. New York: McGraw-Hill.

Miller, S. L. 1953. A production of amino acids under possible primitive Earth conditions. *Science 117*:528–529.

Miller, S. L., and L. E. Orgel. 1974. *The Origins of Life on Earth*. Englewood Cliffs, N. J.: Prentice-Hall.

Misra, S. B. 1969. Late Precambrian (?) fossils from southeastern Newfoundland. *Geol. Soc. Am. Bull. 80*:2133–2140.

Mitchell, P. 1966. Chemiosmotic coupling in oxidative and photosynthetic phosphorylation. *Biol. Rev. Camb. Philos. Soc. 41*:445–502.

Mizukami, I., and J. Gall. 1966. Centriole replication. II. Sperm formation in the fern *Marsilea* and the cycad *Zamia. J. Cell Biol. 29*:97–111.

Moncrief, N. D., M. Goodman, and R. H. Kretsinger. 1990. Evolution of EF-hand calcium-modulated proteins. I. Relationships based on amino acid sequences. *J. Mol. Evol. 30*:522–562.

Monty, C. 1973a. Les nodules de manganese sont des stromatolithes oceaniques. *C. R. Seances Acad. Sci., Ser. D 276*:3285–3288.

Monty, C. 1973b. Precambrian background and Phanerozoic history of stromatolite communities, an overview. *Ann. Soc. Geol. Belg. 95*:585–624.

Monty, C. 1973c. Remarques sur la nature, la morphologie et la distribution spatiale de stromatolithes. *Sci. Terre 18*:189–212.

Moraes, C. T., S. DiMauro, M. Zeviani, A. Lombes, S. Shanske, A. F. Miranda, H. Nakase, E. Bonilla, L. C. Werneck, S. Servidei, I. Nonaka, Y. Koga, A. J. Spiro, A. K. W. Brownell, B. Schmidt, D. L. Schotland, M. Zupanc, D. C. DeVivo, E. A. Schon, and L. P. Rowland. 1989. Mitochondrial DNA deletions in progressive external ophthalmoplegia and Kearns-Sayre syndrome. *New Engl. J. Med. 320*:1293–1299.

Morowitz, H. J., and D. C. Wallace. 1973. Genome size and life cycle of the mycoplasm. *Ann. N. Y. Acad. Sci. 225*:62–73.

Mossman, D. J., and B. D. Dyer. 1985. The geochemistry of Witwatersrand-type gold deposits and the possible influence of ancient prokaryotic communities on gold dissolution and precipitation. *Precambrian Res. 30*:303–319.

Mounolou, J. C., H. Jakob, and P. P. Slonimski. 1967. Molecular nature of hereditary cytoplasmic factors controlling gene expression in mitochondria. In *The Control of Nuclear Activity*. L. Goldstein, ed. Englewood Cliffs, N. J.: Prentice-Hall, pp. 413–431.

Müller, M. 1988. Energy metabolism of protozoa without mitochondria. *Annu. Rev. Microbiol. 42*:465–488.

Myers, C. R., and K. H. Nealson. 1990. Iron mineralization by bacteria: metabolic coupling of iron reduction to cell metabolism in *Alteromonas putrefaciens* strain MR-1. In *Iron Biominerals*. R. B. Frankel and R. P. Blakemore, eds. New York: Plenum, pp. 131–149.

Nagy, B., and L. A. Nagy. 1976. Interdisciplinary search for early life forms and for the beginnings of life on Earth. *Interdiscip. Sci. Rev. 4*:291–307.

Nagy, B., L. A. Nagy, J. E. Zumberge, D. S. Sklarew, and P. H. Anderson. 1977. Indications of a biological and biochemical evolutionary trend during the Archean and early Proterozoic. *Precambrian Res. 5*:109–120.

Nagy, L. A., and J. E. Zumberge. 1976. Fossil microorganisms from the approximately 2800 to 2500 million year old Bulawayan stomatolite: application of ultramicrochemical analyses. *Proc. Natl. Acad. Sci. U. S. A. 73*:2973–2976.

Nakayama, S., N. D. Moncrief, M. Goodman, and R. H. Kretsinger. 1993. Evolution of EF-hand calcium-modulated proteins. II. Domains of several subfamilies have diverse evolutionary histories. *J. Molec. Biol. 34*:416–448.

Nardon, P., V. Gianinazzi-Pearson, A.M. Grenier, L. Margulis, and D.C. Smith, eds. 1989. *Endocytobiology IV:* Proceedings of the Fourth International Colloquium on Endocytobiology and Symbiosis, July 4–8, 1989, Villeurbanne, France. Paris: Institut National de la Recherche Agronomique (INRA).

Nass, M. M. K. 1969. Mitochondrial DNA: advances, problems and goals. *Science* *165*:25–35.

Nature. 1919. *Les Symbiotes* [anonymous review of Portier, p., 1918. *Les Symbiotes.* Paris: Masson]. *Nature 103*:482–483.

Nealson, K., D. Cohn, G. Leisman, and B. Tebo. 1981. Co-evolution of luminous bacteria and their eukaryotic hosts. *Ann. N. Y. Acad. Sci. 361*:76–91.

Nealson, K. H. 1991. Luminous bacteria symbiotic with entomopathogenic nematodes. In *Symbiosis As a Source of Evolutionary Innovation: Speciation and Morphogenesis.* L. Margulis and R. Fester, eds. Cambridge, Mass.: MIT Press, pp. 205–218.

Nealson, K. H., and J. W. Hastings. 1979. Bacterial bioluminescence: its control and ecological significance. *Microbiol. Rev. 43*:496–518.

Nealson, K. H., and C. R. Myers. 1992. Microbial reduction of manganese and iron: New approaches to carbon cycling. *Appl. Environ. Microbiol. 58*:439–443.

Newsom, H. E., and S. R. Taylor. 1989. Geochemical implications of the formation of the Moon by a single giant impact. *Nature 338*:29–34.

Nicklas, R. B. 1989. The motor for poleward chromosome movement in anaphase is in or near the kinetochore. *J. Cell Biol. 109*:2245–2255.

Nisbet, E. G. 1987. *The Young Earth: An Introduction to Archaean Geology.* Boston: Allen & Unwin.

Noguchi, H. 1921. *Cristispira* in North American shellfish. A note on a spirillum found in oysters. *J. Exp. Med. 34*:295–315.

Oishi, K. 1971. Spirochaete-mediated abnormal sex-ratio (SR) condition in *Drosophila*: a second virus associated with spirochaetes and its use in the study of the SR condition. *Genet. Res. 18*:45–56.

Oparin, A. I. 1924. The origin of life. (In Russian; English translation in Bernal, J. D. 1967. *Origin of Life.* Cleveland: World; London: Weidenfeld and Nicolson, pp. 197–234.)

Oparin, A. I. 1968. *Genesis and Evolutionary Development of Life.* New York: Academic Press.

Orgel, L. 1970. [Oral remarks]. In *Origins of Life: Proceedings of the First Conference.* L. Margulis, ed. New York: Published for Interdisciplinary Communication Associates by Gordon and Breach.

Ormerod, W., S. Francis, and L. Margulis. 1976. Delay in the appearance of clamp connections in *Schizophyllum commune* by inhibitors of microtubule protein assembly. *Microbios 17*:189–205.

Oró, J. 1970. [Oral remarks]. In *Origins of Life: Proceedings of the First Conference.* L. Margulis, ed. New York: Published for Interdisciplinary Communication Associates by Gordon and Breach.

Oró, J., G. Holzer, and A. Lazcano-Araujo. 1980. The contribution of cometary volatiles to the primitive Earth. In *Open Meeting of the Working Group on Space Biology of the 22nd Plenary Meeting of COSPAR, Bangalore, India, May 22–Jun 9, 1979.* (Life Sciences and Space Research, vol. 18). R. Holmquist, ed. Oxford: Pergamon.

Oró, J., S. L. Miller, and A. Lazcano. 1990. The origin and early evolution of life on Earth. *Annu. Rev. Earth Planet. Sci. 18*:317–356.

Ourisson, G., P. Albrecht, and M. Rohmer. 1984. The microbial origin of fossil fuels. *Sci. Am. 251*(2):44–51.

Outka, D. E., and B. C. Kluss. 1967. The amoeba to flagellate transformation in *Tetramitus rostratus. J. Cell Biol. 35*:323–346.

Owen, T., K. Biemann, D. R. Rushneck, J. E. Biller, D. W. Howarth, and A. L.

LaFleur. 1977. The composition of the atmosphere at the surface of Mars. *J. Geophys. Res. 82*:4635–4639.

Page, F. C. 1976. A revised classification of the Gymnamoebae (Protozoa: Sarcodina). *Zool. J. Linn. Soc. 58*:61–77.

Palmer, A. R. 1968. Oral communication reported in Siever, R. 1968. Environment of the primitive Earth. *Science 161*:711–712.

Palmer, J. D. 1985. Evolution of chloroplast and mitochondrial DNA in plants and algae. In *Molecular Evolutionary Genetics*. R. J. MacIntyre, ed. New York: Plenum, pp. 131–240.

Panganiban, A. T., Jr., T. E. Patt, W. Hart, and R. S. Hanson. 1979. Oxidation of methane in the absence of oxygen in lake water samples. *Appl. Envir. Microbiol. 37*:303–309.

Parsons, D. F. 1967. Ultrastructure and molecular aspects of cell membranes. In *Proceedings of the Seventh Canadian Cancer Research Conference, Honey Harbor, Ontario*, vol. 7. J. F. Morgan, ed. New York: Pergamon, pp. 193–246.

Parsons, J. A., and R. C. Rustad. 1968. The distribution of DNA among dividing mitochondria of *Tetrahymena pyriformis*. *J. Cell Biol. 37*:683–693.

Pasteur, L. 1857. Mémoire sur la fermentation appelée lactique. *C. R. Acad. Sci. 45*:913–916. (Facsimile in *The Pasteur Fermentation Centennial, 1857–1957*: A Scientific Symposium on the Occasion of the One Hundredth Anniversary of the Publication of Louis Pasteur's "Mémoire sur la fermentation appelée lactique." New York: Charles Pfizer and Co., 1958.)

Patrusky, B. 1990. Heat shock proteins: a biological imperative. *Mosaic 21*:2–11.

Pettersson, M. 1977. Major integrative levels and the *Fo-so* series. *Aslib Proc. 30*:215–237.

Picken, L. E. R. 1960. *The Organization of Cells and Other Organisms*. Oxford: Clarendon.

Pickett-Heaps, J. 1974. The evolution of mitosis and the eukaryotic condition. *BioSystems 6*:37–48.

Pickett-Heaps, J. 1975. *Green Algae: Structure, Reproduction, and Evolution in Selected Genera*. Sunderland, Mass.: Sinauer Associates.

Pickett-Heaps, J., D. H. Tippett, and J. A. Andreozzi. 1978. Cell division in the pennate diatom *Pinnularia*. II. Later stages of mitosis. *Biol. Cell. 33*:79–84.

Pickett-Heaps, J., D. H. Tippett, and J. A. Andreozzi. 1979. Cell division in the pennate diatom *Pinnularia*. V. Observations on live cells. *Biol. Cell. 35*:295–304.

Pirozynski, K. A. 1991. Galls, flowers, fruits, and fungi. In *Symbiosis as a Source of Evolutionary Innovation: Speciation and Morphogenesis*. L. Margulis and R. Fester, eds. Cambridge, Mass.: MIT Press, pp. 364–379.

Pirozynski, K. A., and D. W. Malloch. 1975. The origin of land plants: a matter of mycotrophism. *BioSystems 6*:153–164.

Ponnamperuma, C. 1972. *The Origins of Life*. New York: Dutton.

Ponnamperuma, C. 1977. Cosmochemistry and the origin of life. *Proc. Robert A. Welch Found. Conf. Chem. Res. 21*:137–197.

Ponnamperuma, C., ed. 1981. *Comets and the Origin of Life*. Dordrecht: Reidel.

Ponnamperuma, C. 1992. Cosmochemical evolution and the origins of life. In *Environmental Evolution: The Effects of the Origin and Evolution of Life on Planet Earth*. L. Margulis and L. Olendzenski, eds. Cambridge, Mass.: MIT Press, 17–27.

Ponnamperuma, C., and N. Gabel. 1968. Current status of chemical studies on the origin of life. *Space Life Sci. 1*:64–96.

Pontecorvo, G. 1958. *Trends in Genetic Analysis*. New York: Columbia University Press.

Portier, P. 1918. *Les Symbiotes*. Paris: Masson.

Portnoy, D. A., R. D. Schreiber, P. Connelly, and L. G. Tilney. 1989. Gamma interferon limits access of *Listeria monocytogenes* to the macrophage cytoplasm. *J. Exp. Med. 170*:2141–2146.

Preer, J. R., Jr. 1975. The hereditary symbionts of *Paramecium aurelia. Symp. Soc. Exp. Biol. 29*:125–144.

Preer, J. R., Jr. 1984. Genus II: *Caedibacter*. In *Bergey's Manual of Systematic Bacteriology*, vol. 1. J. G. Holt, ed. Baltimore: Williams and Wilkins, pp. 803–806.

Preer, J. R., Jr., L. B. Preer, and A. Jurand. 1974. Kappa and other endosymbionts in *Paramecium aurelia. Bacteriol. Rev. 38*:113–163.

Prescott, D. M. 1964. Cellular sites of RNA synthesis. *Prog. Nucleic Acid Res. Mol. Biol. 3*:33–57.

Prescott, D. M., and K. G. Murti. 1973. Chromosome structure in ciliated protozoans. *Cold Spring Harbor Symp. Quant. Biol. 38*:609–618.

Proudlock, J. W., L. W. Wheeldon, D. J. Jollow, and A. W. Linnane. 1968. Role of sterols in *Saccharomyces cerevisiae. Biochim. Biophys. Acta 152*:434–437.

Raff, R. A., and H. R. Mahler. 1972. The non-symbiotic origin of mitochondria. *Science 177*:575–582.

Ragan, M. A., and D. J. Chapman. 1978. *A Biochemical Phylogeny of the Protists*. New York: Academic Press.

Raikov, I. B. 1971. Bacteries epizoiques et mode de nutrition de cilie psammophile *Kentrophoros fistulosum* Fauré-Fremier (étude au microscope electronique). *Protistologica 7*:365–378.

Raikov, I. B. 1982. *The Protozoan Nucleus: Morphology and Evolution*. Wien: Springer-Verlag.

Rambler, M., and L. Margulis. 1979. An ultraviolet light induced bacteriophage in *Beneckea gazogenes. Origins Life 9*:235–240.

Rambler, M. B., and L. Margulis. 1980. Bacterial resistance to ultraviolet irradiation under anaerobiosis: implications for pre-Phanerozoic evolution. *Science 210*:638–640.

Raudaskoski, M. 1972. Occurrence of microtubules in the hyphae of *Schizophyllum commune* during intracellular nuclear migration. *Arch. Mikrobiol. 86*:91–100.

Raven, P. H. 1970. A multiple origin for plastids and mitochondria. *Science 169*:641–646.

Reddy, C. A., M. P. Bryant, and M. J. Wolin. 1972. Characteristics of the S organism isolated from *Methanobacillus omelianskii. J. Bacteriol. 109*:539–545.

Reich, E., and D. J. L. Luck. 1966. Replication and inheritance of mitochondria DNA. *Proc. Natl. Acad. Sci. U. S. A. 55*:1600–1608.

Reijnders, L. 1975. The origin of mitochondria. *J. Mol. Evol. 5*:167–176.

Reimer, T. O., E. S. Barghoorn, and L. Margulis. 1979. Primary productivity in an early Archean microbial ecosystem. *Precambrian Res. 9*:93–104.

Reinheimer, H. 1920. *Symbiosis: A Sociophysiological Study of Evolution*. London: Headley Brothers.

Renaud, F. L., and H. Swift. 1964. The development of basal bodies and flagella in *Allomyces arbusculus. J. Cell Biol. 23*:339–354.

Renfro, A. R. 1974. Genesis of evaporite-associated stratiform metalliferous deposits—a sabkha process. *Econ. Geol. 69*:33–45.

Rich, A. 1970. [Oral remarks.] In *Origins of Life: Proceedings of the First Conference.* L. Margulis, ed. New York: Published for Interdisciplinary Communication Associates by Gordon and Breach.

Richmond, M. H., and D. C. Smith, eds. 1979. *The Cell as a Habitat.* London: The Royal Society.

Ris, H. 1975. Primitive mitotic mechanisms. *BioSystems 7*:298–304.

Ris, H., and D. F. Kubai. 1974. An unusual mitotic mechanism in the parasitic protozoan *Syndinium* sp. *J. Cell Biol. 60*:702–720.

Ris, H., and W. Plaut. 1962. Ultrastructure of DNA-containing areas in the chloroplast of *Chlamydomonas. J. Cell Biol. 13*:383–391.

Roberts, J. D., and M. C. Caserio. 1964. *Basic Principles of Organic Chemistry.* Menlo Park: W. A. Benjamin.

Robinson, D. R., and K. Gull. 1991. Basal body movements as a mechanism for mitochondrial genome segregation in the trypanosome cell cycle. *Nature 352*:731–733.

Roodyn, D. B., and D. Wilkie. 1968. *The Biogenesis of Mitochondria.* London: Methuen.

Rossignol-Strick, M., and E. S. Barghoorn. 1971. Extraterrestrial abiogenic organization of organic matter: the hollow spheres of the Orgueil meteorite. *Space Life Sci. 3*:89–107.

Saffo, M. B. 1991. Symbiogenesis and the evolution of mutualism: lessons from the *Nephromyces*-bacterial endosymbiosis in molgulid. In *Symbiosis As a Source of Evolutionary Innovation: Speciation and Morphogenesis.* L. Margulis and R. Fester, eds. Cambridge, Mass.: MIT Press, pp. 410–429.

Sagan, D. 1991. *Biospheres: Metamorphosis of Planet Earth.* New York: McGraw-Hill.

Sagan, L. 1967. On the origin of mitosing cells. *J. Theor. Biol. 14*:225–274.

Sagan, L., Y. Ben-Shaul, H. T. Epstein, and J. A. Schiff. 1965. Studies of chloroplast development in *Euglena.* XI. Radioautographic localization of chloroplast DNA. *Plant Phys. 40*:1257–1260.

Sagan, D., and L. Margulis. 1991. Epilogue: The uncut self. In *Organism and the Origins of Self: Boston Studies in the Philosophy of Science*, vol. 129. A. I. Tauber, ed. Dordrecht, The Netherlands: Kluwer Academic Publishers, pp. 361–374.

Sahagún, Fray Bernadino de. 16th century. *Florentine Codex, General History of the Things of New Spain.* Translated from Nahuatl by C. E. Dibble and A. J. O. Anderson. Salt Lake City: University of Utah Press, 1963.

Salisbury, J. L., and G. L. Floyd. 1978. Calcium-induced contraction of the rhizoplast of a quadriflagellate green alga. *Science 202*:975–977.

Sapp, J. 1987. *Beyond the Gene: History of Cytoplasmic Inheritance.* New York: Cambridge University Press.

Sapp, J. In press. *Evolution by Association.* New York: Oxford University Press.

Satir, P. 1974. How cilia move. *Sci. Am. 231*(4):44–52.

Satir, P. 1984. The generation of ciliary motion. *J. Protozool. 31*:8–12.

Scannerini, S., and P. Bonfante-Fasolo. 1991. Bacteria and bacteria-like objects in endomycorrhizal fungi (Glomaceae). In *Symbiosis As a Source of Evolutionary Innovation: Speciation and Morphogenesis.* L. Margulis and R. Fester, eds. Cambridge, Mass.: MIT Press, pp. 273–287.

Scheer, H., ed. 1991. *Chlorophylls.* Boca Raton, Fla.: CRC Press.

Schein, S. J., M. V. L. Bennet, and G. M. Katz. 1976. Altered calcium conductance in pawns, behavioral mutants of *Paramecium aurelia. J. Exp. Biol. 65*:699–724.

Schell, J., M. Van Montagu, M. DeBenckeleer, M. DeBloch, A. Depicker, M. DeWilde, G. Engler, G. Genetillo, J. P. Hernalsteens, M. Holsters, J. Seurinck, B. Silva,

F. Van Vliet, and R. Villarroel. 1979. Interactions and DNA transfer between *Agrobacterium tumefaciens*, the Ti plasmid and the plant host. *Proc. R. Soc. Lond., Ser. B. Biol. Sci.* 204:251–266.

Schidlowski, M., P. W. U. Appel, R. Eichmann, and C. E. Junge. 1979. Carbon isotope geochemistry of the 3.7×10^9-yr-old Isua sediments, West Greenland: implications for the Archean carbon and oxygen cycles. *Geochim. Cosmochim. Acta* 43:189–199.

Schidlowski, M., R. Eichmann, and C. E. Junge. 1975. Precambrian sedimentary carbonates: carbon and oxygen isotope geochemistry and implications for the terrestrial oxygen budget. *Precambrian Res.* 2:1–69.

Schiff, J. A., and H. Lyman. 1982. *On the Origins of Chloroplasts.* New York: Elsevier.

Schildkraut, C., M. Mandel, S. Levisohn, J. E. Smith-Sonneborn, and J. Marmur. 1962. DNA base composition and taxonomy of some protozoans. *Nature* 196:795–797.

Schneider, S., and P. Boston, eds. 1992. *Scientists on Gaia.* Cambridge, Mass.: MIT Press.

Schopf, J. W. 1967. Antiquity and evolution of Precambrian life. In *McGraw-Hill Yearbook of Science and Technology.* New York: McGraw-Hill, pp. 46–55.

Schopf, J. W. 1972. Precambrian paleobiology. In *Exobiology.* C. Ponnamperuma, ed. Amsterdam: North-Holland, pp. 16–61.

Schopf, J. W. 1975. Precambrian paleobiology: problems and perspectives. *Annu. Rev. Earth Planet. Sci.* 3:213–249.

Schopf, J. W. 1976. Are the oldest 'fossils,' fossils? *Origins Life* 7:19–36.

Schopf, J. W., ed. 1983. *Earth's Earliest Biosphere: Its Origin and Evolution.* Princeton, N. J.: Princeton University Press.

Schopf, J. W., ed. 1992. *Major Events in the History of Life.* Boston: Jones and Bartlett.

Schopf, J. W., and E. S. Barghoorn. 1967. Algal-like fossils from the early Precambrian of South Africa. *Science* 156:508–512.

Schopf, J. W., E. S. Barghoorn, M. D. Maser, and R. O. Gordon. 1965. Electron microscopy of fossil bacteria two billion years old. *Science* 149:1365–1367.

Schopf, J. W., and J. M. Blacic. 1971. New microorganisms from the Bitter Springs Formation (late Precambrian) of the North Central Amadeus Basin, Australia. *J. Paleontol.* 45:925–960.

Schopf, J. W., T. D. Ford, and W. J. Breed. 1973. Microorganisms from the late Precambrian of the Grand Canyon, Arizona. *Science* 179:1319–1321.

Schopf, J. W., and C. Klein, eds. 1992. *The Proterozoic Biosphere*: A Multidisciplinary Study. New York: Cambridge University Press.

Schuster, F. 1963. An electron microscope study of the amoebo-flagellate, *Naegleria gruberi* (Schardinger). I. The amoeboid and flagellate stages. *J. Protozool.* 10:297–313.

Schuster, F. 1968. The gullet and trichocysts of *Cyathomonas truncata. Exp. Cell Res.* 49:277–284.

Schwartz, R. M., and M. O. Dayhoff. 1978. Origins of prokaryotes, eukaryotes, mitochondria and chloroplasts. *Science* 199:395–403.

Schwemmler, W. 1973. Ecological significance of endosymbiosis: an overall concept. *Acta Biotheor.* 22:113–119.

Schwemmler, W. 1984. *Reconstruction of Cell Evolution: A Periodic System.* Boca Raton, Fla.: CRC Press.

Schwemmler, W. 1991. Symbiogenesis in insects as a model for morphogenesis, cell differentiation, and speciation. In *Symbiosis As a Source of Evolutionary Innovation:*

Speciation and Morphogenesis. L. Margulis and R. Fester, eds. Cambridge, Mass.: MIT Press, pp. 178–204.

Science. July 3, 1880. [An investigation of a meteorite.]

Searcy, D. G. 1970. Measurements by DNA hybridization in vitro of the genetic basis of parasitic reduction. *Evolution 24*:207–219.

Searcy, D. G., and R. J. Delange. 1980. *Thermoplasma acidophilum* histone-like protein: partial amino acid sequence suggestive of homology to eukaryotic histones. *Biochim. Biophys. Acta 609*:197–200.

Searcy, D. G., and W. G. Hixon. 1991. Cytoskeletal origins in sulfur-metabolizing archaebacteria. *BioSystems 25*:1–11.

Searcy, D. G., and D. B. Stein. 1980. Nucleoprotein subunit structure in an unusual prokaryotic organism: *Thermoplasma acidophilum. Biochim. Biophys. Acta 609*:180–195.

Searcy, D. G., D. B. Stein, and G. R. Green. 1978. Phylogenetic affinities between eukaryotic cells and a thermophilic mycoplasma. *BioSystems 10*:19–28.

Seckbach, J., J. F. Fredrick, and D. J. Garbary. 1983. Auto- or exogenous origin of transitional algae: an appraisal. In *Endocytobiology II: Intracellular Space as Oligogenetic Ecosystem*. H. E. A. Schenk and W. Schwemmler, eds. Berlin: Walter de Gruyter, pp. 947–1007.

Seilacher, A. 1985. Discussion of Precambrian metazoans. *Philos. Trans. R. Soc. Lond. B. Biol. Sci. 311*:47–48.

Seilacher, A. 1993. Early multicellular life: Late Proterozoic fossils and the Cambrian explosion. In *Early Life on Earth: Nobel Symposium 84*. S. Bengtson, ed. New York: Columbia University Press (in press).

Seilacher, A., W.-E. Reif, and F. Westphal. 1985. Sedimentological, ecological, and temporal patterns of fossil Lagerstatten. *Philos. Trans. R. Soc. Lond. B. Biol. Sci. 311*:5–23.

Sewall, R. B. 1974. *The Life of Emily Dickinson*. New York: Farrar, Straus and Giroux.

Shmagina, A. P. 1948. *Ciliary Movement*. Moscow: Moscow State Publishing House for Medical Literature (MEDGIZ). [In Russian.]

Siegel, S. M., K. Roberts, H. Nathan, and O. Daly. 1967. Living relative of microfossil *Kakabekia. Science 156*:1231–1234.

Silver, S., G. Nucifora, L. Chu, and T. Misra. 1989. Bacterial resistance ATPases: primary pumps for exporting toxic anions and cations. *TIBS* (Feb):129–133.

Silver, W. S., and J. R. Postgate. 1973. Evolution of asymbiotic nitrogen fixation. *J. Theor. Biol. 40*:1–10.

Simpson, G. G. 1953. *The Major Features of Evolution*. New York: Columbia University Press.

Simpson, G. G. 1967. The crisis in biology. *Am. Schol. 36*:363–377.

Simpson, G. G., and W. S. Beck. 1965. *Life: An Introduction to Biology*. 2nd ed. New York: Harcourt Brace and World.

Simpson, L. 1968. Effect of acriflavin on the kinetoplast of *Leishmania tarentolae. J. Cell Biol. 37*:660–682.

Singh, G., M. T. Lott, and D. C. Wallace. 1989. A mitochondrial DNA mutation as a cause of Leber's hereditary optic neuropathy. *New Engl. J. Med. 320*:1300–1305.

Sleigh, M. A. 1962. *The Biology of Cilia and Flagella*. Oxford: Pergamon.

Smillie, R. M. 1976. Temperature control of chloroplast development. In *Genetics and*

Biogenesis of Chloroplasts and Mitochondria. T. Bücher, W. Neupert, W. Sebald, and S. Werner, eds. Amsterdam: North-Holland, pp. 103–110.

Smith, D. C. 1979. From extracellular to intracellular: the establishment of symbiosis. *Philos. Trans. R. Soc. Lond. B. Biol. Sci. 204*:115–130.

Smith, D. C., and A. Douglas. 1989. *The Biology of Symbiosis.* London: E. A. Arnold.

Smith, H. E., and H. J. Arnott. 1974. Epi- and endobiotic bacteria associated with the external surface of *Pyrsonympha vertens*, a symbiotic protozoan of the termite *Reticulitermes flavipes. Trans. Am. Microsc. Soc. 93*:180–189.

Smith, H. E., H. E. Buhse, Jr., and S. J. Stamler. 1975. Possible formation and development of spirochaete attachment sites found on the surface of symbiotic polymastigote flagellates of the termite *Reticulitermes flavipes. BioSystems 7*:374–379.

Smith, H. E., S. J. Stamler, and H. E. Buhse, Jr. 1975. A scanning electron microscopy survey of the surface features of polymastigote flagellates from *Reticulitermes flavipes. Trans. Am. Microsc. Soc. 94*:401–410.

Sogin, M. L. 1991. The phylogenetic significance of sequence diversity and length variations in eukaryotic small subunit ribosomal RNA coding regions. In *New Perspectives on Evolution.* L. Warren and H. Kaproski, eds. New York: Wiley-Liss, pp. 175–188.

Sogin, M. L. 1993. The origin of eukaryotes and evolution into major kingdoms. In: *Early Life on Earth: Nobel Symposium 84.* Stephen Bengtson, ed. New York: Columbia University Press (in press).

Soifer, D., ed. 1975. *The Biology of Cytoplasmic Microtubules. (Ann. N.Y. Acad. Sci. 253)* New York: New York Academy of Sciences.

Sokatch, J. R. 1969. *Bacterial Physiology and Metabolism.* London: Academic Press.

Soldo, A. 1963. Axenic culture of *Paramecium*—some observations on the growth and behavior and nutritional requirements of a particle-bearing strain of *Paramecium aurelia. Ann. N. Y. Acad. Sci. 108*:380–388.

Sonea, S. 1991. Bacterial evolution without speciation. In *Symbiosis As a Source of Evolutionary Innovation: Speciation and Morphogenesis.* L. Margulis and R. Fester, eds. Cambridge, Mass.: MIT Press, pp. 95–105.

Sonea, S., and M. Panisset. 1976. Manifesto for a new bacteriology. *Rev. Can. Biol. 35*:159–161.

Sonneborn, T. M. 1959. Kappa and related particles in *Paramecium. Adv. Virus Res. 6*:229–356.

Sonneborn, T. M. 1974. Ciliate morphogenesis and its bearing on general cellular morphogenesis. *Actualités Protozool. 1*:337–355.

Sorokin, S. P. 1968. Reconstruction of centriole formation and ciliogenesis in mammalian lungs. *J. Cell Sci. 3*:207–230.

Soyer, M.-O. 1972. Les ultrastructures nucléaires de la *Noctiluca* (dinoflagellé libre) au cours de la sporogénesis. *Chromosoma 39*:419–441.

Soyer, M.-O., and O. K. Haapala. 1974. Division and function of dinoflagellate chromosomes. *J. Microsc. 19*:137–146.

Soyer-Gobillard, M.-O. 1974. *Titres et Travaux Scientifiques. Chargée de Recherche au C. N. R. S.* Banyuls-sur-Mer, France: Laboratoire Arago.

Spiegelman, S. 1967. An in vitro analysis of a replicating molecule. 80th Jubilee Lecture. *Am. Sci. 55*:221–264.

Stackebrandt, E., R. G. E. Murray, and H. G. Trüper. 1988. Proteobacteria new class, a name for the phylogenetic taxon that includes the "purple bacteria and their relatives." *Int. J. Syst. Bacteriol. 38*:321–325.

Stanier, R. Y., and G. Cohen-Bazire. 1977. The phototrophic prokaryotes: the cyano-bacteria. *Annu. Rev. Microbiol. 31*:225–274.

Stanier, R. Y., M. Doudoroff, and E. A. Adelberg. 1970. *The Microbial World.* 3rd ed. Englewood Cliffs, N. J.: Prentice-Hall.

Stanley, S. M. 1976a. Fossil data and the Precambrian-Cambrian evolutionary transition. *Am. J. Sci. 276*:56–76.

Stanley, S. M. 1976b. Ideas on the timing of metazoan diversification. *Paleobiology 2*:209–219.

Starr, M. P. 1975. *Bdellovibrio* as symbiont: the associations of bdellovibrios with other bacteria interpreted in terms of a generalized scheme for classifying organismic associations. *Symp. Soc. Exp. Biol. 29*:93–124.

Stebbins, G. L. 1966. *Process of Organic Evolution.* Englewood Cliffs, N. J.: Prentice-Hall.

Steinman, G., A. E. Smith, and J. J. Silver. 1968. Synthesis of a sulphur-containing amino acid under simulated prebiotic conditions. *Science 159*:1108–1109.

Stoecker, D. K., and M. W. Silver. 1990. Replacement and aging of chloroplasts in *Strombidium capitatum* (Ciliophora: Oligotrichida). *Marine Biol. 107*:491–502.

Stolp, H. 1979. Interaction between *Bdellovibrio* and its host cell. *Proc. R. Soc. Lond., Ser. B. Biol. Sci. 204*:211–217.

Strother, P. 1992. Evidence for earliest life. In *Environmental Evolution: The Effects of the Origin and Evolution of Life on Planet Earth.* L. Margulis and L. Olendzenski, eds. Cambridge, Mass.: MIT Press, pp. 87–101.

Stubblefield, E., and B. R. Brinkley. 1966. Cilia formation in Chinese hamster fibro-blasts in vitro as a response to Colcemid treatment. *J. Cell Biol. 30*:645–652.

Stubblefield, E., and B. R. Brinkley. 1967. Architecture and function of the mammalian centriole. *Symp. Int. Soc. Cell Biol. 6*:175–218.

Sueoka, N. 1961. Variation and heterogeneity of basic composition of DNA. A compilation of old and new data. *J. Mol. Biol. 3*:31–40.

Szathmary, E. 1987. Early evolution of microtubules and undulipodia. *BioSystems 20*:115–131.

Takhtajan, A. L. 1973. Four kingdoms of the living world. *Priroda 2*:22–32. [In Russian.]

Tamm, S. L. 1982. Flagellated ectosymbiotic bacteria propel a eucaryotic cell. *J. Cell Biol. 94*:697–709.

Tartar, V. 1961. *The Biology of Stentor.* New York: Pergamon.

Tartar, V. 1967. Morphogenesis in protozoa. In *Research in Protozoology*, vol. 2. T. T. Chen, ed. Oxford: Pergamon, pp. 1–116.

Tartar, V. 1968. Regeneration in situ of membranellar cilia in *Stentor coeruleus. Trans. Am. Microsc. Soc. 87*:297–306.

Taylor, F. J. R. 1974. Implications and extensions of the serial endosymbiosis theory of the origin of eukaryotes. *Taxon 23*:229–258.

Taylor, F. J. R. 1976. Autogenous theories for the origin of eukaryotes. *Taxon 25*:377–390.

Taylor, F. J. R. 1978. Problems in the development of an explicit hypothetical phylogeny of the lower eukaryotes. *BioSystems 10*:67–89.

Taylor, F. J. R. 1979. Symbionticism revisited: a discussion of the evolutionary impact of intracellular symbioses. *Philos. Trans. R. Soc. Lond. B. Biol. Sci. 204*: 267–286.

Taylor, S. R. 1987. The origin of the moon. *Am. Sci. 75*:469–477.

Taylor, S. R., and S. M. McLennan. 1985. *The Continental Crust: Its Composition and Evolution.* Oxford: Blackwell.

Tebo, B. M., D. S. Linthicum, and K. H. Nealson. 1979. Luminous bacteria and light emitting fish: ultrastructure of the symbiosis. *BioSystems 11*:269–280.

Tewari, K. K., R. D. Kolodner, and W. Dobkin. 1976. Replication of circular chloroplast DNA. In *Genetics and Biogenesis of Chloroplasts and Mitochondria.* T. Bücher, W. Neupert, W. Sebald, and S. Werner, eds. Amsterdam: North-Holland, pp. 379–386.

Thomas, L. 1974. *The Lives of a Cell: Notes of a Biology Watcher.* New York: Viking.

Thompson, M. P., ed. 1988. *Calcium-Binding Proteins.* 2 vols. Boca Raton, Fla.: CRC Press.

Thorington, G., B. Berger, and L. Margulis. 1979. Transmission of symbionts through the sexual cycle of *Hydra viridis*. I. Observations on living organisms. *Trans. Am. Microsc. Soc. 98*:401–413.

Thorington, G. U. 1980. The algal and bacterial symbionts of *Hydra viridis*: metabolic relations and transmission through the host sexual cycle. Ph.D. thesis. Boston University, Boston.

Thorington, G. U., and L. Margulis. 1981. *Hydra viridis*: transfer of metabolites between hydra and symbiotic algae. *Biol. Bull. 160*:175–188.

Thorne, S. W., E. H. Newcomb, and C. B. Osmond. 1977. Identification of chlorophyll *b* in extracts of prokaryotic algae by fluorescence spectroscopy. *Proc. Natl. Acad. Sci. U. S. A. 74*:575–578.

Tiffney, B. H., and E. S. Barghoorn. 1974. *The Fossil Record of the Fungi.* (Occasional Papers of the Farlow Herbarium of Cryptogamic Botany, No. 7.) Cambridge, Mass.: Harvard University.

Tilney, L. G., Y. Hiramoto, and D. Marsland. 1966. Studies on the microtubules in Heliozoa. III. A pressure analysis of the role of these structures in the formation and maintenance of the axopodia of *Actinosphaerium nucleofilum* (Barrett.) *J. Cell Biol. 29*:77–95.

Tilney, L. G., and K. R. Porter. 1967. Studies on the microtubules in Heliozoa. II. The effect of low temperature on these structures in the formation and maintenance of axopodia. *J. Cell Biol. 34*:327–343.

Tilney, L. G., and D. A. Portnoy. 1989. Actin filaments and the growth, movement, and spread of the intracellular bacterial parasite, *Listeria monocytogenes. J. Cell Biol. 109*:1597–1608.

Timasheff, S. N. 1981. The biophysical approach to the self-assembly of biological organelles: the tubulin-microtubule system. In *Molecular Approaches to Gene Expression and Protein Structure.* M. A. Q. Siddiqui, M. Krauskopf, and H. Weissbach, eds. New York: Academic Press, pp. 245–285.

To, L. 1987. Are centrioles semiautonomous? *Ann. N. Y. Acad. Sci. 503*:83–91.

To, L., and L. Margulis. 1978. Ancient locomotion: prokaryotic motility systems. *Int. Rev. Cytol. 54*:267–293.

To, L., L. Margulis, D. Chase, and W. L. Nutting. 1980. The symbiotic microbial community of the Sonoran desert termite: *Pterotermes occidentis. BioSystems 13*:109–137.

To, L., L. Margulis, and A. T. W. Cheung. 1978. Pillotinas and hollandinas: distribution and behavior of large spirochaetes symbiotic in termites. *Microbios 22*:103–133.

Todd, N. 1967. A theory of karyotypic fissioning. Genetic potentiation and eutherian evolution. *Mammalian Chromosome Newsletter 8*:268–279.

Todd, N. 1970. Karyotypic fissioning and canid phylogeny. *J. Theor. Biol. 26*:445–480.

Todd, N. 1992. Mammalian evolution: karyotypic fission theory. In *Environmental Evolution: The Effects of the Origin and Evolution of Life on Planet Earth.* L. Margulis and L. Olendzenski, eds. Cambridge, Mass.: MIT Press, pp. 279–296.

Tornabene, T. G., T. A. Langworthy, G. Holzer, and J. Oró. 1979. Squalenes, phytanes and other isoprenoids as major neutral lipids of methanogenic and thermoacidophilic "Archaebacteria." *J. Mol. Evol. 13*:73–83.

Trench, R. K. 1979. The cell biology of plant-animal symbiosis. *Annu. Rev. Plant Physiol. 30*:485–531.

Trench, R. K. 1991. *Cyanophora paradoxa* Korschikoff and the origins of chloroplasts. In *Symbiosis As a Source of Evolutionary Innovation: Speciation and Morphogenesis.* L. Margulis and R. Fester, eds. Cambridge, Mass.: MIT Press, pp. 143–150.

Trench, R. K., R. R. Pool, Jr., M. Logan, and A. Engelland. 1978. Aspects of the relationship between *Cyanophora paradoxa* (Korschikoff) and its endosymbiotic cyanelles *Cyanocyta korschikoffiana* (Hall and Claus). I. Growth, ultrastructure, photosynthesis and the obligate nature of the association. *Proc. R. Soc. Lond., Ser. B. Biol. Sci. 202*:423–443.

Trüper, H. G., and H. W. Jannasch. 1968. *Chromatium buderi* nov. spec., eine neue Art der "grossen" Thiorhodaceae. *Arch. Microbiol. 61*:363–372.

Tucker, J. B. 1968. Fine structure and function of the cytopharyngeal basket in the ciliate *Nassula. J. Cell Sci. 3*:493–514.

Tucker, J. B. 1971. Development and deployment of cilia, basal bodies, and other microtubular organelles in the cortex of the ciliate *Nassula. J. Cell Sci. 9*:539–567.

Turner, F. R. 1968. An ultrastructural study of plant spermatogenesis. Spermatogenesis in *Nitella. J. Cell Biol. 37*:370–393.

Urbach, E., D. L. Robertson, and S. W. Chisholm. 1992. Multiple evolutionary origins of prochlorophytes within the cyanobacterial radiation. *Nature 335*:267–270.

Ursprung, H., K. D. Smith, W. H. Sofer, and D. T. Sullivan. 1968. Assay systems for the study of gene function. *Science 160*:1075–1081.

Uzzell, T., and C. Spolsky. 1974. Mitochondria and plastids as endosymbionts: a revival of special creation? *Am. Sci. 62*:334–343.

Vallee, R. 1990. Dynein and the kinetochore. *Nature 345*:206–207.

van der Velden, W., and A. W. Schwartz. 1977. Search for purines and pyrimidines in the Murchison meteorite. *Geochim. Cosmochim. Acta 41*:961–968.

Verma, D. P. S., D. T. Nash, and H. M. Schulman. 1974. Isolation and in vitro translation of soybean leghaemoglobin mRNA. *Nature 251*:74–77.

Vetter, R. D. 1991. Symbiosis and the evolution of novel trophic strategies: thiotrophic organisms at hydrothermal vents. In *Symbiosis As a Source of Evolutionary Innovation: Speciation and Morphogenesis.* L. Margulis and R. Fester, eds. Cambridge, Mass.: MIT Press, pp. 219–245.

Vidal, G. 1984. The oldest eukaryotic cells. *Sci. Am. 250*(2):48–57.

Vidal, G. 1990. Giant acanthomorph acritarchs from the upper Proterozoic in southern Norway. *Palaeontology 33*:287–298.

Vidal, G., 1993. Early ecosystems—limitations imposed by the fossil record. In *Early Life on Earth: Nobel Symposium No. 84.* S. Bengtson, ed. New York: Columbia University Press (in press), pp. 247–260.

Vidal, G., and A. H. Knoll. 1983. Proterozoic plankton. *Geol. Soc. Am. Memoirs 161*:265–277.

von Wettstein, D. 1967. Chloroplast structure and genetics. In *Harvesting the Sun:*

Photosynthesis in Plant Life. A. San Pietro, F. A. Greer, and T. J. Army, eds. New York: Academic Press, pp. 153–190.

Walker, C. W., and M. P. Lesser. 1989. Nutrition and development of brooded embryos in the brittlestar *Amphipholis squamata*: do endosymbiotic bacteria play a role? *Mar. Biol. 103*:519–530.

Walker, J. C. G. 1977. *Evolution of the Atmosphere.* New York: Macmillan.

Walker, J. C. G. 1980. Atmospheric constraints on the evolution of metabolism. In *Limits to Life*: Proceedings of the Fourth College Park Colloquium on Chemical Evolution. C. Ponnamperuma and L. Margulis, eds. Dordrecht: Reidel, pp. 121–132.

Walker, J. C. G. 1983. *Earth History: The Several Ages of the Earth.* Boston: Jones and Bartlett.

Wallin, I. E. 1923. Symbionticism and prototaxis: two fundamental biological principles. *Anat. Rec. 26*:65–73.

Wallin, I. E. 1925. On the nature of mitochondria. IX. Demonstration of the bacterial nature of mitochondria. *Am. J. Anat. 36*:131–146.

Wallin, I. E. 1927. *Symbionticism and the Origin of Species.* Baltimore: Williams & Wilkins.

Walsby, A. E. 1968. Mucilage secretion and movements of blue-green algae. *Protoplasma 65*:223–238.

Walsh, M. M. 1992. Microfossils and possible microfossils from the early Archean Onverwacht Group. *Precambrian Res. 54*:271–293.

Walsh, M. M., and D. R. Lowe. 1985. Filamentous microfossils from the 3,500 M-yr-old Onverwacht Group, Barberton Mountain Land, South Africa. *Nature 314*:530–532.

Walter, M. R., ed. 1976. *Stromatolites.* (*Dev. Sedimentol. 20*) Amsterdam: Elsevier.

Walter, M. R. 1993. Stromatolites: The main source of information on the evolution of the early benthos. In *Early Life on Earth: Nobel Symposium No. 84.* S. Bengtson, ed. New York: Columbia University Press (in press), pp. 215–228.

Watson, M. E., and P. W. Signor. 1986. How a clam builds windows: shell microstructure in *Corculum* (Bivalvia: Cardiidae). *Veliger 28*:348–355.

Weinrich, D. H., ed. 1954. Sex in protozoa. A comparative review. In *Sex in Microorganisms.* Washington, D. C.: American Association for the Advancement of Science, pp. 134–265.

Weissman, A. 1892. *Das Keimplasma. Eine Theorie der Vererbung.* Jena: Gustav Fischer.

Wenyon, C. M. 1926. *Protozoology.* London: Ballière, Tindall and Cox.

Weyl, P. 1968. Organic particles in the ocean. *Science 162*:587.

Whatley, J. M. 1977. The fine structure of *Prochloron. New Phytol. 79*:309–313.

Whatley, J. M., and C. Chapman-Andresen. 1990. Phylum Karyoblastea. In *Handbook of Protoctista.* L. Margulis, J. O. Corliss, M. Melkonian, and D. J. Chapman, eds. Boston: Jones and Bartlett, pp. 167–185.

Whatley, J. M., P. John, and F. R. Whatley. 1979. From extracellular to intracellular: the establishment of mitochondria and chloroplasts. *Proc. R. Soc. Lond., Ser. B. Biol. Sci. 204*:165–187.

White, H. B., III. 1976. Coenzymes as fossils of an earlier metabolic state. *J. Mol. Evol. 7*:101–104.

Whitehouse, H. L. K. 1973. *Towards an Understanding of the Mechanism of Heredity.* 3rd ed. New York: St. Martin's.

Whitfield, M. 1988. Mechanisms or machinations: is the ocean self-regulating? In *Gaia, the Thesis, the Mechanisms and the Implications*: Proceedings of the First Annual Camelford Conference on the Implications of the Gaia Hypothesis. P. Bunyard and E. Goldsmith, eds. Camelford, Cornwall: Wadebridge Ecological Centre, pp. 79–90.

Whittaker, R. H. 1959. On the broad classification of organisms. *Q. Rev. Biol.* *34*:210–226.

Whittaker, R. H. 1969. New concepts of kingdoms of organisms. *Science 163*:150–160.

Whittaker, R. H., and L. Margulis. 1978. Protist classification and the kingdoms of organisms. *BioSystems 10*:3–18.

Williamson, D. I. 1987. Incongruous larvae and the origin of some invertebrate life-histories. *Prog. Oceanog. 19*:87–116.

Williamson, D. I. 1992. *Larvae and Evolution: Toward a New Zoology*. New York: Rutledge, Chapman, and Hall.

Wilson, E. B. 1959. *The Cell in Development and Heredity*. 3rd ed. New York: Macmillan. [Reprint of 1925 edition]

Wilson, L. 1975. Microtubules as drug receptors: pharmacological properties of microtubule protein. *Ann. N. Y. Acad. Sci. 253*:213–231.

Woese, C. R. 1981. Archaebacteria. *Sci. Am.* 244(6):98–122.

Woese, C. R., and G. E. Fox. 1977. Phylogenetic structure of the prokaryotic domain: the primary kingdoms. *Proc. Natl. Acad. Sci. U. S. A. 74*:5088–5090.

Woese, C. R., G. E. Fox, L. Zablen, T. Uchida, L. Bonen, K. Pechman, B. J. Lewis, and D. Stahl. 1975. Conservation of primary structure in 16S rRNA. *Nature 254*:83–86.

Woese, C. R., O. Kundler, and M. L. Wheelis. 1990. Towards a natural system of organisms: proposal for the domains Archaea, Bacteria and Eucarya. *Proc. Natl. Acad. Sci. U. S. A. 87*:4576–4579.

Wujek, D. E. 1969. Ultrastructure of flagellated chrysophytes. I. *Dinobryon. Cytologia 34*:71–79.

Yamamoto, N., and M. L. Droffner. 1985. Mechanisms determining aerobic or anaerobic growth in the facultative anaerobe *Salmonella typhimurium*. *Proc. Natl. Acad. Sci. U. S. A. 82*:2077–2081.

Yamin, M. A. 1980. Cellulose metabolism by the termite flagellate *Trichomitopsis termopsidis*. *Appl. Environ. Microbiol. 39*:859–863.

Younger, K. B., S. Banerjee, J. K. Kelleher, M. Winston, and L. Margulis. 1972. Evidence that the synchronized production of new basal bodies is not associated with DNA synthesis in *Stentor coeruleus*. *J. Cell Sci. 11*:621–637.

Zheng, Y., M. K. Jung, and B. R. Oakley. 1991. γ-Tubulin is present in Drosophila melanogaster and Homo sapiens and is associated with the centrosome. *Cell 65*:817–823.

INDEX

boldface = indexed term appears in figure or figure legend and may also appear elsewhere in the text on the page indicated.
n = indexed term appears in footnote.